EVALUATING AND PREDICTING DESIGN PERFORMANCE

PRINCIPLES OF COMPUTER-AIDED DESIGN

A Series by

Yehuda E. Kalay
School of Architecture and Planning
State University of New York, Buffalo

Graphic Introduction To Programming
Modeling Objects and Environments
Evaluating and Predicting Design Performance
Computability of Design

on architectural design

EVALUATING AND PREDICTING DESIGN PERFORMANCE

Edited by

Yehuda E. Kalay

School of Architecture and Planning
State University of New York at Buffalo

A WILEY-INTERSCIENCE PUBLICATION

JOHN WILEY & SONS, INC.

New York / Chichester / Brisbane / Toronto / Singapore

1992

Library of Congress Cataloging in Publication Data:

Evaluating and predicting design performance / edited by Yehuda E. Kalay.
p. cm. — (Principles of computer aided design)
A selection of the papers were originally presented at a symposium.
Includes bibliographical references
ISBN 0-471-85385-2
1. Architectural design—Data processing—Evaluation. 2. Computer-aided design—Evaluation. I. Kalay, Yehuda E. II. Series: Kalay, Yehuda E. Principles of computer-aided design.

NA2728.E94 1991 91-9767
721′.0285—dc20 CIP

Printed and bound in the United States of America by Braun-Brumfield, Inc.

10 9 8 7 6 5 4 3 2 1

CONTRIBUTORS

Rachel Becker, National Building Research Institute, The Technion–Israel Institute of Technology, Haifa, Israel

Claude Bédard, Centre for Building Studies, Concordia University, Montreal, Canada

Isabel J. Canete, Parsons Brinckerhoff Quade & Douglas, Inc., New York, New York

Gianfranco Carrara, Department of Architecture and Environmental Controls, University of Rome, Rome, Italy

Paul Fazio, Centre for Building Studies, Concordia University, Montreal, Canada

Steven J. Fenves, Department of Civil Engineering, Carnegie Mellon University, Pittsburgh, Pennsylvania

Ulrich Flemming, Department of Architecture, Carnegie Mellon University, Pittsburgh, Pennsylvania

Gabriela Goldschmidt, Faculty of Architecture and Town Planning, The Technion–Israel Institute of Technology, Haifa, Israel

Krishnan Gowri, SIRICON Inc., Montreal, Canada

Mark D. Gross, College of Environmental Design, University of Colorado, Boulder, Colorado

Eek J. H. Hacfoort, Faculty of Architecture, Building, and Planning, Eindhoven University of Technology, Eindhoven, The Netherlands

Volker Hartkopf, Department of Architecture, Carnegie Mellon University, Pittsburgh, Pennsylvania

Craig Hendrickson, Engineering Design Research Center, Carnegie Mellon University, Pittsburgh, Pennsylvania

Peter R. A. Jockusch, Department of Architecture, University of Kassel, Kassel, Germany

Bharati Jog, School of Architecture and Interior Design, University of Cincinnati, Ohio

Robert E. Johnson, College of Architecture and Urban Planning, The University of Michigan, Ann Arbor, Michigan

Yehuda E. Kalay, School of Architecture and Planning, State University of New York at Buffalo, New York

Loukas N. Kalisperis, Department of Architecture, The Pennsylvania State University, University Park, Pennsylvania

Robin S. Liggett, Graduate School of Architecture and Urban Planning, University of California, Los Angeles, California

Vivian Loftness, Department of Architecture, Carnegie Mellon University, Pittsburgh, Pennsylvania

Mary L. Maher, Engineering Design Research Center, Carnegie Mellon University, Pittsburgh, Pennsylvania

Peter Manning,* College of Architecture, Technical University of Nova Scotia, Halifax, Canada

Samir Mattar, United Nations Centre for Human Settlements, Nairobi, Kenya

Peter Mill, Department of Architecture, Carnegie Mellon University, Pittsburgh, Pennsylvania

William J. Mitchell, Harvard Graduate School of Design, Cambridge, Massachusetts

Rashid A. Mohsini, School of Architecture and Planning, State University of New York at Buffalo, New York

Foster Nichols Jr., Parsons Brinckerhoff Quade & Douglas, Inc., New York, New York

Gabriele Novembri, CARTESIANA Consortium, Rome, Italy

Michael J. O'Neill, Department of Environment, Textiles, and Design, The University of Wisconsin, Madison, Wisconsin

Rivka E. Oxman, Faculty of Architecture and Town Planning, The Technion–Israel Institute of Technology, Haifa, Israel

Gerhard Schmitt, Department of Architecture, Swiss Federal Institute of Technology, Zurich, Switzerland

Edna Shaviv, Faculty of Architecture and Town Planning, The Technion–Israel Institute of Technology, Haifa, Israel

Melvin Siegel, Robotics Institute, Carnegie Mellon University, Pittsburgh, Pennsylvania

Edward Steinfeld, School of Architecture and Planning, State University of New York at Buffalo, New York

Mitch Steinman, Department of Architecture, The Pennsylvania State University, University Park, Pennsylvania

Luis H. Summers, Department of Civil, Environmental and Architectural Engineering, University of Colorado at Boulder, Colorado

Milton Tan, Harvard Graduate School of Design, Cambridge, Massachusetts

Sagun Tuladhar, Parsons Brinckerhoff Quade & Douglas, Inc., New York, New York

K. Jan Veldhuisen, Faculty of Architecture, Building, and Planning, Eindhoven University of Technology, Eindhoven, The Netherlands

Avi Wiezel, Department of Civil Engineering, The Technion–Israel Institute of Technology, Haifa, Israel

Craig Zimring, College of Architecture, Georgia Institute of Technology, Atlanta, Georgia

**Present address:* 1470 Summer St. No 1601, Halifax, Nova Scotia, Canada B3H 3A3.

SERIES PREFACE

Computers have made their debut as design tools in many engineering disciplines, providing designers with flexible means to represent design products. In that capacity they have already demonstrated their utility in improving the productivity and the quality of the production end of the design process, much as word processors have improved the production of documents in offices.

A growing number of researchers, developers, and users of computer-aided design (CAD) systems, however, have reached the conclusion that the utility of computers in design is far from what it might become if their use were extended over the design process as a whole, including the complete, accurate, and efficient representation of the designed artifacts and the processes that are employed in their conception and creation.

This series of four books is intended to help realize the potential of CAD, through the introduction—in an integrative form—of principles, methodologies, and practices that underlie CAD. It is intended to be used by people who wish to engage in the process of research, development, and maintenance of the new generation of CAD systems. Since the current cadre of people involved with R&D in CAD is relatively small, the series assumes that the reader has no particular knowledge of the field and thus it can be used as a text for beginners. The topics it introduces progress, nevertheless, rapidly toward the frontiers of CAD and thus make the series useful as a text for advanced courses and as reference for professionals.

The first volume, *Graphic Introduction to Programming*, introduces the basic concepts of computing that are needed to master the tool—namely, the computer. These concepts include programming, structured problem solving, and interactive computer graphics. Programming is the means through which computers are instructed how to perform the desired tasks. Structured problem solving, through algorithm design and analysis, encourages a disciplined approach to the design process as a whole and to computer-aided design software development in particular. Given the established manner of communicating design between humans, and because of the ease with which graphic and pictorial information can be disseminated compared to text and numerals, computer graphics have become the standard means of interaction with CAD systems. These three topics have been integrated and are covered through learning the programming language Pascal. This language has certain features that qualify it for this task, which include ease of comprehension, enforcing good programming practices, availability on most computers, and provision of dynamic memory management facilities.

The second volume, *Modeling Objects and Environments*, introduces the concepts of modeling real-world phenomena in the computer's memory. It covers a set of methods and techniques for representing the physical environment as symbol structures that are understood and can be manipulated by computers. The book includes the study of data structures that are particularly suitable for representing two- and three-dimensional artifacts, and operating on these structures in a way that preserves their semantic integrity. The book covers the formative and other attributes of objects, such as topology, geometry, transformations, assemblies, and general database query/update operations.

The third volume, *Evaluating and Predicting Design Performance*, presents an assortment of task-specific applications that draw upon the model of the designed artifact and provide the designer (and knowledge-based-CAD systems) with a variety of evaluative, simulative, and tabulative measures of the artifacts' expected performances. Written by noted experts in their respective disciplines, the applications present both state-of-the-art knowledge in specific fields relating to CAD and an integrative approach to employing computers as design consultants.

The fourth volume, *Computability of Design*, completes the series by reviewing the key topics that were presented, and by raising the fundamental issues of design computation: How can design processes be mathematically modeled? What is design knowledge, and how can it be computed? How can the design process as a whole be computed? Alternatively, how can particular design tasks be computationally assisted? This volume is based on a symposium that was held in Buffalo, New York, In December 1986, where noted experts discussed the feasibility, utility, and desirability of various approaches to the computation of design.

The series as a whole is concerned with the principles of CAD rather than with an exhaustive survey of techniques, practices, and existing systems. Its goal is to educate students, researchers, and professionals, who will develop the CAD systems of the future and who will maintain them. It is intended to be used by designers in many disciplines rather than by computer scientists alone. It may, however, also be of considerable interest to computer scientists, by exposing them to the computational concerns of design professionals.

It is recommended that the books be used in sequence, but this is not a prerequisite for their utility. Neither is adherence to the programming language Pascal that is used to exemplify the theory. It is the concepts and principles presented in the series that are of primary importance, and they transcend the technologies and techniques used for their implementation.

YEHUDA E. KALAY

Buffalo, New York
October 1991

PREFACE

Predicting is risky business, especially about the future
—ALVIN TOFFLER

Design, in its broadest sense, is a discipline that formally recognizes our ability to influence the future and our responsibility to do so in the "proper" way. It is, therefore, a purposeful and conscious specification of the actions that must be taken at the present to attain some desired future conditions.

Characterized by discovery, learning, and judicious decision making, the design process can be likened to exploratory search, where alternative courses of action are hypothesized and their effects are predicted and evaluated against the predefined set of desired conditions. The evaluation may find that the predicted effects of the actions match the desired conditions and thus validate the hypothesis. More often, however, the predicted effects of the proposed actions do not match the desired conditions, or they conflict with each other. In such cases the actions must be modified to achieve the desired conditions, or the desired conditions must be revised to match the predicted effects of the proposed actions. This is done through iterative adjustment of the proposed actions and the desired conditions until a match is achieved.

The design process can thus be said to comprise three major tasks:

1. Defining a set of desired conditions that comprise the objectives to be achieved.
2. Specifying actions that, in the opinion of the designer, will achieve the desired objectives.
3. Predicting and evaluating the effects of the specified actions to verify that they are consistent with each other and that they achieve the desired objectives.

It is the last task in this triad that is the focus of this book. Its centrality to the process of design stems from the fact that prediction and evaluation comprise the "glue" that connects and binds the other two components of the triad. Specifying actions that achieve a desired set of conditions cannot be separated from predicting and evaluating the expected effects of these actions, nor can the desired conditions be fully understood without such evaluation. Prediction and evaluation comprise processes whereby the expected future behavior of environments, artifacts, and people are estimated and compared to a desired state of affairs. The evaluative process guides the generative process toward achieving the stated objectives and uncovers opportunities to be had and trade-offs that will improve the overall quality of the solution. It also produces insights into the particular design problem that were not apparent at its inception, in terms of compromises that must be made between conflicting objectives. These insights not only produce new design parameters and define new objectives to be achieved, but may also radically alter previously stated objectives and beliefs.

In general, it is very difficult to define precisely the objectives to be achieved without actually engaging in the search for actions that will achieve them. Many desired conditions are simply unknown, or the particular values they ought to have cannot be determined prior to knowing the effects of actions that attempt to achieve them. Furthermore, specifying the desired conditions that should be attained by complex systems, such as environments, artifacts, and groups of people, is difficult, because it involves political, psychological, social, economical, environmental, technical, and many other variables. It is difficult to specify the desired performances of each one of these variables alone and even more difficult to specify a combination of their effects and interactions. Their combined effects may cancel each other, be impossible to achieve due to internal contradictions, or they may be totally undesirable.

Predicting and evaluating the expected performances of design actions and the objectives they aim to achieve comprise three distinct, yet related tasks:

1. Evaluating a proposed set of design actions for internal consistency.
2. Predicting the expected effects and side effects of these design actions and comparing them to the desired objectives.
3. Evaluating the desired objectives themselves.

These tasks are further complicated by the different kinds of representations that we use to express the actions and objectives. The actions we specify to attain the desired conditions are typically represented as discrete operations involving concrete objects or people. The desired conditions, on the other hand, are often represented in terms of their functionalities. To evaluate the correspondence between the effects of the proposed actions and the desired functionalities we must translate between the two different representation schemes. This translation is error prone, because the functional behavior of discrete objects and people is often context dependent, time dependent, and subject to interpretation. These difficulties put great mental strains on designers, demanding their creative talents, capacity for judgment, experience, and most important, abilities to predict the expected performances of the proposed actions and to draw operational conclusions from these evaluations to make subsequent design decisions that will bring the specified actions closer to achieving the desired objectives.

To alleviate these mental strains, the search for methods of predicting the effects of design decisions having began in the 1960s, seeking a synthesis between the designer's creative and rationale faculties. These methods, comprising a series of logical and mechanical operations, would free designers from the tedious evaluative aspects of their work, to the great advantage of their creative powers. Use of the methods would provide instant feedback on the effects and side effects of decisions made by the designers' creative skills and thus allow them to adjust these decisions to achieve the desired objectives. Furthermore, a rational and deductive approach to the processes of predicting and evaluating design decisions would make it possible to identify a certain number of operations that could be automated. Consequently, proponents of such evaluative and predictive methods tried to model them in a way that would also take full advantage of the potential offered by computers. Needless to say, three decades of research failed to realize this dream. In fact, the considerable knowledge gained in predicting and evaluating specific types of design decisions remains largely unutilized in everyday architectural design practice.

Three major developments over the past decade have made computer-aided evaluation of design alternatives and prediction of their expected performances a topic of current interest:

1. Explosive growth of knowledge concerning various aspects of building materials and construction technology
2. Advancements in computer-aided representation and modeling of buildings and their performances
3. Growing understanding of knowledge-based and computer-aided design techniques

The explosive growth of knowledge is evident in the multitude of manuals and handbooks that summarize the results of many research projects and the accumulated experiences of designers and builders. The advancements in computer-aided representation and modeling of buildings make it possible to evaluate the expected performance of buildings more accurately than before. The continuing development of knowledge-based and computer-aided design techniques, such as expert systems and semantic networks, make it possible to apply the evaluation at the earliest design phases, allowing designers to consider the effects of design decisions from many points of view when they can still be modified.

Still, many problems persist, limiting the aid that computers can provide to designers as far as evaluating and predicting the performance of design alternatives is concerned. First, although certain areas of performance modeling have become fine tuned (e.g., energy, lighting, and cost estimation), others are still based on highly subjective and inaccurate measures. They include such important areas as privacy control, usability and habitability, and environmental impact. Second, many side effects that accompany all design decision are hard to foresee and evaluate. Third, the need to trade off certain qualities to gain others, and thus achieve a better overall balance of performances, is difficult, if not impossible, to model and compute.

This book explores recent developments concerning computer-aided evaluation of design alternatives and prediction of their expected performances. In particular, it examines six major issues:

1. Theoretical issues concerning the roles of evaluation and prediction in the design process
2. General frameworks for integrating evaluation and prediction in the overall design process
3. Multicriteria evaluation of design alternatives

4. Multilevel evaluation of design alternatives
5. Evaluation of physical factors concerning the built environment
6. Evaluation of human factors concerning the built environment

These topics are highly interrelated. Theory is the basis of the exploration. It explains why evaluation and prediction are needed in the first place, how they fit in the overall design process, and why they are difficult to perform. General frameworks describe specific computational, manual, and combined approaches to integrating evaluation and prediction in the overall design process. Multicriteria evaluation is a direct derivative of general computational design frameworks. It strives to automate some of the functions currently assumed exclusively by designers and to present them with a basis for making trade-off decisions. Multilevel evaluation recognizes the varying informational characteristics of the design solution as it evolves through the hierarchical phases of the design process.

The book examines each of these topics, as well as related issues, by presenting three or four different views on each subject. It considers the roles of evaluation and prediction in the design process in light of the growing understanding of both the design process itself and the needs its products must satisfy. The discussions of frameworks for integrating evaluation in the overall design process present various methodologies for embedding design criteria in the generative process, by constraint propagation, derivation of optimized solutions, and reliance on precedents. Multicriteria evaluation explores various attempts to present designers with feasible chocies to be made regarding trade-offs between competing qualities and values offered by the emerging design solution. Discussion of multilevel evaluation explores the possibility of partitioning the design process vertically or horizontally, in an attempt to better match the informational tools at the designers' disposal. Evaluation of the physical and human factors concerning the built environment considers two specific areas: energy and wayfinding. These admittedly narrow areas demonstrate the wealth of tools available to designers and represent evaluative and predictive measures that have already been codified and embedded in knowledge-based computational systems.

This book and the symposium that preceded it are the culmination of efforts by many individuals. I thank the contributors, both those whose papers were selected for inclusion in the book and those whose papers were not. I also thank the following reviewers for generously donating their time and expertise to thoughtfully review and critique the many papers that were submitted for publication: Dennis Andrejko, John Archea, Michael Brambley, Michael Brill, Collin Davidson, Steven Fenves, Ulrich Flemming, Mark Gross, Anton Harfmann, Robert Johnson, Robin Liggett, Vivian Loftness, Bruce Majkowski, Peter Manning, Raymond McCall, Rashid Mohsini, Marcos Novak, Michael O'Neill, Gunter Schmitz, Edna Shaviv, Edward Steinfeld, Lucien Swerdloff, James Turner, John Wade, and Robert Woodbury. I also thank the State University of New York at Buffalo, for supporting the symposium through a Conferences in the Disciplines grant. Many thanks to the administration of the School of Architecture and

Planning for their support and encouragement. Most thanks are due to Bruce Majkowski, who arranged for and coordinated all the logistics of mounting the symposium, and to Debi Plotnik for her assistance.

YEHUDA E. KALAY

Buffalo, New York
October 1991

CONTENTS

EVALUATING AND PREDICTING DESIGN PERFORMANCE

INTRODUCTION: THE USES OF INCONSISTENCY IN DESIGN

William J. Mitchell

Harvard Graduate School of Design
Cambridge, Massachusetts

In this paper two of the central dogmas underlying most current theories of design evaluation are challenged: that the representations used by a designer must be well formed, and that a designer must have a consistent belief framework within which to make judgments about design proposals. The crucial roles in design of ambiguous and inconsistent representations and provisional beliefs are examined, and a model of design exploration based on nonmonotonic modes of reasoning is sketched.

INTERPRETED FORMAL SYSTEMS

The study of natural language has been dominated for several decades by two theses, the first put forward by Noam Chomsky and the second by Richard Montague.* Chomsky's thesis, essentially, is that natural languages can be described as formal systems specified by generative grammars. Montague's thesis is that natural languages can, furthermore, be described as interpreted formal systems with semantics specified by model-theoretic means. Design products and processes can usefully be studied within a closely analogous framework. That is, designs can be treated as expressions within a language of two- or three-dimensional shapes specified by a shape grammar, and we can go on to establish rules for interpretation of designs. Design processes can then be seen as computations directed at producing shapes with satisfactory interpretations as depictions of physical systems performing certain functions.

I have developed this approach in considerable detail in my book *The Logic of Architecture* [Mitchell 1990]. In this paper I extend it to cover designs with ambiguous

*Introductory texts on formal semantics develop these theses in detail. See Allwood et al. [1977], Bach [1989], Dowty et al. [1981], and Martin [1987].

and inconsistent interpretations. I begin by showing how designs within languages specified by shape grammars can be interpreted, then I consider uninterpretable, ambiguous, and inconsistent designs, and finally, I discuss the identification and resolution of inconsistencies in evolving designs.

SHAPE GRAMMARS

Grammars are defined within algebras. In particular, shape grammars are commonly defined within an algebra of two-dimensional straight-line shapes under the operations of shape addition and shape subtraction [Stiny 1980a, 1986, 1990]. But shape grammars can also be defined within algebras of three-dimensional shapes [Stiny 1980b], algebras of curved-line shapes, algebras of colored surfaces [Knight 1989], and so on. By taking the Cartesian product of a shape algebra and an algebra of labels, an algebra of labeled shapes can be produced. The labels might specify material properties, and so on. Shape grammars can also be defined within algebras of this sort [Stiny 1989].

A shape grammar is defined by specifying a starting shape and a set of shape rules. Shape rules associate conditions and actions. Each shape rule specifies that if the shape on its left can be matched to a subshape in an evolving design, then that subshape can be replaced by the shape on the right. Shapes in the language specified by the grammar are derived by recursive application of the shape rules to the starting shape. Figure 1 shows a simple example.

A shape grammar can be thought of purely as an abstract computational mechanism for specifying a set of design possibilities. Alternatively, if we wish, we can give a shape grammar some psychological content by using the left sides of rules to describe the conditions that a designer is on the alert for, and using the right sides to describe characteristic responses to those conditions. Nothing in the nature of the shape grammar formalism itself determines whether a particular grammar will be abstract or descriptive. Often the

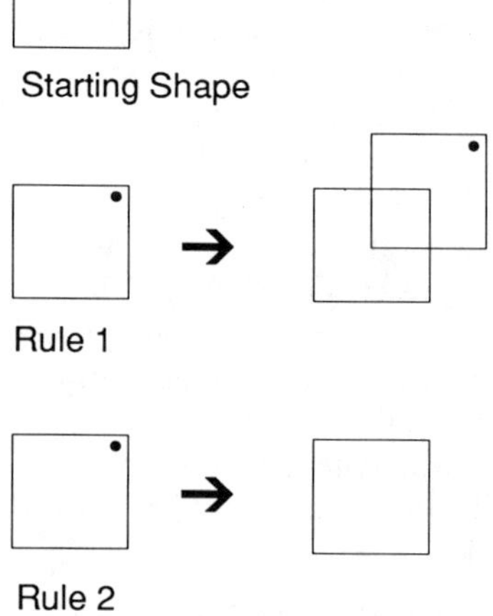

Figure 1. Simple shape grammar.

objective in writing a shape grammar is to produce an efficient and elegant specification of some interesting set of design possibilities, but knowledge of how a designer actually reasons is used heuristically to suggest appropriate rules.

FUNCTIONAL INTERPRETATION

In general, expressions in a language are interpreted by assigning other kinds of things to them. Thus French sentences may be interpreted as English sentences, Pascal statements may be interpreted as sequences of computer operations, expressions written in musical notation may be interpreted as audible melodies, and sentences in logical notation may be interpreted as true or false. In design, we usually want to interpret shapes as depictions of physical structures that function in various ways: as load-bearing elements, as protection against the elements, and so on. Interpretation thus becomes a matter of classifying shapes by expected roles (as columns, beams, windows, rooms, etc.), measuring various geometric properties, and deriving inferences from these measurements about adequacy in performance of these roles.

A first step in formalization of the semantics of shape languages in this way is to establish a set of functional labels, for example *{column, beam, arch, . . .}*, together with a set of shape classification rules that associate such labels with subshapes in an evolving design. The left side of a shape classification rule is an unlabeled or labeled shape, and the right side is the same shape with at least one functional label associated to it. Application of shape classification rules interprets an abstract design as a set of spatially related functional components—a system.

In practice, shape classification may be carried out in two different ways. First, a designer may proceed by assembling prelabeled shapes—column shapes, beam shapes, and so on—in Tinkertoy fashion. Many computer-aided design systems provide for this. Second, a designer may instantiate unlabeled shapes and later apply classification rules. This second approach may seem perverse, but it has the important advantage of allowing for formation of interpretations in terms of emergent form. Consider, for example, the floor plans of Palladio's villas (Figure 2). The underlying grid is most naturally constructed with traditional drafting instruments, or by hand sketching, by overlaying evenly spaced parallel lines. From this, a grid of small rectangles emerges. Furthermore, less obvious closed shapes—T-shapes, +-shapes, and multicell rectangles—also emerge. Interpretation of these closed shapes as rooms yields various villa layouts. When closed shapes are interpreted as rooms, fragments of the original gridlines are variously interpreted as walls and as lines of ceiling vaulting spanning unobstructed floor space. Contrast this process of emergence and interpretation with one of laying out prelabeled room, wall, and vault elements.

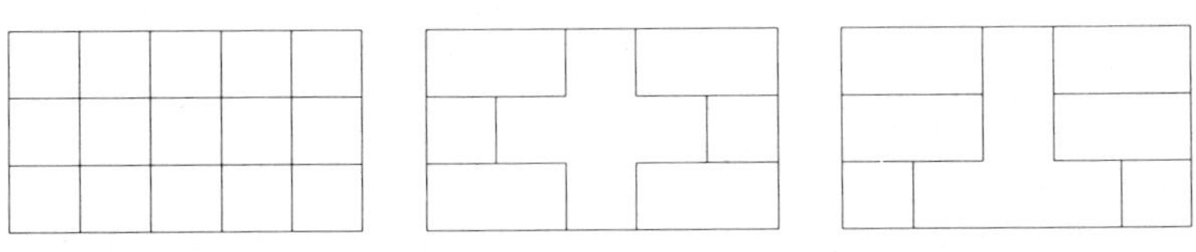

Figure 2. Palladian villa plan grid and some emergent shapes within it.

Generally, shape classification rules will encode knowledge of what things of various shapes are good for: Palladio knew that long narrow rectangles, or squares, or T-shapes, or +-shapes on the central axis were good for central rooms. In different contexts, the same shape might be seen as good for different things. Where shapes are interpreted as structural elements rather than rooms, for example, a long rectangular element might be labeled as a beam (Figure 3).

Functional classification rules can, however, be quite arbitrary: you might use a thing of any shape as a beam, although it might not be very good for this purpose. And you might use a brick as, say, a writing instrument. Under normal circumstances this use of a brick would be perverse. But if you were Robinson Crusoe on a desert island and nothing more suitable were available, you might think that a brick served quite well for writing in the sand. In general, functional classification implicitly asserts a hypothesis that an object will, in the given context, exhibit behaviors that serve the specified purpose adequately well. The hypothesis may be verified or disproved by appeal to precedent, by analysis, or by testing a prototype (you cannot fool nature). Skilled designers produce good hypotheses—they know what things with particular shape and material properties are likely to be good for in various contexts.

Analysis of behavior to determine functional adequacy begins with taking observations of shapes. This amounts to evaluation of mathematical functions such as *span (beam)*, and *depth(beam)*, and of predicates such as *under(column, beam)*. Then these observed values can be used, in standard ways, as arguments in other functions and predicates that predict behavior. Predicted behavior can then be compared with a specification of desired behavior to determine functional adequacy. Usually, we expect designers not only to know what things are good for, but also to be able to *show* what they are good for by analysis of behavior.

INTERPRETABLE AND UNINTERPRETABLE DESIGNS

Interpretation procedures are, in general, functions mapping from things of one kind (such as strings of English words) to things of another kind (such as values in the set *{true,false}*). These procedures cannot take just anything at all as arguments. A random string of English words is not true or false, it is just uninterpretable. Similarly, a symbol string that does not comply with the rules of Pascal syntax cannot be translated into actions by a Pascal interpreter. And we can hardly expect to be able to produce functional interpretations of random shape compositions: we will not know what some of the shapes are, or the functional significance of their spatial interrelationships. Thus shape grammars of practical interest in design usually specify sets of shapes that we know how to interpret as artifacts that function in particular ways: as villas [Stiny and Mitchell 1978a,b], as teahouses [Knight 1981], as Queen Anne houses [Flemming 1986], as eave details [Mitchell and Radford 1987], as structural frames [Mitchell 1990], and so on. We know what the parts do, and we know the functional significance of the spatial relationships that can be formed—placement of a portico on the front of a villa, or a column under a beam. In general, the rules of grammar tell how to form significant relationships of recognizable parts: word strings that mean something, or designs that can be understood in functional terms.

Shapes in a shape vocabulary have many possible functions, just as words listed in a dictionary have many possible senses. A rectangular prism, for example, might function

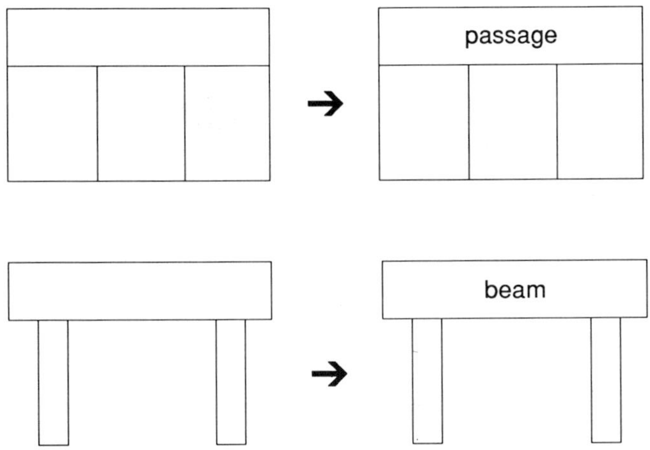

Figure 3. The same shape is seen as different things in different contexts.

as a paperweight, as a column, as a handrail, or as a device for smashing windows. But sense and function are more narrowly determined in context. A shape grammar for domestic construction might allow such a prism to be instantiated in vertical orientation under a beam (where it functions as a column) or in horizontal orientation beside a balcony (where it functions as a handrail), but would probably make no provision for use as a paperweight or window smasher. A shape grammar for Palladian villa plans [Stiny and Mitchell 1978a] allows T-shapes and +-shapes to serve as central rooms but not as side rooms, it allows rectangular shapes to serve as rooms anywhere, and it allows off-axis interior rectangles to serve only as stairs or storage rooms—uninhabited spaces because they get no light or ventilation.

A grammatical, interpretable expression in some language may or may not be appropriate in a specific context: a declarative English sentence may be true or false with respect to a particular situation, and a design may or may not work under the given conditions. The task of a designer is to produce a shape that is, first, interpretable, and second, has an interpretation that matches the requirements of the situation. Given that a rectangle, a T-shape, or a +-shape might serve as the central room of a villa, which is appropriate in a particular design?

AMBIGUOUS DESIGNS

Sometimes a shape in a design is interpretable in several different ways. Consider, for example, the shape shown in Figure 4.* You can easily interpret it as a depiction of a rabbit or as a depiction of a duck. If you initially see it as a duck, you can reread it as a rabbit, and vice versa. Figure 5 shows a set of classification rules that apply in bottom-up fashion to produce these possible readings.

*Wittgenstein [1968] makes this shape the focus of a penetrating discussion of the interpretation of drawings.

Figure 4. Rabbit/duck figure.

Shapes usually have competing structural descriptions, and this can become an important source of ambiguity.† Figure 6, for example, shows a shape and some different ways of describing it as a relationship of simpler shapes. Such shapes are analogous to English sentences, such as *Time flies like an arrow*, that can be parsed in different ways to yield alternative meanings. The most interesting and appropriate way to decompose a shape into subshapes may be far from obvious: as the song says, breaking up is hard to do.

Decomposition of a shape in a new way not only yields a new set of objects, but new values for the predicates and functions that interest us. In development of a floor plan, for example, a designer might be interested in the predicates *symmetrical(shape)* and *overlap(shape, shape)*, where *shape* is a closed planar shape, *symmetrical* specifies bilateral symmetry about a vertical axis, and *overlap* specifies the relation of overlap between two shapes. The function *area(shape)* might also be of interest. Now consider the composition shown in Figure 7. If we read it in the most natural way, as two overlapping closed shapes, we can describe it by saying that *symmetrical(left)* and *symmetrical(right)* are both false, by saying that *overlap(left, right)* is true, and by giving numerical values for *area(left)* and *area(right).* But if we read it in a somewhat less obvious way, as adjacent shapes touching only at two vertices, we must redescribe it by saying that *symmetrical(left)* and *symmetrical(right)* are now both true, that *overlap(left, right)* is now false, and that *area(left)* and *area(right)* now have new values.

IMPOSSIBLE OBJECTS

Sometimes a shape initially seems to make sense—to be a well-formed representation of an artifact in some domain—but on closer inspection turns out to be contradictory. Figure 8, for instance, seems to depict a three-dimensional, plane-faced solid that might function as some sort of structural frame. Credence is given to this interpretation by the fact that each vertex, considered in isolation, can be seen as a joint formed from square

†Competing structural descriptions and related issues are discussed by Finke [1989], Hinton [1979], Kosslyn et al. [1983], Palmer [1977], and Reed [1974].

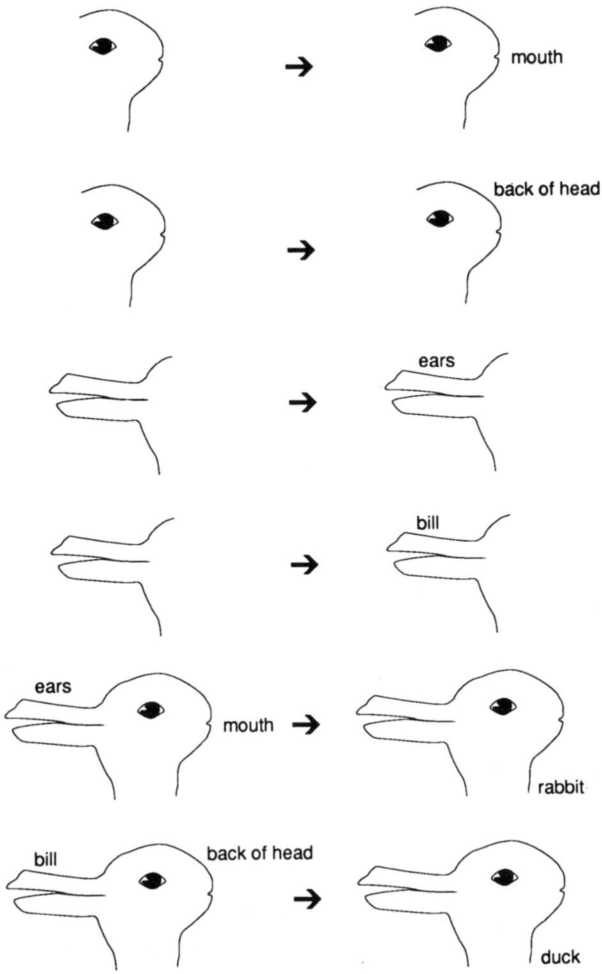

Figure 5. Classification rules that produce alternative readings.

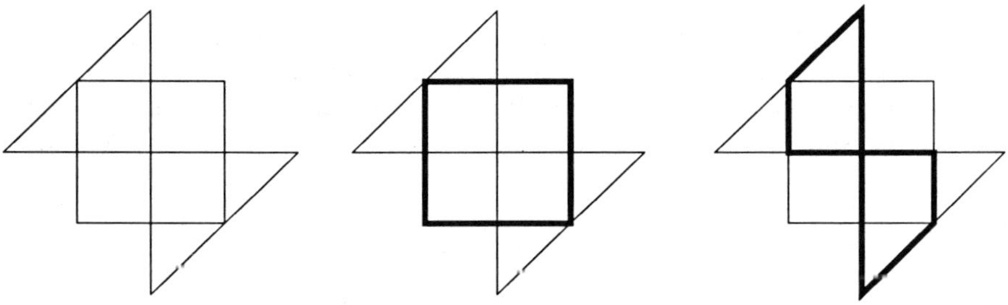

Figure 6. Alternative decompositions of a shape.

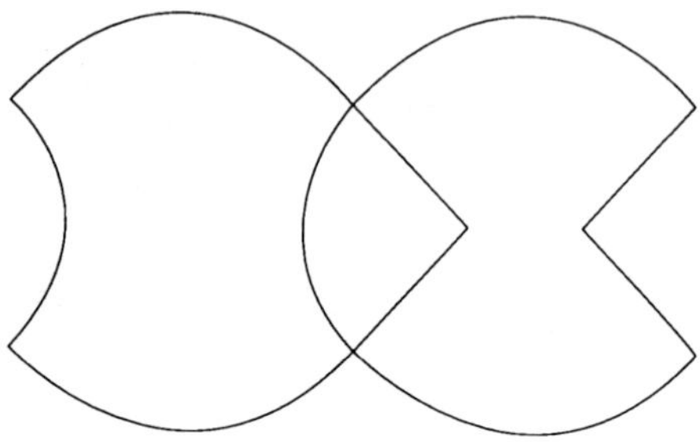

Figure 7. Are the left snape and the right shape both symmetrical?

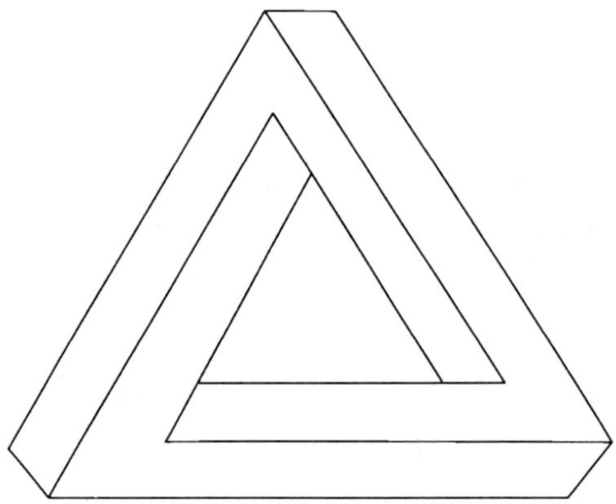

Figure 8. Impossible figure.

prisms. But there is no consistent way to assign orientations to all the faces such that this can be so. Thus the mind cannot settle on a reading, and the shape seems to oscillate in a disturbing way. This shape turns out to be an "impossible object"—a graphic object that complies with many of the syntactic rules for line depiction of prismatic solids, but contains contradictions such as the nonsense sentence, *The round square cupola of Berkeley College is pink.**

Impossible objects and nonsense sentences are not meaningless (although some philosophers have held that they are). We can think of them as conjectures that objects with the properties that they specify exist in the domain that concerns us. Such conjectures can be investigated directly, by production of objects in the domain in an effort to match the description. Or they can be investigated indirectly, by seeking to demonstrate contradictions. It is easy to see the contradictions in the examples given here, but in general it may require considerable intellectual effort to determine whether a design specifies an impossible object.

*The analogy between impossible objects and nonsense sentences was explored originally by Huffman [1971]. See Ballard and Brown [1982] for a survey of subsequent work on the interpretation of line drawings of prismatic solids.

EMERGENCE OF CONTRADICTIONS

Our behavior when confronted with ambiguous and inconsistent designs illustrates the extraordinary resourcefulness and flexibility of the human mind in attempting to make sense of situations. We quickly leap to conclusions based on whatever evidence initially presents itself, but we are prepared to retract them if necessary in the face of new evidence. We draw default conclusions in the absence of information to the contrary, but will abandon or revise these if contrary information does show up later.

There are several common ways for contrary information to show up in designs and force reinterpretation. The nonsense figure, for example, can be partitioned spatially into consistent parts, as shown in Figure 9. It is only when these parts are put together that a contradiction emerges. Figure 10 shows a design for the Villa Poiana from Palladio's *Four Books of Architecture.* The plan presents an internally consistent view of a villa, and so does the elevation. But when we put plan and elevation together we can see that they depict different patterns of openings. In this case a contradiction emerges when different views of the same object are combined.

Figure 9. Partitioning the impossible object into consistent parts.

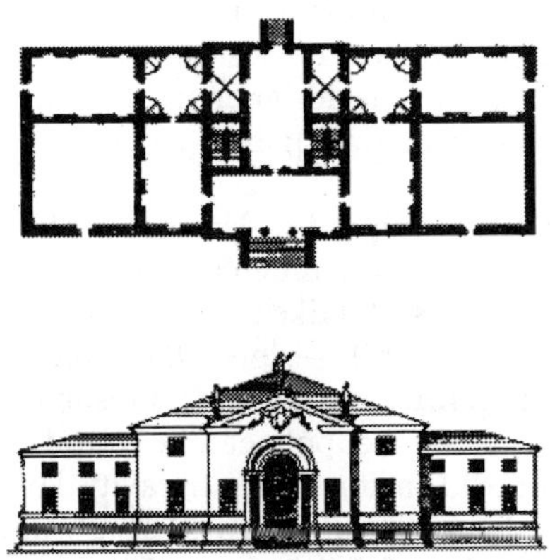

Figure 10. Palladio's Villa Poiana, from the *Four Books of Architecture.* Openings to the side courts show in plan but not in elevation.

Now consider the structural frame shown schematically in Figure 11*a*. Unless we have reason to be suspicious, we will usually see this as a structurally stable component of a building design. If the design is subsequently elaborated to show pin joints, as in Figure 11*b*, this default conclusion is confirmed. But if subsequent elaboration shows pin jointing as in Figure 11*c*, we are forced to revise the default conclusion. Here a contradiction emerges when further detail is added to a design.

Finally, consider the possibility of wholesale changes in attitude. An architect might begin a project with little sensitivity to energy issues and later, as a result of criticism, become more energy conscious. It might then emerge that decisions about orientation and fenestration were not consistent with meeting established constraints on operating cost.

RESOLUTION OF AMBIGUITIES AND INCONSISTENCIES

In the end, a design must be made sufficiently consistent and unambiguous to provide a basis for confident prediction of performance and efficient construction: the inconsistencies and ambiguities that emerge during the course of a design process must somehow be ironed out. The designer of a cupola presumably has to make up his or her mind whether it is round or square before acoustic or structural performance can be predicted and before construction can proceed.

Basically, an inconsistency exists when a fact derived by interpretation of a design and a belief used in the production of that interpretation cannot be true simultaneously. So the inconsistency can be resolved by altering the design or by abandoning the belief. Thus with the inconsistent Palladian villa we can alter the plan to correspond with the elevation, we can alter the elevation to correspond with the plan, or we can ask whether we are reading the drawings in the right way (perhaps the plan and elevation are not meant to be of the same three-dimensional object after all). With the impossible object we can change the joint details to yield a consistent reading (Figure 12). Or we can challenge the belief that all the prisms are connected at their ends: perhaps we have a twisted shape with an open corner, viewed from a vantage point that conceals the open corner (Figure 13).

An inventive design can challenge widely accepted beliefs, as the Wright brothers' flying machine challenged the belief that heavier-than-air machines could not fly. Successful demonstration of a cold-fusion device, or a gyroscopic antigravity machine, would have a similar effect. Perhaps the designer of a cupola for Berkeley College could, inventively, have it both ways—by showing how a cupola may be square in plan and round in section (Figure 14).

The beliefs that we employ in interpreting designs vary in their rootedness: some are tentative and readily abandoned without much regret, but others are harder to dislodge. We might believe that a shape looks a lot like a rabbit's ears, but we are easily persuaded that it might just as easily be a duck's bill—especially when this new belief is found to be equally consistent with the other information that is presented in a drawing. It takes only a simple sketch to convince us that a cupola can be round and square at once. We are very suspicious of designs for cold-fusion and antigravity devices, though, because these challenge belief in well-established physical laws. These laws are deeply consistent with the entire fabric of our knowledge of the world, and abandoning belief in them would

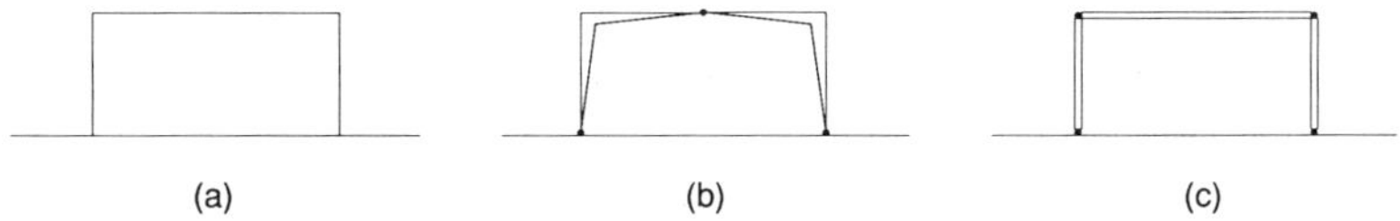

Figure 11. Elaborations of a schematic structural frame.

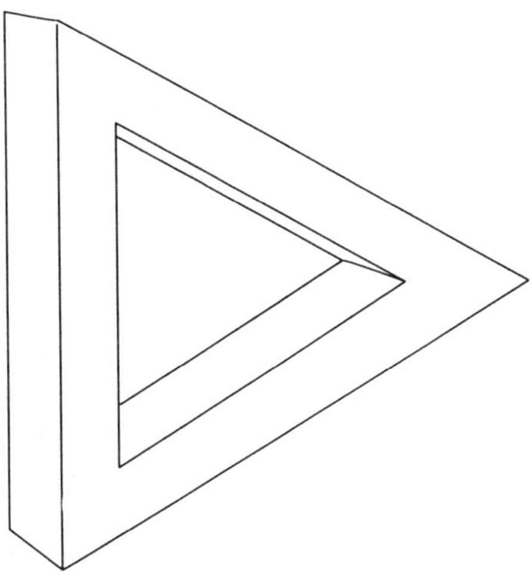

Figure 12. Impossible object made into a plane frame.

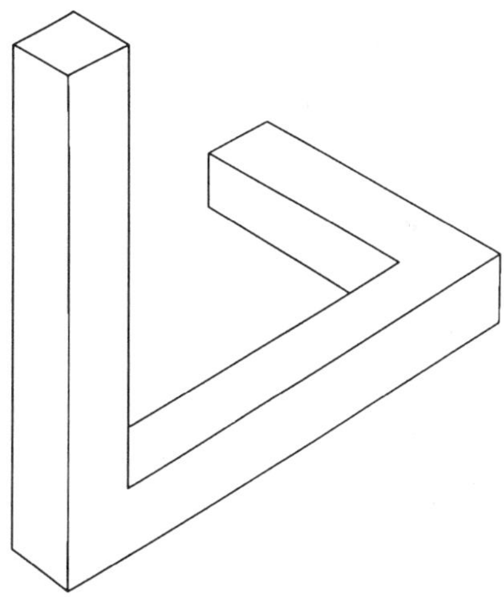

Figure 13. Impossible object reinterpreted.

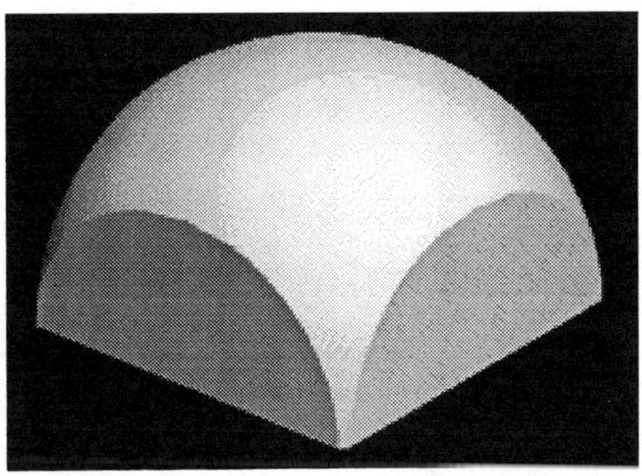

Figure 14. Cupola that is round in section and square in plan.

entail major revisions of our world view. Only the most watertight reasoning demonstrating the workability of a design proposal, or (better) a working prototype will be likely to convince us that we should abandon deeply rooted beliefs.

CONCLUSION: NONMONOTONIC DESIGN EXPLORATION

Traditional approaches to prediction and evaluation of design performance are based on the assumptions that design representation is a matter of progressively accumulating a consistent set of assertions about a building's geometry and materials (hence the emphasis on well-formedness and semantic consistency in the representation schemes used in computer-aided design), that evaluation of performance can be based on a well-established set of facts and rules about the real world (hence the use of fixed analysis procedures and the notion that we can just go on adding analysis procedures to a computer-aided design system), and that as the knowledge base about the proposed building and relevant aspects of the real world grows so does the set of conclusions that can be drawn from it by application of suitable inference procedures. In other words, monotonicity is assumed.* But as I have demonstrated, this is not always realistic. Designers construct shapes and try to make sense of them. Interpretation of shapes is not merely mechanical: the effort after sense is a probing, speculative, imaginative one. It requires alertness to emergent objects and relationships that can form the basis for fruitful new interpretations, the capacity to draw tentative conclusions on the basis of incomplete information, and willingness to revise beliefs that have formed the basis of an interpretation.

This suggests revision of the traditional view of design as a rather straightforward form of instrumental reasoning—trial-and-error search for consistent proposals that can be shown, by means of fixed analysis procedures, to satisfy predefined sets of performance requirements. We should regard this merely as a boundary case. More generally, and of more cultural importance, design is exploration of the interrelationships between beliefs and possibilities. Our beliefs establish what proposals we will accept as realistic and useful possibilities, but production of new proposals can also challenge our beliefs. The most creative designs are those which demonstrate that surprising, previously unimagined possibilities are consistent with deeply rooted beliefs, or conversely, which successfully challenge such beliefs.

REFERENCES

Allwood, J., L.-G. Andersson, and O. Dahl (1977). *Logic in Linguistics*, Cambridge University Press, Cambridge.

Bach, E. (1989). *Informal Lectures on Formal Semantics*, State University of New York Press, Albany, N.Y.

Ballard, D. H., and C. M. Brown (1982). *Computer Vision*, Prentice Hall, Englewood Cliffs, N.J.

Dowty, D. R., R. E. Wall, and S. Peters (1981). *Introduction to Montague Semantics*, D. Reidel, Dordrecht, The Netherlands.

*Distinctions between monotonic and nonmonotonic reasoning, formalizations on nonmonotonic reasoning, and applications of nonmonotonic reasoning are discussed in Ginsberg [1987].

Finke, R. A. (1989). *Principles of Mental Imagery*, MIT Press, Cambridge, Mass.

Flemming, U. (1986). "More Than the Sum of Parts: The Grammar of Queen Anne Houses," *Planning and Design,* **14:** 323–350.

Ginsberg, M. L. (ed.) (1987). *Readings in Nonmonotonic Reasoning*, Morgan Kaufmann, Los Altos, Calif.

Hinton, G. E. (1979). "Some Demonstrations of the Effects of Structural Descriptions in Mental Imagery," *Cognitive Science,* **3:** 231–250.

Huffman, D. A. (1971). "Impossible Objects as Nonsense Sentences," *Machine Intelligence,* **6:** 295–323.

Knight, T. W. (1981). "The Forty-One Steps," *Environment and Planning B*, **8:** 97–114.

Knight, T. W. (1989). "Color Grammars: Designing with Lines and Colors," *Planning and Design*, **16**(4): 417–450.

Kosslyn, S. M., B. J. Reiser, M. J. Farah, and S. L. Fliegel (1983). "Generating Visual Images: Units and Relations," *Journal of Experimental Psychology: General,* **112:** 278–303.

Martin, J. N. (1987). *Elements of Formal Semantics*, Academic Press, Orlando, Fla.

Mitchell, W. J. (1990). *The Logic of Architecture*, MIT Press, Cambridge, Mass.

Mitchell, J. R., and A. D. Radford (1987). "EAVE: A Generative Expert System for Detailing," *Planning and Design*, **14:** 281–292.

Palmer, S. E. (1977). "Hierarchical Structure in Perceptual Representation," *Cognitive Psychology*, **9:** 441–474.

Reed, S. K. (1974). "Structural Descriptions and the Limitations of Visual Images," *Memory and Cognition,* **2:** 329–336.

Stiny, G. (1980a). "Introduction to Shape and Shape Grammars," *Environment and Planning B*, **7:** 343–351.

Stiny, G. (1980b). "Kindergarten Grammars: Designing with Froebel's Building Gifts," *Environment and Planning B*, **7:** 409–462.

Stiny, G. (1986). "A New Line on Drafting Systems," *Design Computing*, **1:** 5–19.

Stiny, G. (1989). "What Is a Design?" in *ADADIA '89 Proceedings: New Ideas and Directions for the 1990s,* C. Yessios (ed.), University of Florida, Gainsville, Fla.

Stiny, G. (1990). "What Designers Do That Computers Should," in *The Electronic Design Studio*, M. McCullough, W. J. Mitchell, and P. Purcell (eds.), MIT Press, Cambridge, Mass.

Stiny, G., and W. J. Mitchell (1978a). "The Palladian Grammar," *Environment and Planning B*, **5:** 5–18.

Stiny, G., and W. J. Mitchell (1978b). "Counting Palladian Plans," *Environment and Planning B*, **5:** 189–198.

Wittgenstein, L. (1968). *Philosophical Investigations,* Basil Blackwell, Oxford.

PART ONE

THEORY

1

COMPUTATIONAL FRAMEWORK FOR SUPPORTING CREATIVE ARCHITECTURAL DESIGN

Gianfranco Carrara

Dipartimento di Techniche dell'Edilizia e del Controllo Ambientale
"La Sapienza" University
Rome, Italy

Yehuda E. Kalay

School of Architecture and Planning
State University of New York at Buffalo
Buffalo, New York

Gabriele Novembri

Cartesiana Consortium
Rome, Italy

Design can be considered a process leading to the definition of a physical form that achieves a certain predefined set of performance criteria. The process comprises three distinct operations: (1) definition of the desired set of performance criteria (design goals); (2) production of alternative design solutions; and (3) evaluation of the expected performances of alternative design solutions, and comparing them to the predefined criteria. Difficulties arise in performing each one of the three operations and in combining them into a purposeful process. First, it is difficult to define the desired performance criteria prior to and independent of the search for a design solution that achieves them, since many aspects of the desired criteria can be discovered only through the search for an acceptable solution. Furthermore, the search for such a solution may well alter the definition of these criteria, as incompatibilities between criteria are discovered, necessitating compromises and relaxation of constraints. Second, the generation of design solutions is a task demanding creativity, judgment, and experience, all three of which are difficult to define, teach, and otherwise capture in an explicit manner. Third, it is difficult to evaluate the expected performances of alternative design solutions and to compare them to the predefined criteria. Design parameters interact with each other in complex ways which cause effects and side effects. Predicting the expected performances of even primary effects relies on assumptions (physical, sociological, psychological, etc.) and hence is seldom a reliable measure. A fourth problem arises from the need to coordinate the three operations in an iterative process that will converge on an acceptable design solution in reasonable time.

Computational techniques were developed to assist each of the three operations comprising the design process, with varying degrees of success. It now appears that a comprehensive and successful computational design assistant will have to recognize the limitations of current computational techniques and incorporate a symbiosis between the machine and the human designer. This symbiosis comprises allocating design tasks between the designer and the computer in a manner that is most appropriate for the task at hand. In general, it will allow the designer to establish gross design criteria and adapt exiting solutions or create new solutions that meet them. The computer will elaborate the gross design criteria established by the designer, check them for consistency and for compliance with design codes, provide the designer with all pertinent information needed to carry out the design task, evaluate the solutions proposed by the designer, monitor the design process as a whole, and guide it to a successful conclusion.

In this paper we analyze the computational needs of architectural design processes and propose a framework for a symbiotic computational partnership which comprises four major components: (1) user interface and design process control, (2) design goals, (3) evaluators, and (4) database. The general content of each component, and the manner in which they interact, are discussed.

PRESENT STATE OF BUILDING DESIGN

In most developed countries, difficulties in the process of designing buildings have been increasing in recent years due to the growing complexity of buildings themselves and to the growing complexity of the processes leading to their construction and management. Underlying these difficulties is the increasing complexity of integrating and coordinating building-related functions with social, technological, economical, political, and other factors in each phase of the design process [Archea 1987]. Moreover, growing complexities are evident not only in the design of new buildings, but also in reconstruction, extension, and transformation of existing buildings (the relative importance of which increases constantly in comparison to new construction). Each of the building design, construction, and management phases requires the constant support and integration of different disciplinary skills. Yet the scientific, technological, and organizational content of these disciplinary skills evolves very rapidly, through specialization and through the development of new knowledge. The need for constant and extensive professional updating, and the growing difficulty of communication between experts in different fields (even those that are seemingly closely allied), further complicate the building design process.

As a consequence, various disciplinary institutions involved in the building process, as well as many professionals, have expressed the need for a clearly defined building design process and a readily accessible and up-to-date building knowledge base [Cross 1977, Jones 1980]. The structured process will help coordinate and harmonize the various design activities and allow for the orderly generation of comprehensive design solutions. The knowledge base will permit rapid access to comprehensive design knowledge, answer queries of the participants in each design phase, and facilitate comparative evaluation of alternative solutions and verification of their compliance with predefined design criteria.

Organizing the processes of design, production, management, and adaptive reuse of building is not contradictory to design innovation, as proven in other design disciplines (e.g., in the aerospace, automotive, and electronic industries) [Akin 1978, Jones 1985, Kalay 1985]. Rather, it provides methods, techniques, and instruments that assure:

1. Suitability of the response to the stated need (e.g., new construction versus retrofit)
2. Effectiveness of the design process (i.e., its convergence on an acceptable solution)
3. Quality of the proposed solution (i.e., meeting predefined criteria and achieving design objectives)
4. Up-to-datedness of technological and organizational solutions
5. Recording of previously devised solutions, and the performance levels they have achieved, to serve as a basis for future developments (i.e., the continuing development of a disciplinary knowledge base)

The conceptual tasks facing any design professionals are the definition of design objectives, the creative generation of solutions, and their judicious evaluation to assure their satisfactory response to the defined objectives [Schon 1983]. In recent years, an attempt was made to assist designers in performing these tasks through the use of computational methods. However, no means have yet been established to help the designers effectively in their creative efforts [Kalay 1987, Carrara and Novembri 1986, Mitchell 1977]. It is the objective of this paper to outline a computational framework that will assist designers in pursuing their creative endeavors.

ARCHITECTURAL DESIGN AND COMPUTATIONAL ASSISTANTS

The use of numerous advanced graphic programs, and programs that support numerical calculation for structures, energy, and other quantitative aspects of a building, have been growing steadily in the last three decades. These tools, designated by the acronym CAD (computer-aided design), have helped in the development of architectural design solutions, particularly in cases where standardized building elements or standardized building practices could be employed, such as in industrial construction [ARU 1971, ARC 1975]. However, valuable as these tools may be for substituting and augmenting traditionally manual operations, they actually provide very little support to the creative and evaluative tasks comprising the bulk of the design process, where choices are made and conflicts are resolved [Tzamir and Churchman 1984].

Stated more specifically, all the techniques and computerized tools developed to date for supporting the processes of design, construction, and management of buildings do not support the five needs outlined earlier. They suffer from limitations that are inherent in the concept of "program" itself, which involves the carrying out in sequence of a predetermined number of instructions in order to obtain a result that is always and in all cases defined by the formulation of the problem corresponding to the data provided. Although this programmatic model works well for cases where the problem can be perfectly formulated, it fails where architecture is concerned. Architectural problems are neither well understood, nor can they be strictly formulated a priori. Rather, they tend to be formulated in reaction and in response to parameters that emerge gradually during the design process. These parameters often vary from case to case, and even within the same case at different times [Akin 1978, Rittel and Webber 1984, Schon 1983]. As a result, the use of current data processing tools in the building design and construction fields, although widespread, is extremely fragmented, often with incompatible or even conflicting aims.

Nevertheless, it would be wrong and counterproductive to place all the blame for this situation on the data-processing tools that are currently available to the participants in the building process. In fact, from a purely technical point of view, programs for calculating and automating design activities exhibit remarkable levels of compactness, efficiency, sophistication, and power, unthinkable only a few years ago. The causes for the present situation lie deeper than the current state of data processing. They are due in part to the context where the first experiments in using computers for architectural design took place, and partially due to commercial motivations.

Computer-aided design was developed through experiments and studies, conducted in the early 1960s, aimed at defining a method of "systematic design" [Alexander et al. 1977, Cross 1977]. The search for a systematic method of design sought to identify a logical deductive process, through which its proponents intended to achieve a synthesis between the designer's creative and rational faculties. This method of design, comprising a series of logical and mechanical operations, would free designers from the most tedious aspects of their work, to the great advantage of their creative powers. Furthermore, a rational and deductive approach to the process of design would make it possible to identify a certain number of operations that could be automated. Consequently, proponents of the new design method tried to model it in a way that would also take full advantage of the potential offered by computers [Buffa et al. 1969, Whitehead and Eldars 1964]. The conviction that it was possible to isolate certain operations and identify specific phases of the design process, and subsequently translate them, once generalized, into programs, has led architectural CAD researchers to adopt the classical approaches of industrialized automation, exhibited in systems theory and operations research [Hashimshoni et al. 1978, Shaviv and Gali 1979]. Unfortunately, these engineering methods did not suite the architectural design process, failing to support intuition, creativity, discovery, and judgment.

This failed quest for a systematic design method has, nevertheless, produced a series of computerized operations that comprise the present generation of CAD software. Unfortunately, they were seized upon and amplified by software production companies that offer CAD systems of the most general applicability, with the aim of covering the broadest market with the least amount of products. Consequently, the architectural profession, which is less adaptable than other professions to CAD instrumentation, is being led by software producers rather than leading the manufacturers to produce software that is specific and appropriate for architects' true needs. Architects are therefore forced to use the more precise, systematic, and more easily mechanizable model of problem solving that follows the logic and methodology offered by available software, rather than a model more suitable for the nondeductive, often irrational, and not easily codifiable architectural design process.

This situation can be overcome only by reconsidering the logic and methodology of design, and by discovering the activities within the process that could and should be supported by computers. To achieve this goal, we must come back to the studies and experiments that have attempted, since the early 1960s, to analyze the design process, in order to discover not only a better method of design, but also all the aspects that characterize design itself, and the roles that computational tools can and should play in it [Carrara et al. 1989, Novembri 1989, Kalay 1985].

THE ARCHITECTURAL DESIGN PROCESS

Other than the fruitless attempts to define exactly what should be understood by *design*, the experiments and research into the nature of design (by architects, design theorists, and others) have also revealed some of its gross characteristics that can guide the development of CAD systems more suitable for architectural design [Mitchell 1975]. Specifically, we consider architectural design *an extremely complex set of operations aimed at the definition of a built object that achieves a predefined set of required functional performances.* The design process can be said to comprise three main operations:

1. Defining a set of required functional characteristics that comprise the objectives to be achieved by the designed artifact
2. Making design decisions that in the opinion of the designer are (or should) be capable of achieving the predetermined objectives
3. Verifying that these choices are internally consistent and that they achieve the required functional characteristics

These three operations are thoroughly intertwined: it is extremely difficult to define precisely the proper objectives of the design process, without actually engaging in the search for solutions that achieve them. Since each architectural design problem is unique, at least in terms of its physical and temporal contexts, many design parameters are simply unknown, or the particular values they must achieve cannot be determined. The search for a solution produces insights into the particular design problems that were not apparent prior to the search, in terms of compromises that must be made, opportunities to be had, and trade-offs that will improve the overall quality of the solution. These insights not only produce new design parameters and define new criteria to be achieved, but may also radically alter previously stated objectives [Rittel and Webber 1984].

The generation of design solutions cannot be separated from their evaluation and from the process of predicting their expected performances, since these evaluative processes guide the solution generation process toward achieving the stated objectives [Simon 1969]. Furthermore, the unpredictability of the consequences of any design decision, due to its multifaceted effects and side effects, precludes preplanning of the design process except at its most abstract level. Rather, designers must assess the accomplishments and drawbacks of each design decision before they can take the next decision. This dynamic decision-making process compares the current state of the emerging design solution with the stated objectives and draws operational conclusions from the comparison. These conclusions may affect the development of the design solution itself or the statement of the objectives to be achieved. Either way, the evaluation process requires that the designer check the most recently made decision for consistency with decisions made previously and predict the future performances of the design alternative from many aspects, only some of which can be stated explicitly. Hence a threefold difficulty in evaluating a design decision:

1. It must be evaluated for consistency with previous design decisions.
2. Its expected implications (effects and side effects) must be predicted and compared to stated design objectives.
3. It requires evaluation of the objectives themselves.

Both generative and evaluative operations put great mental strains on the designer. The creative generation of design solutions demands the creative talents, capacity for judgment, and experience of the designer. The evaluation of design alternatives requires prediction of the expected performances of the solutions, and the ability to draw operational conclusions from these evaluations to make subsequent design decisions that will bring the solution closer to achieving the desired objectives.

DESIGN PROCESS PARADIGMS

In the scientific literature, two fundamentally different approaches to design have been investigated that can be adopted for the purposes of developing computational systems to support design [Aksoylu 1982]. The first approach is often called *prototype refinement* [Oxman and Gero 1988]. It is based on the adaptation and placement in context of a previously developed design solution that has been proven successful according to some criteria. This previously developed solution thus attains the status of a prototype, from which similar design solutions may be derived. In the absence of universal design rules, this approach provides an empirically validated corpus of "successful" design solutions, whose adaptation to particular spatiotemporal contexts eliminates much of the uncertainty concerning performance prediction and evaluation, and provides a holistic framework, or direction, for searching for a pertinent solution. In architecture, this approach has been most clearly manifested in the Beaux Arts tradition and in the work of Christopher Alexander [Norman 1987].

The second approach, called *prototype generation,* is based on the derivation of new design solutions from scratch by attempting to find the most appropriate answer to predefined objectives and constraints [Gross et al. 1987]. This approach follows the well-known paradigm of problem solving, which embodies the means–ends analysis [Newell and Simon 1972]. In effect, it is the paradigm underlying the functional traditions in architecture, as expressed by the Bauhaus [Mitchell 1987]. This approach requires the designer to begin with a general definition of the objectives he or she wishes to achieve, and to find design solutions that meet them. The formulation of these objectives, however, is itself a design task, requiring generation and testing of compatible, precise, and analytically verifiable goals. Such definition is possible only in strictly theoretical cases. It is difficult or even impossible to achieve in practice. Therefore, designers often rely upon prototypical sets of objectives adapted from building codes, common practices, experience, and the ubiquitous "average occupant" stereotype.

It is often difficult to find an appropriate prototype for adaptation from which to begin one's own design exploration. Furthermore, the consequences of selecting the wrong prototype may be disastrous. Once a prototype has been selected, its adaptation to specific spatiotemporal conditions may alter it in a manner that will eliminate its designation as a prototype and qualify it as a new design solution. The two methods described above therefore actually coexist in practice and are both applied alternately by designers in the course of their work. If the designer starts from an existing prototype, situations may arise in which he or she must develop alternative solutions. At the same time, once a new design solution has been defined, it necessarily requires perfection and completion in all its elements to arrive at the final design. The two approaches are thus complementary, practically indivisible, and often interchangeable.

Given these two different design paradigms, different methods have been proposed to carry out the design search process. The proponents of the *problem-solving* process, where alternative solutions are generated and tested until a solution that meets the stated objectives is found [Simon 1969, Woodbury 1987], hold that the search for a "satisficing" solution is goal directed, and can therefore employ means–ends analysis to guide it toward the desired solution. Thus the skills that are employed when following the problem-solving paradigm are mainly analytical: the ability to compare the current state of the designed artifact to its desired state, and the ability to draw operational conclusions from this comparison so that the differences can be reduced. Such *goal-driven* approaches have been computationally represented as *backward-reasoning* search strategies: operators are applied to the goal statement to convert it to a set of subgoals that are easier to solve [Swerdloff and Kalay 1987]. This method can be applied recursively, until a set of subgoals that can be solved without further reduction is found.*

Proponents of the prototype refinement paradigm, on the other hand, hold that the design search process may be compared to *puzzle making*: the search for the most appropriate effects that can be attained in unique spatiotemporal contexts through the manipulation of a given set of components, following a given set of combinatoric rules [Archea 1987]. Thus, rather than rely on a goal-driven strategy, they propose reliance on precedents, symbols, and metaphors for guidance [Mitchell 1987]. The main skills employed when following this paradigm are synthetical: the ability to compose given parts into a new, unique, whole. This *data-driven* approach has been represented computationally as *forward-reasoning* search strategies: operators are applied to the current state with the aim of transforming it according to predefined rules.†

*For example, the design of a house can be divided into a set of design subgoals that include programming, schematic design, and design development. The programming subgoal can be further subdivided into the subgoals of selecting spaces, fixing the budget, and setting the desired quality of the house, while the schematic design subgoal can be further subdivided into the subgoals of designing the plan and the massing, and so on.

†Shape grammars are well known systems of this form. A shape grammar consists of an initial shape and a set of rules that are iteratively applied to transform successive shapes into a final form, thus generating a design. Shape grammars have been developed for a variety of objects, including Frank Lloyd Wright prairie houses [Koning and Eizenberg 1981], Palladian villas [Stiny and Mitchell 1978], Pittsburgh Shadyside district houses [Flemming 1987], and more.

COMPUTER SYSTEMS TO ASSIST DESIGN

Both problem-solving and puzzle-making design methods have been implemented as computational design systems [Flemming 1987, Shaviv 1987, Stiny 1980]. Each system demonstrates certain achievements and many drawbacks. Built upon radically different paradigms, no system has yet been successful in incorporating both problem solving and puzzle making in one computational design system. Thus although they represent significant achievements in automating partial or specific design processes, the development of an integrated computational system to support the overall design process, with its alternating problem-solving and puzzle-making modes of search, is yet to come.

The difficulty in automating design as a whole is due, in part, to a lack of understanding of the *learning*, *creative*, and *judgmental* processes that comprise design. These processes are, and for the foreseeable future will continue to be, the prerogative of human beings. There are, however, many design processes that have already been computed successfully, such as a host of analyses, visual representations, and even certain solution-generating algorithms [Mitchell 1977]. The first step toward the future computation of design is to recognize the necessary distinction between processes and objects that can be represented as overt knowledge (in some formal environment), and hence can be computed, and those that cannot.

Recognition of this dichotomy of knowledge (in light of the current state of computation science), and at the same time the real need to assist designers computationally, has lead to computational approaches that *facilitate*, but do not fully automate the design process. These approaches are based on the observation that designers are able to cope with and manage complex design processes and have for centuries achieved outstanding results doing so without the aid of computers. Perhaps it is not necessary to fully automate the process of design to improve design productivity and quality significantly. Perhaps it may be more prudent to develop a practical symbiosis between the capabilities of designers and machines.

The development of computational design assistants is still a formidable task. It requires the identification of those tasks that are computable, and making decisions concerning which of the computable tasks should or should not be computed. Furthermore, the problem of combining computational and manual design practices into a single design process must be resolved. Attempts to achieve such symbiosis range from instilling in designers first principles of design technology, to assisting designers in the creative phases of design and to the appreciation of the emerging product through visually realistic images and detailed analytical methods. It has been postulated that computers can help design students learn and retain design knowledge as a result of simulation and experimentation with a large number of alternatives [Bryan 1987]. Others have proposed that computers can enhance intuition by generating alternative solutions that nurture the intuitive leap and by providing an extensive knowledge base at the point of conception, on which the judgmental decisions can be based [March and Steadman 1971]. That is, computers can be used to generate large numbers of obvious, logically predictable design solutions, which can act as catalysts to the designer in generating more novel solutions. Realistic visual representation of the designed artifact lets the designer infer semantic information directly from the image, as one would from a physical scale model or the real artifact [Cohen 1987].

The major problems in integrating computational and manual design practices have been, so far, rigidity and loss of design continuity: the designer had to decide early in the design process which design operations would be done with the aid of the computer and which would not. Once so decided, the designer was committed (due to cost and time constraints) to the predefined allocation of tasks and could not change it on the basis of new information gathered later. The continuity of the design process was hampered by the need to relinquish either manual or computational methods when the time had come to shift from one to the other.

Solutions to the rigidity and discontinuity problems have been suggested in the form of *design partnership* between the designer and the computer [Negroponte 1975, Swerdloff and Kalay 1987]. In such a partnership, the role of the computer can be shifted dynamically between passive representation–evaluation and active generation–evaluation of design solutions. Such dynamics would allow the designer and the system to respond to changing requirements, unforeseen problems, and emerging opportunities as they arise during the design process.

A FRAMEWORK TO SUPPORT CREATIVE ARCHITECTURAL DESIGN

To support creative architectural design, a software system that can assist in the intuitive and judgmental operations of the process must incorporate both prototype refinement and prototype generation modes of design. The system should make it possible to define and specify design objectives, to adapt or create design solutions, to evaluate the functional performances achieved by these solutions, and to compare them with the stated objectives.

Prototype refinement can be achieved by defining an appropriate database of prototypes and the particular design solutions derived from them. The prototypes will include all the information pertinent to generic building objects, such as kitchens, bathrooms, schools, and so on. The particular solutions derived from earlier adaptations of prototypes to particular contexts will include specific dimensions, relationships, orientations, and the performance measurements achieved by them, thus serving as a basis for assessing the quality of these adaptations.

To support prototype generation, the system must be capable of representing design objectives, or goals, and include a host of evaluators for comparing the emerging solutions with the objectives. Together with the database, the goals and the evaluators comprise the system's knowledge base.

Additionally, a user interface, which facilitates the dialogue between the system and the designer, must be provided. The user interface will include a mechanism to control the design process task allocation and the means–ends planning component that guides the process to a closure. The functional scheme of the proposed system is illustrated in Figure 1.1.

Goals

The design objectives are represented in the system by a hierarchical list of design criteria and their relative weighing indexes, which describe the desired performance of the solution. These criteria include such aspects as cost, functional requirements, and

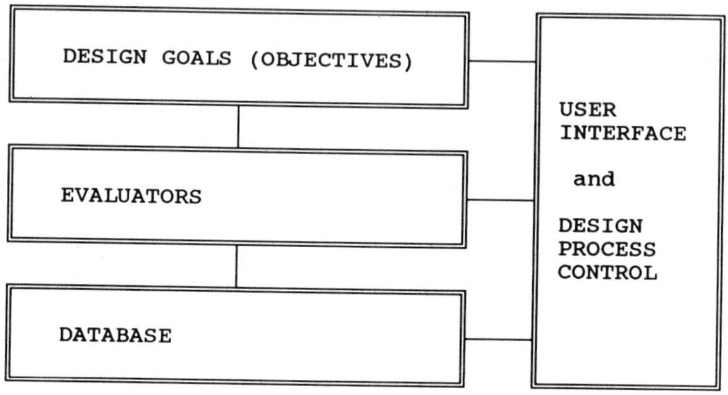

Figure 1.1. Framework for knowledge-based computer-aided design.

social implications. They are grouped in context-sensitive clusters called *goals*, where the level of accomplishing one criterion may be traded off against the levels of accomplishing others by changing the relative weighing indexes such that the overall performance of the solution is maximized [Swerdloff and Kalay 1987]. Each goal thus represents an acceptable combination of performances of the design solution. For example, the following two goals represent alternative combinations for achieving the PREDESIGN conditions of a bridge:

```
goal (g020,
      predesign,
           [(span,80), (depth,80), (aesthetics,50),
            (redundancy,85), (constructability,90),
            (maintenance,60), (cost,95)]
      ).
goal (g021,
      predesign,
           [(span,80), (depth,80), (aesthetics,50),
            (redundancy,85), (constructability,90),
            (maintenance,90), (cost,80)]
      ).
```

Two types of criteria must be supported [Maher and Fenves 1985]:

1. *Hard criteria*, which must be satisfied (of the pass/fail kind)
2. *Soft criteria*, which can be satisfied at different levels

Criteria of the first type are nonnegotiable. Only soft criteria may be traded against each other. For example, in the second alternative (g021), the threshold for satisfying the cost criterion is lower than in the first case (i.e., the initial cost is higher). This difference is counterbalanced by a requirement for better satisfaction of the maintenance criteria (i.e., life-cycle costs are higher). Hence alternative goals allow for alternative design solutions and account for performance trade-offs (this does not mean that the system itself is capable of determining the actual trade-off of performances, only that it facilitates and supports them).

The conditions under which various criteria are considered satisfied and the minimal acceptable levels of satisfaction for the soft criteria are established by defaults at the time the criteria are introduced, and are modifiable by the designer during the design process. The system allows the designer to define the goals either prior to initiating the design process, or dynamically as the process unfolds. Their dynamic creation may be prompted by new information and insight discovered through design exploration, or by the discovery of unresolvable conflicts. Some situations may arise in which the design solution requires modification of the goals already defined, the introduction of new goals, or the redefinition of priorities or trade-offs among the relative weighing indexes.

Alternative sequences of establishing individual goals form different design solution strategies. A particular sequence represents a *design plan.* Due to the exploratory nature of design processes, plans cannot always be developed prior to and independent of the process of accomplishing the goals that comprise them (especially since new goals may be added dynamically). A method for the dynamic construction of plans is therefore an important part of the proposed control structure. Still, certain generalized design strategies can be preplanned, reflecting the goal-oriented nature of design. Accordingly, two distinct classes of design plans can be identified: *preplanned* strategies and *dynamic* strategies. Preplanned strategies involve the mapping out of a sequence of goals before actually executing them. Preplanning is intended to optimize the problem-solving process by most appropriately sequencing actions, resolving high-level conflicts, and discovering strategic errors. Dynamic strategies typically use generate-and-test techniques to accomplish goals and to observe their effects before selecting subsequent goals for accomplishment. This method better supports design exploration and is a prerequisite for supporting discovery; however, it is less directed than the preplanned approach and is likely to require a wider search.

Goals are often defined at a high level of abstraction. It may not always be possible, therefore, to directly generate a solution that achieves them. In such cases, goals may be decomposed into sets of more specific subgoals, whose combined achievement will constitute achievement of a higher-level goal. Subgoals therefore represent a decomposition hierarchy of design objectives. The process of accomplishing goals through decomposition can be considered hierarchical planning, as depicted in Figure 1.2 for the design of a bridge. For example, a plan for achieving the PRE_DESIGN goal may include examination of alternative bridge types, estimation of their costs, and determination of the overall appearance of the bridge:

```
(bridge_type --> cost_est <-> appearance)
```

(The notation <-> implies that the COST_EST goal and the APPEARANCE goal may be achieved in any order, whereas --> implies strict ordering.)

A plan for achieving the MAIN_DESIGN goal can include detailed structural design and the production of drawings and specifications:

```
(structural_design --> drawings <-> specs)
```

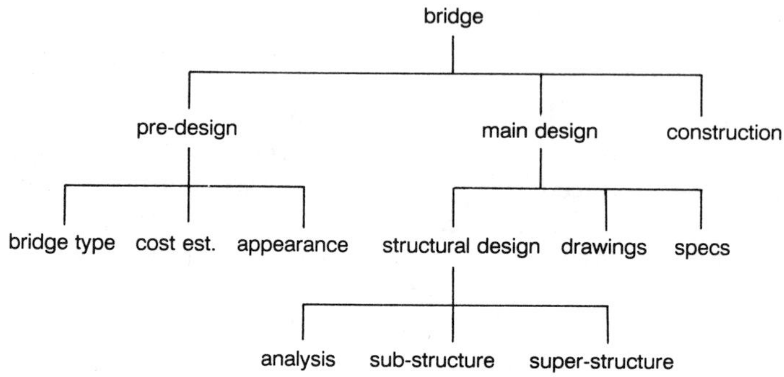

Figure 1.2. Goal decomposition hierarchy for a bridge.

The STRUCTURAL_DESIGN goal can be further divided into ANALYSIS, SUBSTRUCTURE, and SUPERSTRUCTURE design:

```
(analysis --> substructure_design <-> superstructure_design)
```

Database

The database is the software component of the system that stores models of both prototypical and specific building objects, including the particular building currently being designed. The database consists of three major components (depicted in Figure 1.3):

1. Generic building objects, called *prototypes*, ranging from doors and walls to rooms and entire buildings
2. Particular design solutions, called *instances*, where the prototypical building objects have been dimensioned and placed in specific contexts
3. Aggregation matrices, called *relations*, which provide access to groups of building objects that share some common attribute (e.g., all furniture on the first floor)

The building objects represented by the database include both descriptive and functional characteristics (Figure 1.4). The descriptive characteristics represent the physical attributes of objects, and include form, material, color, texture, thermal properties, and so on. The functional characteristics represent nonphysical attributes, such as the behavior and performance of building objects under various conditions. For example, a CHAIR could be represented as an object with certain shape and material attributes, and with functional attributes such as stability, durability, and support.

The descriptive and functional information associated with an object is represented by the FRAME formalism. Descriptive attributes are represented as SLOT–FACET couplets, whereas functional attributes are represented as rules and procedures. In the frame formalism, each SLOT represents a particular feature of the prototype (e.g., color), while the FACET represents the range of values that can be assigned to that attribute (e.g., red, blue, or green). SLOTS and FACETS make it possible to represent particular information, as well as default values or maximum–minimum values.

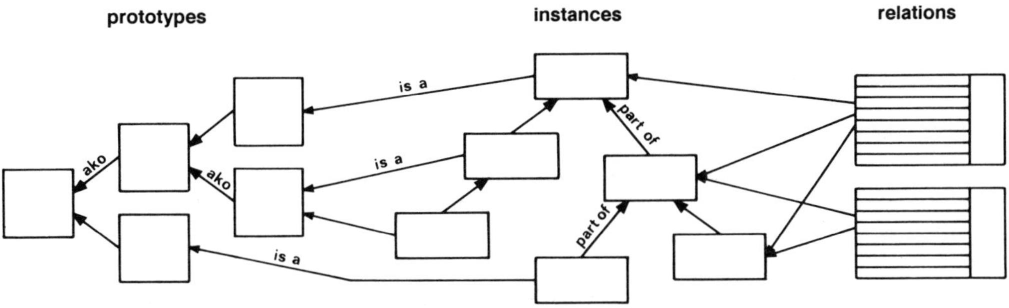

Figure 1.3. Database schema.

Object CHAIR

Descriptive charac.		Functional charac.
construction	steel, wood	stability
color		durability
number of legs		support
type	office, rocker	stackability
angle of back		
arms	yes, no	

Figure 1.4. Object frame.

Rules represent condition–action couplets, using IF–THEN premise–action tuples. For example, they could represent the fact

```
IF the chair has two legs
THEN it is not stable.
```

Procedures represent elaborate actions, to be taken under certain conditions. For example:

```
IF the size of a window has been reduced
THEN check level of lighting and if necessary
     add more artificial lighting.
```

The functional description provides the system with knowledge related to certain characteristics of the objects themselves and establishes the relationships between objects. This type of information is very important, making it possible to assess the appropriateness of choices made by the designer.

To reduce the amount of space needed to store attributes of many objects, a *generalization hierarchy* is used. Generalization provides a means to associate individual objects with classes of objects of the same kind, such that shared properties need not be stated separately for each object. For example, KITCHEN and BEDROOM are particular entities of a class of objects called ROOM, which is known to be an indoor space enclosed by walls. It is not necessary, therefore, to state separately that KITCHEN is an indoor space enclosed by walls and that BEDROOM is an indoor space enclosed by walls. Rather, these properties, which characterize all rooms, can be associated with the generalization class ROOM and known to be true for KITCHEN and for BEDROOM by way of *inheritance* (Figure 1.5). On the other hand, COOKING and SLEEPING, which are particular properties of KITCHEN and of BEDROOM, respectively, are not class properties. Therefore, they must be associated with each object separately.

Inheritance is a logical deductive mechanism, which by traversing the generalization hierarchy assembles the information that can be derived from each class in the hierarchy. A prototype inherits all the characteristics of the more general class and may add to it specialized information not shared by other entities in that class. This information, in turn, will be inherited by objects that are even more specific than the one that added the information. For example, a BEDROOM is a kind of room; therefore, it inherits the properties of ROOM (i.e., an indoors space surrounded by walls). It is, however, a

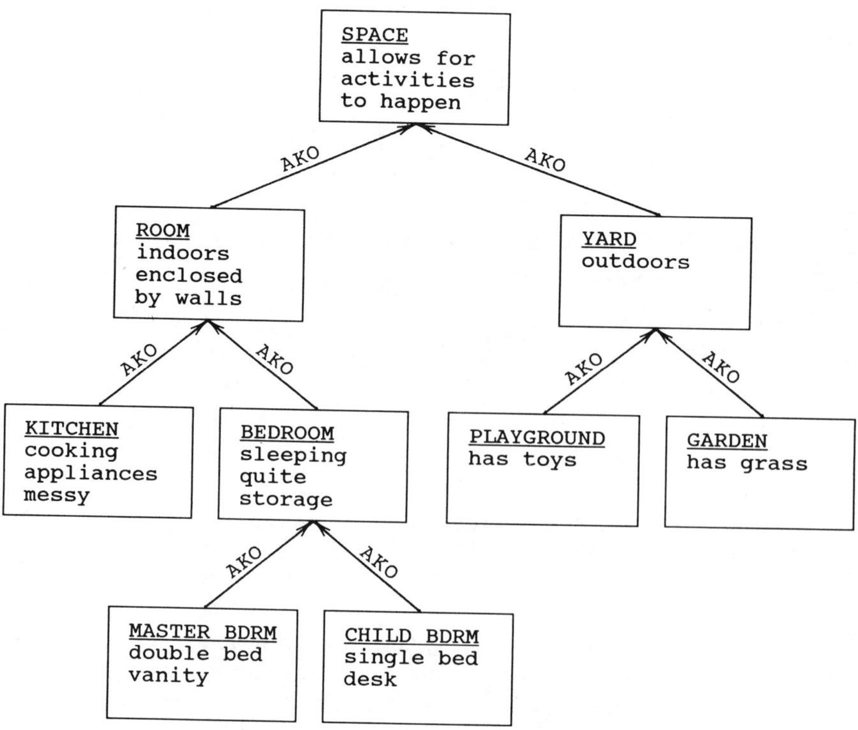

Figure 1.5. Generalization hierarchy and inheritance.

general class as far as MASTER BEDROOM and CHILD BEDROOM are concerned, which inherit in addition to the ROOM attributes, the attributes of BEDROOM (i.e., a quiet place used for sleeping, and the need for storage). The generalization relationship is denoted by AKO ("a kind of") links that connect a more specialized entity with a more general one. Inheritance helps maintain database integrity by eliminating redundant information that may not be updated when database changes occur.

Instances are particular examples of objects defined by a prototype. An instance inherits all its characteristics from the prototype, through so-called IS_A links. In the process of creating instances, known as *instanciation*, actual values are substituted for the default values established for the prototype, thus creating a concrete copy of the prototype. This copy is placed at a specific location in space, established through a transformation matrix. The unique location distinguishes the instance from other instances of the same prototype.

Instances are connected to each other through PART–WHOLE relationships, establishing *assembly* hierarchies between parts of a building. A window, for example, is considered part of a wall, which in turn is considered part of a room. Assembly hierarchies allow propagation of changes from parent objects to their dependent children. For example, when the wall containing the window is relocated, so is the window. This automatic propagation of changes helps maintain the semantic integrity of the database by eliminating inconsistencies due to oversight.

Aggregation matrices establish relationships between various instances, of types that do not fall under the part–whole or the master–instance categories. These relationships bind objects that share a common property, or objects that must be considered together when database changes occur. Furthermore, they provide for bidirectional change propagation, a feature that the other types of relationships lack. For example, an aggregation relationship can express the *perpendicularity* constraint between two walls in a symmetrical way, so when either one of the two walls is rotated, the other will be adjusted accordingly. Aggregation relationships also differ from the other two relationship types in that they require explicit definition of the nature of the link. This explicit definition makes aggregation more general than part–whole and master–instance relations, where the nature of the relation is implied by the type itself. By adding a conditional component to the definition (in the form of a set of rules), the aggregation relation can exercise the link selectively, depending on the nature of change and the nature of the objects affected. For example, the perpendicularity constraint may be waived for external walls but enforced for internal partitions.

Every type of entity in the database is accompanied by operators that create, delete, and modify the entities and that associate them in the various relationships. Additionally, the database includes certain knowledge that maintains the consistency of the information it contains and propagates changes due to variations introduced or caused by outside operators (e.g., by the designer).

Evaluators

Evaluation is an integral part of the design search process. It determines how well proposed design solutions meet the goals and abide by the constraints. Evaluation is performed by specific programs, each determining the level of expected performance of a certain building object characteristic. These programs are typically written by experts in the particular field of their expertise. The information provided by the evaluators can be used by the design process controller to determine what action must be taken next, or by the designer personally.

Typical architectural design problems involve many different goals and constraints. Therefore, specific evaluators that specialize in one narrow domain each, must be employed, such as energy analysis, cost estimation, structural analysis, and so on. However, because each evaluator is a specialist in one area of expertise, the results of all the evaluations must be combined and traded-off against each other to determine the overall quality of the design alternative. Trading-off dissimilar qualities is a task that involves personal preferences and is project specific. Project independence of the evaluators can be maintained by assigning the task of trading off the merits and drawbacks they report to the goals the design process is to achieve, which are project specific by nature. Furthermore, since many of the trade-offs are highly personal, designers must exercise their own preferences and judgment.

The hierarchical nature of the design process requires that evaluation be performed at many different levels of abstraction, with different levels of information. This can be achieved by augmenting the information available in the database with defaults, or by designing the evaluators to work with less information [Antony 1987].

User Interface and Design Process Control

The user interface is responsible for communicating between the user and the system and for controlling its operation. It allows the user to monitor the progression of the design process and to affect it such that it reflects the user's preferences. The user interface is also the means through which the system's knowledge base is developed and augmented, modified, and updated, either explicitly, or implicitly by monitoring the transactions initiated by the designer. The ability to acquire new knowledge and to update existing knowledge is a necessary feature to prevent early obsolescence of the system and to customize it to the designer's particular preferences.

Computer-aided design research has advanced the frontiers of user interfaces much more than has any other application, by introducing computer graphics as the standard means of communication between the user and the machine. However, the nature of all user interfaces is still governed by technical considerations influenced by the constraints of the tool rather than by the needs of the user. All user interfaces are based either on the *operator–operant–data* paradigm or on the *operant–operator–data* paradigm. The first forces the user to select an operator prior to selecting the operant to which the operation will be applied. For example, the user must select a word processor, spreadsheet, or database program before he or she can select the document file, projection, or data file on which he or she wishes to operate. The second paradigm forces the user to select an operant before the operation can be selected. For example, the user must first select a polygon, solid, or dimension line, and later select an operation, such as translation, copy, or deletion of that operant. In both cases the data are supplied last (e.g., delete a particular word, move the polygon by 5 feet, etc.). Both user interface paradigms are based on sound technical reasons, and both have merits and drawbacks. The most significant drawback of both is *rigidity*: they enforce one way, and only one way, of performing each task.

The system proposed will allow the user to select operandi, operators, and data in any sequence, thereby increasing the bandwidth of human–machine communication. For example, the user will be free to select a polygon first, then indicate that he or she wishes to move it, then provide the displacement data. Alternatively, the user can select the MOVE operator, then indicate the displacement, and finally, select the object (or objects) to be moved. The latter example allows the user to apply the same translation to more than one operant and to operandi of different types (e.g., to polygons, solids, and to dimension lines), without changing "modes," by applying the same translation to different types of objects (e.g., from 2D to 3D mode, or to Dimensioning mode). It is not clear that eliminating the rigidity of the user interface will benefit the design process. However, only by experimenting with different modes of interaction will we be able to endorse the most productive means of human–machine communication.

The interaction with the user may be graphical or alphanumeric. *Graphical* interaction can be achieved by conventional methods of user interface management systems. *Alphanumeric* interaction may be based on a language for which a suitable interpreter must be provided.

Controlling the design process is a means to cope with and manage the complexity of searching a huge space filled with millions of alternative combinations of possible solutions to the given problem. This space is much too large for systematic search and much too complex to depend on the successful application of educated guessing (although this is a commonly used practice). Thus the critical stage in the design process is not the mutual adjustment of solution subcomponents until they fit each other, but rather, pruning the search space in some quick and efficient manner that will not exclude reasonable potential solutions. In traditional design practices, this pruning leads to rapid convergence upon a single solution that will eventually be elaborated in every detail. It is often called the "creative leap" by which the brain of a sufficiently informed person can select a set of promising solution components in the first place.

Controlling the process of design is also a process of allocating the responsibility for handling each task between the design participants (whether human or machine). The proposed knowledge-based design environment will facilitate control of the design search process, where design goals are selected for accomplishment, and where the tasks of accomplishing them are allocated between the designer and the system.

The control subsystem will invoke the necessary evaluators, which will determine if the design decisions abide by the constraints and achieve the goals. It will comprise a series of conditional statements of the IF–THEN kind, which can be invoked by the system's inference engine. It will analyze the properties of partial design solutions and by invoking the rules, will devise appropriate actions to improve them. Inference will help the designer recognize a solution state once it has been found. Furthermore, by comparing what the designer proposes with the model stored in the system's knowledge base, inference should make it possible to point out errors, missed opportunities, and directions for further elaboration.

CONCLUSION

The knowledge-based system described above has been conceived to cooperate interactively with the architectural designer in defining design objectives and solutions, as well as evaluating their mutual consistency. But what is the meaning of "cooperation" or "symbiosis" in this context? Given that creative design is an extremely complex activity, deeply rooted in human subjectivity, and that it involves the designer's culture, intelligence, and experience, is it actually useful that a computational system will carry out this task on our behalf? Probably not.

An expert system that could deal with a complex design problem would not be any different from another designer, albeit mechanized, but nevertheless very similar to the actual designer. Hence, instead of endeavoring to develop a system capable of autonomous creative design, it appears more prudent to define a context in which creative design could be made easier and its results better. The system described above is intended to be such a context.

By means of defining design constraints, the system will allow designers to explicate and make operational their goals and evaluate the impact of their solutions. Since the knowledge base can hold previous design solutions, designers will be able to utilize them at any stage of the design process, modifying and refining them as needed. Alternatively, they will be able to establish new solutions. But most important, they will be able to shift between these two modes of design continuously and interactively, as necessitated by the nature of the evolving design process itself and its products.

The verification of design choices, with respect to the established set of constraints, will help the designer judge results, and possibly suggest new and better solutions. A system that will do all this without designing anything directly will play an important role in the design process, assisting the designer in developing design solutions, predicting their expected performances, and comparing them to predefined design objectives.

REFERENCES

Akin, O. (1978). "How Do Architects Design?" in *Artificial Intelligence and Pattern Recognition in Computer-Aided Design* J. C. Latombe (ed.), IFIP, North-Holland, New York.

Aksoylu, Y. (1982). *Two Different Systematic Approaches to Design*, Technical Report, University of California, Berkeley.

Alexander, C., S. Ishikawa, and M. Silverstein, with M. Jacobsen, I. Fiksdahl-King, and S. Angel (1977). *A Pattern Language*, Oxford University Press, New York.

Antony, S. (1987). "ARCADE: Architectural Computer Aided Design Evaluator," M.Arch. thesis, School of Architecture and Planning, State University of New York at Buffalo, Buffalo, N.Y..

ARC (1975). *OXSYS: Computer-Aided Design, Typical Automatic Documentation*, Applied Research of Cambridge, Cambridge.

Archea, J. (1987). "Puzzle Making: What Architects Do When No One Is Looking," in *Computability of Design*, Y. E. Kalay (ed.), Wiley-Interscience, New York.

ARU (1971). *Application of Computer Graphics to Architectural Practice*, Project A25/SSHA-DOE, Architectural Research Unit, Edinburgh University, Edinburgh, England.

Bryan, H. (1987). "A Study in the Use of Qualitative Methods in Design," *AIA/ACSA Council on Research Annual Conference*, Boston, November.

Buffa, E. S., G. C. Armour, and T. E. Vollman (1969). "Allocating Facilities with CRAFT," *Harvard Business Review*, **42**, November.

Carrara, G., and G. Novembri (1986). "Constraint Bounded Design Search," in *Computer-Aided Architectural Design Futures*, A. Pipes (ed.), Butterworths, London.

Carrara, G., G. Novembri, and A. Fioravanti (1989). "Towards a New Generation of Computer

Assistants for Architectural Design: An Existing Scenario," in *Proceedings of the ECAADE Conference*, Aarhus, Denmark, September.

Cohen, M. F. (1987). "Radiosity Based Lighting Design," in *Computability of Design*, Y. E. Kalay (ed.), Wiley-Interscience, New York.

Cross, N. (1977). *The Automated Architect*, Pion, London.

Flemming, U. (1987). "The Role of Shape Grammars in the Analysis and Creation of Designs," in *Computability of Design*, Y. E. Kalay, (ed.), Wiley-Interscience, New York.

Gross, M., S. Ervin, J. Anderson, and A. Fleisher (1987). "Designing with Constraints," in *Computability of Design*, Y. E. Kalay (ed.), Wiley-Interscience, New York.

Hashimshoni, R., E. Shaviv, and A. Wachman (1978). "A Decomposition Model As a Tool for Evaluation of Multicell Solutions: A Case Study," in *Proceedings of CAD '78, 3rd International Conference on Computers in Engineering and Building Design*, IPC Science and Technology Press, Guildford, England.

Jones, J. C. (1980). *Design Methods*, Wiley, London.

Jones, G. (1985). "Can Creativity Be Automated?" *Computer Graphics World*, April, pp. 73–77.

Kalay, Y. E. (1985). "Redefining the Role of Computers in Architecture: From Drafting/Modeling Tools to Knowledge-Based Assistants," *Computer-Aided Design*, **17**(7): 319–328.

Kalay, Y. E. (ed.) (1987). *Computability of Design*, Wiley-Interscience, New York.

Koning, H., and J. Eizenberg (1981). "The Language of the Prairie: Frank Lloyd Wright's Prairie Houses," *Environment and Planning B*, **8:** 295–323.

Maher, M. L., and S. J. Fenves (1985). "HI-RISE: An Expert System for the Preliminary Structural Design of High-rise Buildings," in *Knowledge Engineering in Computer-Aided Design*, J. S. Gero (ed.), North-Holland, Amsterdam, pp. 125–135.

March, L., and P. Steadman (1971). *The Geometry of Environment*, MIT Press, Cambridge, Mass.

Mitchell, W. J. (1975)."The Theoretical Foundation of Computer-Aided Architectural Design," *Environment and Planning B*, **2:** 127–150.

Mitchell, W. J. (1977). *Computer-Aided Architectural Design*, Petrocelli-Charter, Princeton, N.J.

Mitchell, W. J. (1987). "Reasoning About Form and Function," in *Computability of Design*, Y. E. Kalay (ed.), Wiley-Interscience, New York.

Negroponte, N. (1975). *Soft Architecture Machines*, MIT Press, Cambridge, Mass.

Newell, A., and H. A. Simon (1972). *Human Problem Solving*, Prentice Hall, Englewood Cliffs, N.J.

Norman, R. B., "Intuitive Design and Computation," in *Computability of Design*, Y. E. Kalay (ed.), Wiley-Interscience, New York.

Novembri, G. (1989). *Progettazióne Edilizia e Tecniche délla Knowledge Engineering*, Technical Report of the CAD-LAB, Dipartiménto di Tècniche dell'Edilìzia e del Contròllo Ambièntale, "La Sapienza" University, Rome.

Oxman, R., and J. S. Gero (1988). "Designing by Prototype Refinement in Architecture," in *Artificial Intelligence in Engineering Design*, J. S. Gero (ed.), Springer-Verlag, New York.

Rittel, H. W., and M. M. Webber (1984). "Planning Problems Are Wicked Problems," in *Developments in Design Methodology*, N. Cross (ed.), Wiley, New York.

Schön, D. A. (1983). *The Reflective Practitioner: How Professionals Think in Action*, Basic Books, New York.

Shaviv, E., and D. Gali (1979). "A Model of Space Allocation in Complex Buildings," *Build International*, June.

Shaviv, E. (1987). "Generative and Evaluative CAAD Tools for Spatial Allocation Problems," in *Computability of Design*, Y. E. Kalay (ed.), Wiley-Interscience, New York.

Simon, H. A. (1969). *The Sciences of the Artificial*, MIT Press, Cambridge, Mass.

Stiny, G., and W. J. Mitchell (1978). "The Palladian Grammar," *Environment and Planning B*, **5:** 5–18.

Stiny, G. (1980). "Introduction to Shape and Shape Grammars," *Environment and Planning B*, **7:** 343–351.

Swerdloff, L. C., and Y. E. Kalay (1987). "A Partnership Approach to Computer-Aided Design," in *Computability of Design*, Y. E. Kalay (ed.), Wiley-Interscience, New York.

Tzamir, Y., and A. Churchman (1984). "Knowledge, Ethics, and Environment: Behavior Studies in Architectural Education," *Environment and Behavior* **16**(1): 111–126.

Whitehead, B., and M. Z. Eldars (1964). "An Approach to the Optimum Layout of Single Storey Buildings," *Architects Journal* **17:** 1373–1380, June.

Woodbury, R. F. (1987). "Strategies for Interactive Design Systems," in *Computability of Design*, Y. E. Kalay (ed.), Wiley-Interscience, New York.

2

HOW CAN WE ACHIEVE A GOOD BUILDING?

Peter R. A. Jockusch

Department of Architecture
University of Kassel
Kassel, Germany

This paper is concerned with the reasons and purposes for which we evaluate and predict building performance. The discussion is based on the author's experience, gained through the preparation and evaluation of more than 50 major architectural competitions. An attempt is made to discover for whom and in what respect a building can be considered a "good building," by asking the following questions:

- What can prediction and evaluation of building performance achieve? How well can we assess the performance and value of an existing building within its sociotechnical context?
- For what purposes and with what degree of confidence can the eventual performance of a designed and specified building be predicted?
- How do these evaluations compare to actual postoccupancy performance?
- To what extent do the roles and motivations of assessors, evaluators, and decision makers affect the value-stating process?

INTRODUCTION

Our activities focus on empirical and sometimes very pragmatic evaluation and prediction of design performance, based on many years of applied multidisciplinary POE (pre- and postoccupancy evaluation) practice, which form part of our briefing, assessment, and feasibility studies. Our experience lies predominantly in the predesign planning of such public buildings as offices, banks, public service facilities, hospitals, cultural institutions, and educational facilities, and covers the following areas:

- Evaluating the existing organizational and operational status of an institution, its objectives, and its institutional performance criteria
- Evaluating the residual values (usability, manageability, rentability, acceptability, etc.) of existing physical facilities as they are now and as they could be (if refitted, upgraded, modernized, rebeautified) in the context of eventual future uses
- Predicting the expected performances of alternative future uses of an institution, as they relate to both the organizational and the physical structure
- Comparing and evaluating, in the predesign feasibility phase, the anticipated performance of planning alternatives
- Organizing and managing town planning and architectural competitions, or finding adequate courses of action to assist a client in finding the best solution for problems
- Developing action strategies that can be implemented in institutions characterized by dynamic growth and change under situations of uncertain circumstances

COMPREHENSIVE UNDERSTANDING OF INSTITUTIONAL RESOURCES

An enterprise can function smoothly in a totally inadequate building. In contrast, a luxurious and comfortable building may sometimes house a badly managed enterprise that shows very poor performance. Therefore, we must admit that performance is by no means simply a reflection of the physical environment itself. Rather, it comprises a complex sociotechnical set of criteria. Built assets, people, information, objectives, methods, tools, materials, good reputation, and rules of conduct, behavior, and operation must all be present and available in a well-balanced mixture and congruence to create an effective overall institutional performance. Consequently, methods for successful and reliable performance prediction and assessment, which are to be used in the design stage of buildings, must be multidisciplinary, as answers to the questions posed above can be given only in a multidimensional way.

The methods of assessment that we use in our practice are, accordingly, mainstream state of the art, conforming to the professional rules of competition management. They extend to multidisciplinary efforts, including empirical building sciences (Gebäudekunde), operational and organizational analysis and diagnosis (methods established by the labor sciences and psychology), and many empirical social science methods, where the assessment covers opinion and behavior issues (Table 2.1).

In the following sections I address several issues that we have encountered in our evaluation practice. I integrate theory and practice, which are inseparable, proceeding through the whole sequence of operations that we normally cover in our work.

INSTITUTIONAL ANAMNESIS: DOCTOR/EVALUATOR AND PATIENT/CLIENT

An important moment in any evaluative action is the first touring interview, when the new external expert is shown the existing institution and is introduced to the people involved in the project. Investigators must be attentive and open to impressions and to responses, like a physician during the first anamnesis of a new patient. One may encoun-

TABLE 2.1 Methods for Evaluation and Prediction, and Their Relevant Sciences

Professional Domain of EPBP	Method Established By
Preparing and organizing architectural competitions	Professional rules of competition management
Evaluation of residual values (technical, functional, recurrent cost, etc.)	Empirical building sciences (Gebäudekunde)
Operational and organizational analysis and diagnosis	Organizational Sociology and Psychology, Labor Sciences Social Psychology
Opinion and behavior analysis	Empirical Social Sciences, Statistics
Health risk analysis	Social Medicine, Chemistry and Physics
Acceptance of working environment Industrial Design	Labor Science, Ergonomics, Environmental and
Expected performance of alternative futures of an institution	Scenario Techniques
Action strategies for implementation in institutions with dynamic growth and change, development under uncertain circumstances	Action Theory, Praxeology, Management Techniques, Social Psychology

ter in the patient/client a strange mixture of pride, attempts to maintain self-respect, or conversely, self-accusation for fear of discovering what should not become overt. Yet the most important tensions between the people involved, weaknesses of the institution, concepts of its future, and thus motives for calling the outsider in the first place are unveiled in the first interview. Even a deep institutional and facilities diagnosis cannot come closer than these first impressions to discovering the real problems of an institution. It can only reveal their relevant facts in a more detailed and reliable manner.

Understanding Critical Appraisals of Existing Facilities in Relation to the Evaluator's Aims and Set of Value Axioms

The values placed by the evaluator on different aspects of the institution being evaluated depend on the evaluator's own belief structure. An evaluator who is motivated to appreciate and/or discover the reusability potential of a building or an institution (e.g., ". . . it could be done!") will arrive at different assessments from those reached by one who is paid by people who want to buy cheap or to demolish the building, and who need an expert's assessment to support their aim. This phenomenon can be observed in the case of two asset evaluators, one commissioned by a buyer, the other by a seller of a property, each having an entirely different view of the building's residual value (see Jockusch [1986]).

Understanding "Value" Issues in Evaluation from a Historical Perspective

Many of our feasibility studies start with an assessment of existing facilities. We notice that the value and usability potential are determined not only by the technical or financial residual value, but also by factors such as historic significance and public emotions. Such emotions are particularly evident in preserving and reusing monuments. If there is high-ranking compensation (see Jockusch [1986]), the client may accept that some requirements will not be properly fulfilled. Yet the importance given to historic heritage issues may diminish as the costs of monument preservation become known.

What to Do if Decision Makers Have No Vision Regarding Their Own Future

Senior officers of a corporation sometimes cannot visualize the future of their own company. We found it necessary to invent scenarios that suggested alternative futures, and to present their operational consequences visually, to stimulate decision makers' imagination and decisiveness. Our imaginatively presented alternatives helped make the debate more objective and less emotional. In analyzing the reasons for lack of perspective and lack of courage by senior management, we found evidence of the sociopsychological dimensions of evaluation, and could attribute indecisiveness to the enormous strain that decision making under uncertainty may create.

Dissonance About the Project Objectives May Be Caused by Power Plays Among Decision Makers

In big organizations with many board members, planning consultants may encounter conflicts if individual board members specify separately their objectives and demands for future developments. An understanding of organizational psychology may help in interpreting the motives of such board members, such as conflicts arising between different generations in a family business.

Premature Planning Results That Draw Precise Conclusions: The Political Consultant's Dilemma

Significant mistakes may result from trying at a premature stage of the project to answer questions very precisely. This applies especially to cost predictions made on schematic design and feasibility study schemes, which may be brought into the political debate too early and have the awkward effect of becoming fixed points in the minds of political opponents.

Give Any Problem to an Architect and a Building Is Created as the Solution

It is a general misconception of the architect's duties that any performance failure of an institution can be cured by building. Most of our jobs started with great doubts about whether the problem really was one of the physical environment. Ever so often we provide help by changing something in the management or workflow of the institution, or by encouraging proper utilization of the potential already in place. Thus some users are satisfied just to learn how better to utilize their buildings for the benefit of the institution.

This procedure is similar to family therapy. The family may pretend that something is wrong with their physical environment, when in fact, faulting the environment hides the fact that severe interpersonal conflicts or managerial failures exist. We have even seen buildings built that were not needed. Although they were used as remedies for a sick institution, they usually failed to cure the malady; they harmed the institutions even more. The question is: How can we utilize performance evaluation in such a case?

COMPETITIONS: ARCHITECTS' PROMISES OF FUTURE PERFORMANCE

Important sources of our awareness and knowledge have been the 50 architectural competitions that we have programmed, organized, and assessed and for which we assisted the juries in finding the optimum solutions. In most of these cases we performed basic diagnoses of existing institutions and their facilities. We usually tested the programs in feasibility studies prior to setting up competition procedures (see Jockusch et al. [1984]).

Assessing Competition Entries Means Evaluative Prediction of Performance

Competition procedures are prime examples of predicting building performance. Entry drawings and descriptions are like the performance promises given by the competitors. In most cases we were able to judge how these promises compare to the real buildings after occupancy. It is necessary to investigate the reasons why there are vast differences between competition documents and jury-praised schemes on the one hand, and between these schemes and the final buildings on the other. I will use our experience in this domain as an empirical source for the arguments that follow.

Can Competition Entries Harmonize Program Conflicts?

Competition programs are sometimes the result of extremely difficult debates among conflicting parties. Conflicting requirements may not be harmonized before distributing the competition documents to the competitors, because promoters may hope that resolution of such conflicts will emerge through design. This hope, our experience tells us, is usually not justified. Having struggled through the entire process of implementation, the designer or builder often cannot overcome performance deficiencies that result from initial conflicting conditions. Furthermore, the designer or builder, who had no input in the debate that created these conflicting requirements, may be held responsible for not having been able to resolve them.

Lack of Evaluative Quality Due to Arrogance, Routine, and Time Pressure

Some juries act under extreme time pressure. In a day and a half they attempt to review more than 80 entries. The relatively short time (few minutes) necessarily devoted to reviewing each entry may be construed by nonwinners as the jury's lack of competence, and cast doubt on its reputation. Yet experienced senior jury members are typically able to identify good projects quickly, a practice that may appear arrogant to competitors who want a lengthy explanation as to why their entries were not selected.

The architectural quality sought by jurors is often reduced to external appearance alone. They may very arrogantly deny the importance of something like functional criteria. The promoters of the project, who normally are deeply concerned with good functional performance, have to accept the jury's decisions and the winner's recommendations for further project development. But when it comes to commissioning the winner for further work, the promoter may resist these recommendations and ask someone else to continue the project, resulting in poor visual quality but acceptable functional performance.

Furthermore, the sole purpose of a competition may not be architectural quality. In many cases, competitions are organized shortly before political elections to enhance the image of senior politicians. These, in turn, may conceal the fact that the budget available for realizing the project is actually much more limited than suggested during the competition itself.

Disharmony and Disagreement Among Jury Members

Professional architects who are members of the jury may dominate the lay members, who are limited in their ability to read plans and to visualize the quality and performance of a scheme correctly and realistically. Ever so often the lay jury members, who represent the political promoters of the project and who ultimately have to sign the contracts and to pay the bills for its construction, come to realize the practical implication of their decisions only after the jury has disbanded following publication of the competition results.

Dilemmas in Value Theory: "Optimum" Cannot Be Achieved

Horst Rittel, in his pessimistic planning ethos (see, e.g., Rittel and Weber [1969]), labels "best" solutions as those that are the "least bad." We should be aware that it is not possible to meet all the requirements in one scheme equally well. We must acknowledge that there are always trade-off to be made, where some demands must be met less optimally than others, so that other requirements can be complied with more fully.

Rittel and the Berkeley approach to planning theory are also concerned with the issue of creating new problems while solving existing ones. This should be kept in mind for all evaluative statements and at all levels of evaluation, such as:

1. Physical versus nonphysical criteria
2. Project cost versus social cost
3. Project impact on or toward the context area

DYNAMIC CHANGES IN PERFORMANCE CRITERIA

Users learn during the briefing and design process. According to the planning rules for public buildings in Germany, user institutions are asked only once, at the beginning of the process, for their needs and wants, mostly in an affirmative, decisive way. If they are prevented from participating in the design and implementation processes, they are likely to be more critical of the final result and will tend to accuse the designers of not fulfilling the brief. But if they are consulted frequently during longer projects developments, the users' opinions can develop into more precise and useful statements about their own needs. They learn on the job, as discussions with the experts foster improvement of their programming skills. As a result, they may utilize their former experiences and apply them more strategically. In one example, our office prepared an in-house newsletter for a major client, publishing information on work environment conditions during the entire project, to familiarize the staff with the new building long before they actually moved into it.

Sometimes briefing is subject to serious in-house discussions about the most desirable future. Inertia and resistance against any change in the built environment may be a factor for some people, particularly elderly staff members, who may oppose the change as such and object to replacing old furniture with new in the new building. To encourage these staff members to accept their new work environment, a full-scale mock-up with test furniture might be built, and the staff encouraged to discuss and test it, even over weekends with their families.

The Relative Importance of Some Objectives May Be Altered by the Changing Influence of the Partners

We must not neglect the fact that competition jury decisions often do not reflect full agreement by all jury members. Furthermore, the decisions may represent a compromise reached by unequal consideration of all the objectives. The jury may give more emphasis to town planning and external appearance than to function, running cost, and serviceability. Yet during the detailed design phase, the latter criteria become increasingly important, to the extent that the winner may be replaced by someone more suited to the client's needs, or the winner may find it necessary to redesign his or her scheme completely.

Changing Interest of Senior Decision Makers in the Project

Some board members who were initially intensively active in promoting a project may lose their interest in it during the prolonged planning period. A board member may decide to leave the organization, a decision that may not be made known officially but may become apparent from changes in the decision-making behavior of that person. Another reason may be that a decisionmaker comes to realize that the organization will not be able to afford the cost of the planned new investment, but is unable to convey this fact to the planning team.

Institutes Without Directors/Directors Without Institutes

In some major building projects for newly founded institutions, senior decision makers, directors, or board members may be appointed only as the building nears completion. Although others have given their best to planning and erecting the building, the new senior management may introduce new ideas and criticize the project for lack of support for these new ideas. This is frequently the case with large hospitals, university buildings in the applied sciences, and with museums, where planning and construction take so many years that the director is appointed only at the final stages of the project. Early appointment may mean that someone of high caliber must act for too long a period without appropriate facilities, which causes other kinds of complications.

Changing Relevance of Evaluation Criteria During the Project Process

Not only do the evaluation criteria as such contribute to a final value statement, but also the relative weight of each criterion in the total value. In the dynamics of project processes, and during the time needed for its realization, some issues and design criteria may lose or gain importance due to changes in general political emphasis. The oil crisis, or the flow of refugees from the east after the opening of the Iron Curtain, for example, could bring changes for which a building cannot sufficiently adapt.

Staff Representative May Deliberately Fail to Use Potentials of Intervention

In matters concerning the quality of, and changes to, the working environment, staff representatives have mandatory rights of intervention. These rights are sometimes not used, because representatives, even if they are well informed about issues and design proposals, may not know what to criticize, avoid, and request. They may have to rely on special training efforts from outside to enable them to perform their important roles. It also happens that staff representatives feel powerless and may prefer not to get involved.

Dynamic Change of Project Constraints and Contextual Design Criteria Caused, for Example, by Changes in Town Planning Policy

Planning issues subject to debates in local councils may lose their functional logic and appropriateness as the dynamics of majorities and party strategies impose different political logic or opportunities. Lack of continuity in the pursuit of project objectives may also occur after majority changes in the local elections and as a result of the need to include elements of the political coalition partners' platform in the new official program.

Lack of Political Guidance Continuity, and Obscurity of Decision Responsibility

Projects often last longer than politicians are in power. After elections and political change, a project may encounter difficulties. In the maze of complex authority systems, one may not know exactly which department or which governmental office really is responsible for the project, and who really is empowered to act as the client. Often, responsibilities and authority are obscured deliberately, because certain projects are subject to a political power play.

Performance Problems That Result from Budget Problems

Architects, in their professional arrogance, may deliberately exceed the cost limits of a project and may attract the sympathy of a jury. For political reasons, an architecturally convincing but too expensive winning scheme usually will not be rejected at the time the jury's results are published. The conflict will emerge later, behind the scenes. Ultimately, the brilliant architecture may be curtailed, and resulting in an ordinary, provincial "feasible solution."

How to Preserve the Initial Quality of a Good Scheme

The initial quality of a scheme, which is praised at the sketch design phase, is often not retained and developed during further phases of design and execution. Architects bear the blame for that if they artificially modify their basic design approach by overrefining the design. Clients may weaken the structural integrity and clarity of the project by pressing an ambitious program onto a good scheme on a tight site, or by demanding substantial changes too late in the process. Especially in Germany, where custom-built, owner-occupied buildings are popular, and where speculative "container" buildings are offensive, the functional and formal distinction is part of the corporate image that is expressed by and through a building. Reputable architects may reject later changes in requirements because that would spoil the quality of their architecture, only to realize a couple of years later that the absence of the rejected facilities paralyzes the operation of the entire enterprise. Or they may structure the cooperation with the client's representatives in such a way that clients have fewer chances to express requirements for change.

COMPENSATORY ABILITIES OF HUMAN–ENVIRONMENT SYSTEMS

Inadequacies of buildings often are not even noticed, because users have adjusted their behavioral and functional attitudes toward the building in such a way that they achieve normality, substituting other rewards and positive stimuli to compensate for building satisfaction. External visitors often wonder why the staff members do not object. Compensatory and adaptation abilities of people, in fact, help a great deal to overcome deficiencies of building performance. Commonly observed examples are self-help remedies in cases of inadequate daylight, sun glare, drafts, limited adjustability of furniture, and adjustment of in-house traffic routes.

When Decision Makers Fall in Love with a Suboptimal Solution

For fear of powerlessness or lack of influence, participants may become resigned to an adaptation that is suboptimal. In other cases, fear of change or lack of ability to conceptualize may be the reason why users accept inappropriate facilities. Elaborate comparison of pros and cons of improved solutions may help to instill courage to seek a better solution; however, persuasive argumentation should be avoided.

The peculiar combination of suboptimal solutions with high satisfaction of users can be found in large administrative institutions. When users were asked to criticize a proposed solution, we discovered that they found it satisfactory only because their limited set of criteria were positively met. Only assessors who have a full set of evaluation criteria are able to evaluate a proposed solution in a comprehensive way.

We also noticed that there may be a common agreement about a building, but various people may come to such a value statement from different perspectives and for a variety of reasons. This leads to several conclusions:

1. Limited scope of, and interest in, the environment may lead to partial judgments that are taken as the whole.
2. In the total range of job satisfaction, acceptance of the physical environment ranks fairly low.
3. High values in some aspects of performance may not be useful as long as a building is deficient in other criteria (e.g., highly valued private homes on a contaminated site).

OBJECTIVITY IN EVALUATION STATEMENTS

A group as a whole may come to different value and acceptance ratings than those that would be reached by the individual members of a group. Consensus of many people about acceptability should not be confused with objectivity of value statements. Given the very different perspectives of people in diverse functions, roles, and status, their statements about the quality of a particular built environment may vary widely.

Understanding Value Statements About Performance and Usability in Light of the Evaluator's Skills and Motivation

A review of building diagnoses done by other people demonstrated to us that evaluation as a skill is not yet fully developed. For this reason, our ASTM subcommittee on overall building performance has attempted to establish some standard evaluation procedures. Also worthy of mention are Public Works Canada, NIOSH, and other organizations, as well as such authors as Volker Hartkopf, Vivian Loftness, and Jacqueline Vischer, the latter stressing the importance of a behavioral science approach to building evaluation.

Conflicts Between Hard and Soft Criteria in Evaluation

When the debate about evaluation and performance ratings escalates, we notice a tendency to undervalue what cannot be expressed in objective, quantifiable data. For example, acceptability, well-being, likes and dislikes, meaning and image, or the contribution of a building to social coherence, group or corporate identity, chances for appropriation, and expression of self are issues that tend to be neglected.

Acceptance Problems with Optimized Solutions

Some reticent people dislike making decisions and prefer to leave all possibilities open. They are afraid to be responsible for bad decisions and want to find an escape by stating that something is not in accordance with expectations. Others do not interpret correctly the drawings and verbal descriptions that are the basis for their decisions, and are therefore opposed to the final result when it materializes. They either lack imagination or they draw distinct but incorrect conclusions about reality from its descriptions and do not accept the real thing as complying with the conditions on which they based their decisions. Some remedies to these problems are full-scale mock-ups, video films from endoscopic models, computer animation, or visits to buildings with similar specifications. Such experience with decision makers confirms that there should be agreement not only about performance criteria of an item to be realized in a building, but also about the mode of measuring performance and of checking compliance with the brief.

Dynamic Change of Objectives During the Evaluation Process

Dynamic change of objectives occurs when evaluators are linked closely with the institution to be evaluated and when the success of a project must be clearly demonstrated. Many performance issues are subject to interpretation. Problems occur when the evaluator is not objective and feels a need to prove that the result is an unqualified success. People who have vested interests in an obviously inadequate or poorly performing result will defend it to maintain self-respect and will be reluctant to acknowledge deficiencies.

ARE PERFORMANCE PREDICTION PROCEDURES DIFFERENT FROM EVALUATION PROCEDURES?

Postoccupancy evaluation (POE) of an existing building has certain advantages over prediction and assessment of expected results. POE has the opportunity to include many of the nonphysical criteria in the judgmental process and is therefore more comprehensive and precise.

The objective of performance prediction is not the physical result but only the written, drawn, and tabulated potentials of an eventual building, even if this may be developed imaginatively. Typically, organizational and operational details are not yet at the same level of precision as that of designed details of the physical structure, so they cannot be included in the performance prediction. However, predictors still have a chance to intervene in the forthcoming process of realization, at least by specifying under what circumstances a proposed scheme might become successful, meet objectives, satisfy expectations, and stay within cost or other limits.

Validity and reliability of such predictive value statements are obviously lower than postoccupancy evaluations. However, in combination with a dynamic action concept (and sufficient political and managerial support), the predicted values may be realized. Even so, it cannot be known in advance whether the end user will like and accept the final result. Thus ex-post evaluation is mostly analytic, diagnostic, and retrospective, whereas ex-ante prediction is (or may be) normative, action oriented, and anticipatory.

In the design stage, the operational details of an enterprise may as yet be defined only vaguely, and thus the criteria for good overall performance may still be obscure. Good building performance is difficult and only vaguely predictable from drawings and descriptions at the design stage. At this stage evaluative statements are restricted to congruence assessment between program items and design or project specifications. Operational criteria can only be determined, gauged, and adjusted by experience and by drawing analogical assumptions from precedents, and must remain at a rough preliminary level. They must exclude the ever-important operational and managerial details and the human–machine–physical environment interactions that make the evaluation so clear, effective, and precise in a POE assessment.

METHODS OF EVALUATION AND PREDICTION OF BUILDING PERFORMANCE: AN APPLIED IMPLEMENTATION THEORY

Let us turn our attention to the theoretical background of how we evaluate and predict building performance successfully. I define success by using the terminology of implementation theory (see the literature on analytical models regarding the evidence of success of program implementations: e.g., Etzioni [1973]). We need to draw distinctions between three types of evaluative actions:

1. Objectives-oriented evaluation
2. System-oriented evaluation, with two alternatives:
   a. Evaluation according to the "survival model"
   b. Success-oriented ("dynamic") evaluation

Objectives-Oriented Evaluation

The key question is: Does the result after implementation comply with the initial brief? This type of evaluation can apply only in cases where the time span from the setting of objectives to the finished implementation (i.e., the design results or the finished building after occupancy by users) is relatively short, where projects are not too complex and dynamic, where program specifications and objectives are extremely clear, and where the framing (contextual) conditions are not too controversial. This type of evaluation turns all people working on project implementation into compliers of prefixed objectives. Inventiveness and deviance are not rewarded. This is the legalistic pattern of evaluation, well known to all competition contributors and to all contractors who are required to stick to the specifications and give evidence of their compliance.

The objectives-oriented model is labeled "traditional." Efficiency of a program is assessed only according to criteria that emanate from the program. Criteria that are originated by the program implementors or actors are not taken into account. The attainment of these program objectives neglects the complexity of implementation processes. This approach may lead to distorted and biased results, where success is achieved outside the frame of contextual factors.

System-Oriented Evaluation

The system-oriented evaluation model (according to Etzioni [1973]) is based on the assumption that any implementation means change of complex sociotechnical systems and that implementors are in search of an optimal combination of means and tools. Etzioni draws a distinction between a "survival" model and a "success" model. While the system survival model is more existential in searching for the conditions under which any existing social and/or physical structure can survive, the success-oriented model asks whether the conditions that were successful in the implementation of a program and in the achievement of its objectives are acceptable.

Evaluation According to the Survival Model

The key question to be answered by this model is whether the final result fulfills the requirements set by the evaluators and by the end users at the time when the evaluation takes place, and whether the context of implementation has been accounted for sufficiently well. This type of evaluation applies where a value analysis of existing facilities is desired, or where new users want to know the suitability and adequateness of a facility to their needs. Of less importance is how this building came into being and the original concepts and objectives when it was conceived and built. Current usability potential and technical performance are of importance. This may apply, for example, when the residual value of existing assets or their usability potential for a certain anticipated future use is to be assessed. This type of evaluation demands the setting of objectives and evaluation criteria prior to, or simultaneous with, the evaluation process.

Success-Oriented Evaluation

The key question to be answered by this model is whether decisions made during the design and implementation process are in accordance with, or appropriate for, the ultimate achievement of preestablished objectives. Are the originally stated objectives still in accordance with the objectives actually respecified by the people in charge at the moment of assessment? Are the actual users happy with the performance of the building after they move into it? This approach to evaluation implies a dynamic, permanent adjustment of actions to changing, developing objectives and requirements, and also a perpetual reassessment of changing constraints during the entire process of design and implementation. It takes into account the circumstances and conditions of the implementors, as well as the "value climate" at the moment of evaluation.

This type of evaluation is the normal case in big and complex building projects with prolonged implementation time and conflict-ridden contextual conditions. This type of evaluation also applies where the objectives must change during project implementation because decision makers learn during the process. Initial objectives that were set under some initial constraints will change as the project proceeds. Budget cuts, political discontinuity, and changes in the legal framework during implementation may be causes for such dynamics, and the decision makers must adjust accordingly.

CONCLUSION

Viewed from a European perspective, building performance prediction and evaluation is a dynamic, argumentative, transsubjective business, but is nevertheless indispensable for improving quality. Evaluators must first define their role and motivation in regard to the assessed institution, and then apply accepted evaluation methods. They should be aware of the multidimensionality of their work and should give equal emphasis to hard and soft criteria.

Appraisal of building performance must always be seen as an intervention that has the potential for enormous impact on the social dynamics of an enterprise, and therefore we conclude that the object of evaluation may be changed in significant ways by the process of evaluation. It is as if evaluation were part of a sociopsychological diagnosis. The purpose of a client in asking an outside expert to investigate an institution is similar to investigating institutional illness. We should be aware of what must follow. Is design a therapeutic action, and as in treatment of some diseases, does the diagnosis help patients to help themselves?

ACKNOWLEDGMENTS

The author is grateful for the editorial assistance received from Catherine A. Keppel (Kassel, Germany) and Lois Beery (Buffalo, New York), who reviewed this contribution for correct English.

REFERENCES

Etzioni, A. (1973). *Soziologie der Organisationen* (German translation of *modern organizations*, translated by J. Baetge), 4th ed., Juventa Verlag, Munich.

Jockusch, P. (1986). "Developmental History of Institutions: Life History of Their Buildings," in *Building Performance: Function, Preservation, and Rehabilitation*, ASTM STP 901, G. Davis (ed.), American Society for Testing and Materials, Philadelphia, pp. 69–85.

Jockusch, P. (1989). "Post-occupancy Evaluation as a Tool for the Preparation of Architectural Competitions," Chapter 4 in *Building Evaluation*, W. Preiser (ed.), Plenum, New York, pp. 35–57.

Jockusch, P., M. Hegger, and B. Ettinger-Brinckmann (1984). "German Lesson II: Three German Competition Advisors Discuss Their Role in Different Types of Public and Private Sector Competitions," *Architects' Journal*, **179**(22): 39–50.

Rittel, H., and H. Weber (1969). *Dilemmas in a General Theory of Planning*, Reprint 86, Institute of Urban and Regional Development, University of California, Berkeley, Calif., pp. 160–167.

3

DESIGN INQUIRY AND RESOURCE ALLOCATION

Robert E. Johnson

College of Architecture and Urban Planning
The University of Michigan
Ann Arbor, Michigan

In this paper we propose that the primary role of resource allocation in design is to assist design decision makers in ordering preferences and exploring trade-offs. Most existing cost evaluation paradigms focus on assessing costs after design decisions are made. This view unnecessarily restricts the active participation of economic knowledge in design decision making. The approach described in this research suggests that the exploration and definition of values and preferences should be the major focus of economic analysis within the design process. A conceptual framework for this approach is presented along with several examples that illustrate the use of this framework. Computational approaches are suggested which play a central role in clarifying preferences and exploring trade-offs during design.

INTRODUCTION

This research starts with the assumption that resource allocation decisions are an important consideration for design theory. Resource allocation is important not only because conservation of scarce resources is one of the criteria for a successful design [Simon 1981], but because decisions about how we use resources today will significantly affect the quality of life for future generations. A weakness of current resource allocation approaches is that they focus on the cost evaluation of completed designs and are not integrated into the thinking processes associated with generating designs. Historically, cost evaluation has been most effective after most of the major design decisions have been made.

The separation of cost evaluation from design decision making has resulted in a tendency to treat cost estimation as if it were an end in itself. Designers delegate cost evaluation to specialists and are not usually trained to consider resource allocation as an integral part of design thinking. The objective of this paper, therefore, is to redefine the resource allocation task from one that is concerned primarily with cost estimation and prediction to one that assists directly in the process of design inquiry. To achieve this objective, a design process model for resource allocation is formulated in which preference theory and choice-structuring knowledge representations play a central role. Within this model, design inquiry is conceived of as a mechanism to help decision makers identify and distinguish between "wants and needs" (values) that are fundamental to making resource allocation decisions. Several examples from recent research projects are used to illustrate the use of this framework. The research reported in this paper builds on previous work by helping to clarify and define the role of "computational assistants" to the design process [Kalay and Swerdloff 1987, and Schön 1988].

It should be noted that this paper focuses exclusively on problems associated with value issues as modified by initial capital cost concerns. The exclusion of longer-term resource allocation concerns does not mean that they are not important. Indeed, life-cycle considerations may become one of the most dominant issues of the next century. However, life-cycle costs and notions of building as an investment have been excluded because they are not central to the argument being presented, because they add a level of complexity to this discussion that would substantially increase the scope of this paper, and because the framework that is presented provides a mechanism for investigating these concerns at a later date.

A BRIEF REVIEW OF COST EVALUATION APPROACHES

Cost estimating is generally thought of and used as a method for design evaluation. At an appropriate point in the design cycle, the design concept will be communicated to a cost estimator. The cost estimator uses the information conveyed by the drawings and specifications to provide an estimate of the cost of constructing the building at a particular location at a particular time. As indicated in Figure 3.1, this information is generally (although somewhat simplistically) classified into two categories: (1) building size, and (2) building specification and configuration [Brandon 1984, p. 116]. If not enough information is available, either the designer will be asked for additional information or design assumptions will be developed by the estimator to fill the gaps. The more complete the design, the more accurate is the estimated cost of construction.

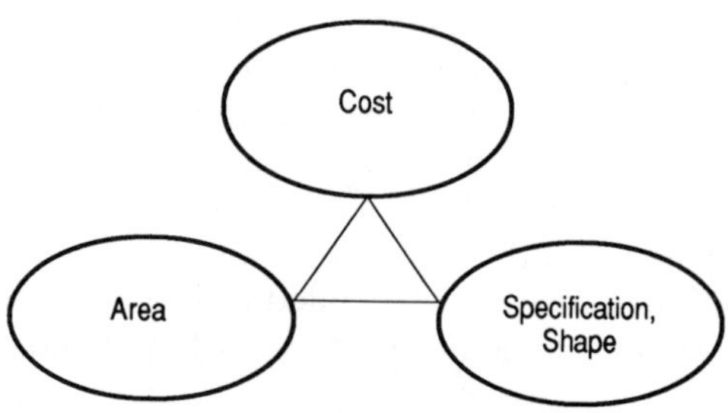

Figure 3.1. Design–cost relationships.

Generally accepted rules of thumb suggest that at the end of schematic design an accuracy of approximately ±10 percent can be expected.*

Research issues in cost estimation have generally focused on improving the predictive quality of the estimate. Various approaches have been proposed to offset some of the inherent reliability and validity problems associated with predicting building costs. These approaches have been described as falling into three "generations": (1) the floor area model, (2) regression, and (3) Monte Carlo and expert systems [Raferty 1987, pp. 51–53]. Recently, more sophisticated approaches have attempted to improve on predictability by treating estimating as a stochastic process. These stochastic approaches inherently require sophisticated computing power. However, despite the apparent suitability of the largely numeric domain of cost estimation and a computer technology that is well suited to numeric computation, computers have yet to make a significant contribution to cost modeling in practice [Newton 1987, p. 42].

The reason for this inadequacy may be that cost evaluation tends to focus on the development and improvement of a cost model as if this were an end in itself. However, a cost model is useful only to the extent that it contributes in some way to helping to make decisions about how to improve the design of a building. Therefore, the point of departure for exploring the usefulness of cost assessment methods must be first to develop an understanding of the design process itself.

A CONCEPTUAL MODEL OF DESIGN INQUIRY

Architectural design has been characterized as a two-stage process of conjecture and analysis [Hillier et al. 1984]. In this approach the conjectural approximation of a design solution is evaluated against increasingly detailed and specific problem data. The design process begins with a framing or prestructuring of the problem. Decision makers reduce the problem to manageable proportions by paring away what are thought to be insignificant elements and focusing on what are thought to be the most important aspects of the problem.

This decision process is not deterministic, but instead is guided by normative approaches that Rowe [1987] refers to as "enabling preconceptions." Schön describes this process in the following manner. "Design situations are always partly indeterminate. In order to formulate a design problem to be solved, the designer must frame a problematic situation: set its boundaries, select particular things and relations for attention, and impose on the situation a coherence that guides subsequent moves" [Schön 1988 p. 182]. Put another way, "In plain language, first decide what you think might be an important aspect of the problem, develop a crude design on this basis and examine it to see what else you can discover about the problem" [Lawson 1980, p. 34].

The crude design concept is refined, elaborated, and tested through this inquiry process. As each specific design refinement is presented, more is learned about the problem. This learning process feeds back into further elaborations of the design. Occasionally (although less frequently in the case of experienced designers), a conclusion is reached

*Cox and Horsley [1983, p. viii] indicates a variance of ±10 percent, while Swineburne [1980, p. 85] indicates a variance of about ±8 percent at this stage of design.

that the original design concept is inadequate. This causes the designer to reframe the problem from a different perspective in order to conjecture a new or substantially revised design concept (see Figure 3.2).

To summarize, design inquiry can be thought of as a form of deliberation that is both judgmental and analytical in nature. It is judgmental in that deciding on a course of action is a normative process that requires deliberation and resists quantification. Judgment is more heavily depended upon during the early stages of design because it clarifies the decision task and isolates preferred courses of action. Analysis, on the other hand, is more quantifiable in nature and confirms the wisdom of our prior judgments. A major problem with economic evaluation approaches is that they focus on evaluation and do not respond to the style of judgmental "conjecturing" that dominates early design thinking.

A MODEL FOR INTEGRATING RESOURCE ALLOCATION AND DESIGN INQUIRY

It should be apparent that although the domain of resource allocation is often thought of as a "well-structured problem," it has more similarities to ill-structured problems such as strategic planning. One obvious reason is that it is impossible to predict accurately the economic performance of a building when not all design decisions have been made. Any attempt to impose an overly systematic evaluation process on a design project that is still in a situation of irresolution is bound to be ineffective.

Approaches to economic evaluation should reconsider the role of economic factors within a more comprehensive framework of design as outlined above. The primary focus of economic evaluation should be the development of a process that enhances a designer's ability to clarify the decision task and isolate the preferred course of decision. Within the specific domain of resource allocation, what is needed is a model for design deliberation that appropriately includes economic assessment approaches within this broader context.

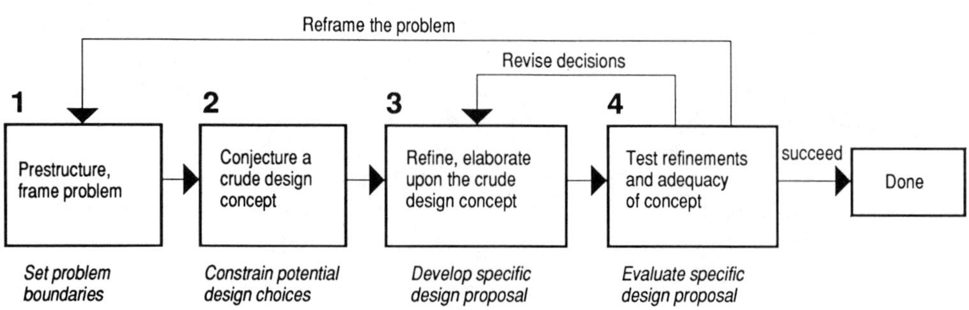

Figure 3.2. Conceptual model of the design process.

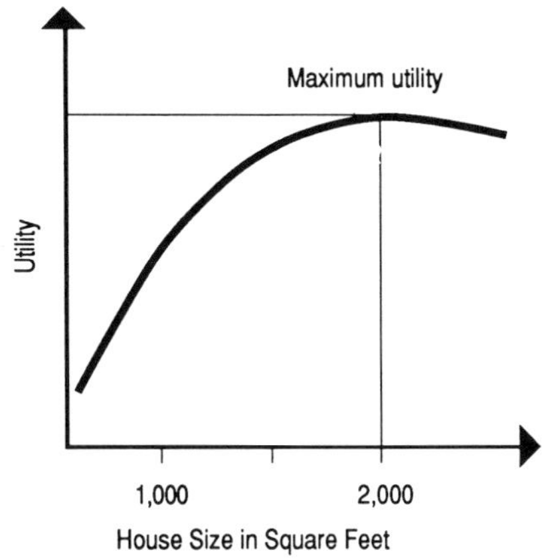

Figure 3.3. Diminishing marginal utility.

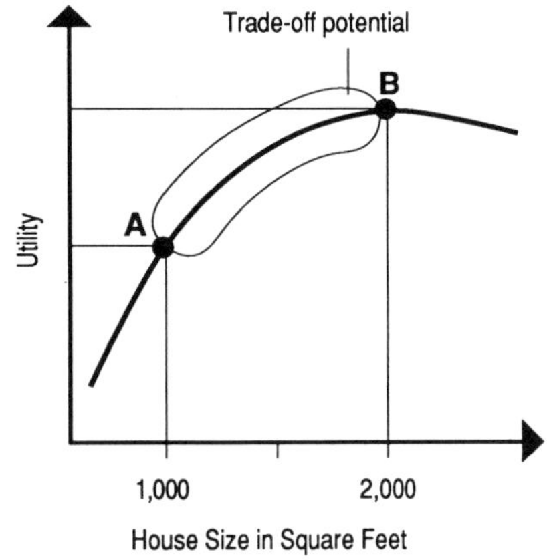

Figure 3.4. Range of acceptable limits for building size.

Preference: A Key Missing Factor

A missing factor in economic evaluation methods (and design theory) is an adequate understanding of the role of preference judgments in design inquiry. The theory of preference goes back to the work of Fechner [1871] on experimental aesthetics. Simply stated, this theory proposes that the pleasure (or utility) derived from a commodity varies with the quantity of that commodity, but increases less rapidly as the quantity of the commodity increases. As an example, Figure 3.3 presents a hypothetical graph of one person's preference for space. This graph shows that utility (or satisfaction) increases rapidly for the first increment of space, but beyond a house size of 2000 square feet (ft^2), satisfaction actually declines. For this individual, a house that is less than 1000 ft^2 is too small; a house that is over 2000 ft^2 is too large.

This characteristic of "diminishing marginal utility" is important to include in any design decision theory. What this says is that for any given design attribute, there is a range of choices that are acceptable to the decision maker and that this range of choices can be identified. Figure 3.4, for example, suggests that there is a minimum value below which the decision maker will not accept any smaller house (A) and a maximum value above which overall utility will decrease (B). The definition of this range as well as the slope of the curve is conditioned by a whole host of social, cultural, and functional considerations that are determined by how the decision maker has framed (or prestructured) the design problem.

Design can then be thought of as making decisions that are partly a function of this utility characteristic and partly a function of cost. Depending on cost and utility assessments, a decision maker may decide to choose more or less of a given attribute. Because decisions must be made about many attributes, there is a considerable opportunity to trade off or sacrifice a certain amount of one attribute in order to obtain more of another attribute.

A CONCEPTUAL FRAMEWORK FOR IMPLEMENTING PREFERENCE THEORY

It has been argued that a weakness of both design theory and, by implication, computer-aided design systems is that they do not consider the implications of decision theory. More specifically, a mechanism to allow for selecting choices and investigating trade-offs is an important, critical element to help decision makers order their preferences and focus on developing courses of action that will help them achieve these preferences. Such a decision process cannot be "computed," because the determination of values and preferences are inherently choices that must be made by people. However, computation can play a central role in clarifying potential choices, assisting the discovery of potential trade-offs and helping the design team to weigh the relative merits of choices they make.

Figure 3.5 presents a modification of the earlier conceptual model of the design process. This model is revised to suggest a means for integrating preference assessment and trade-off analysis into the design process. It is organized into (1) an object database, (2) the prestructuring of design into a set of potential choices, (3) the refinement and elaboration of design through a series of decisions, and (4) a reconsideration of those decisions through evaluation which leads to an improved understanding of the design problem and a revision of previous decisions. It allows the design process to be guided by normative judgments that may be modified by the results of more quantitative analysis.

Object Database

Theoretically, the database consists of all objects that may be used in the development of the building design. Within the domain of cost estimating, this database is referred to as a unit price catalog. Table 3.1 presents an example of a part of this unit price catalog. A complete implementation would contain a separate table for each classification of building component. Typically, the only attribute included in a unit price catalog is the initial capital cost and, for structural systems, the superimposed load associated with the selection of that system. However, a more comprehensive economic analysis (e.g., life-cycle cost evaluation or real estate feasibility analysis) would require the inclusion of

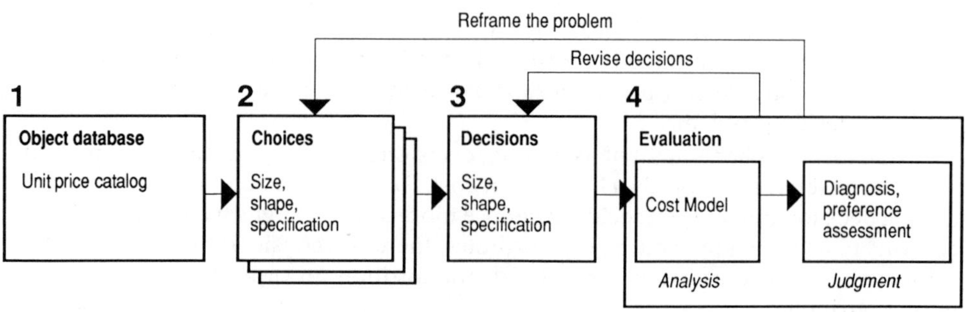

Figure 3.5. Conceptual framework for cost evaluation.

other attributes such as maintenance, cost, frequency of maintenance and U value. This information is easily represented in a table format and therefore can be readily implemented as a series of relations within a database management system or as a series of tables within a spreadsheet.

Choices

The conceptualization and organization of design choices used in this paper draw their inspiration partly from multiattribute utility theory* or the theory of subjective expected utility. The SEU theory of choice behavior is a mathematical theory that explicitly accounts for an individual's preference structure. It assumes that decision makers are completely rational (i.e., will always seek to maximize utility), that all alternatives and the outcomes from selecting an alternative are known, and that any complex problem can be reduced to a number of independent, smaller, and more manageable issues. These assumptions are very rigorous and cannot be expected to be satisfied reasonably for even moderately complex decisions [Simon 1986]. This theory is difficult to apply directly to design decisions because design is characterized by ambiguity and complexity. However, some of the major organizational concepts can be effectively utilized to fit the previously proposed model of design if one relaxes and sometimes ignores these rigorous assumptions.

TABLE 3.1 Partial Unit Price Catalog

No.	Object Description	Cost (per ft²)	Load (lb/ft²)
	Exterior Walls		
1	6-in. Tilt-up concrete panels	$ 4.72	
2	6-in. Concrete block	5.07	
3	8-in. Split-ribbed concrete block	6.84	
4	Standard brick, running bond, 8-in. CMU backup	15.60	
5	2-in. Indiana limestone, 8-in CMU backup	29.30	
6	2-in. Granite, metal studs	30.08	
7	Bronze curtain wall, aluminum framing	32.66	
8	4-in granite, 8-in CMU backup	45.55	
	Steel Joist Superstructure		
1	Steel joists, beam and slab on columns (20-ft bay)	$ 6.66	119
2	Steel joists, beam and slab on columns (25-ft bay)	7.65	120
3	Steel joists, beam and slab on columns (30-ft bay)	8.11	
4	Steel joists, beam and slab on columns (35-ft bay)	9.71	121
	Composite Beam Superstructure		
1	Composite beam and deck, lightweight slab (20-ft bay)	$ 7.98	115
2	Composite beam and deck, lightweight slab (25-ft bay)	8.27	118
3	Composite beam and deck, lightweight slab (30-ft bay)	8.32	116
4	Composite beam and deck, lightweight slab (35-ft bay)	9.52	121

*Edwards and Newman [1986] presents a straightforward, relatively simple and nontechnical introduction to the basics of decision theory.

The first concept to be utilized is the view that design is a complex decision problem that can be configured as a hierarchically organized series of smaller decision problems. The selection and definition of these smaller decision problems is determined partly by how the problem was framed and partly by the inherent structure of inflation necessary to define a building. Table 3.2 outlines a portion of this hierarchy. The building portion of the hierarchy reflects the size–shape–specification cost-estimating framework. The specification part of the hierarchy is organized along the lines of a generally recognized system for classifying cost information known as the UNIFORMAT system. This hierarchy presents a structure whereby design decisions can be made at different levels. That is, design decision making can proceed in either a bottom-up or top-down fashion. It also points out the interactive nature of decisions. For example, a decision about building type will determine occupancy requirements, which, in turn, will determine exit requirements.

A second concept borrowed from decision theory is that a range of acceptable choices can be determined for any decision in the hierarchy (see the "trade-off potential" in Figure 3.4). The acceptable range is determined by how the problem was framed. A design problem that was framed to be a warehouse would probably suggest very different types of exterior wall systems compared to those that would be considered in the design of a corporate headquarters building. Table 3.3 presents one possible set of

TABLE 3.2 Partial Project Information Hierarchy

- Project
  - Neighborhood
  - Building
    - Type (residence, office, retail, warehouse, school, industrial, etc.)
    - Occupancy (number of people)
    - Size
      - Gross floor area
    - Shape
      - Perimeter
      - Floor-to-floor heigh
      - Number of floors
    - Specification (UNIFORMAT)
      - 01 Foundations
      - 02 Substructure
      - 04 Exterior closure
        - 041 Exterior walls
        - 042 Exterior doors and windows
        - 0421 Windows
        - 0422 Curtain walls
        - 0423 Exterior doors
          - Number of required exits
        - 0424 Store fronts
      - 05–12 Other building systems
  - Financing
    - Market capialization rate
    - Mortgage financing terms
    - Construction financing terms
  - Other project attributes

choices for these two different building types. In general, choices are arranged in order of cost, with low-cost items appearing first in the table.

Metaphorically, what is being suggested is the definition of a "palette" from which subsequent design thinking is shaped. In addition to specification choices, the palette can also be conceptualized for size and shape decisions. Table 3.4 illustrates a range of alternative perimeter configurations that is one of the factors that determine building shape. The computed value of the perimeter is a function of both the footprint area and the configuration of the building. A more complex configuration results in a larger perimeter and ultimately will be translated into a larger cost for the exterior wall system. The formulas in Table 3.4 also show that a larger footprint area will increase the size of the perimeter and thus result in a larger cost for the exterior wall system. As with the exterior wall choices, these perimeter choices are arranged from low to high cost.

A more complicated illustration of interaction among design decisions is presented in Figure 3.6. This example also suggests the need to integrate a variety of different knowledge types to determine an appropriate foundation design. For example, the decision to design an office building type should trigger the inheritance of characteristics common to office buildings, such as the required live load. In addition, the decision about bay size will require engineering knowledge and calculations in order to size the structural system for the type of superstructure desired.

TABLE 3.3 Possible Exterior Wall Choices for Two Building Types

No.	Description	Cost (per ft^2)
	Warehouse Exterior Walls	
1	6-in. Tilt-up concrete panels	\$ 4.72
2	8-in. Concrete block	5.07
3	8-in. Split-ribbed concrete block	6.84
	Corporate Headquarters Exterior Walls	
1	4-in brick, 8-in block	\$15.03
2	2-in Granite, metal studs	30.08
3	Bronze curtain wall, aluminum framing	32.66

TABLE 3.4 Possible Perimeter Configuration Choices

No.	Description	Configuration	Formula
1	Rectangle	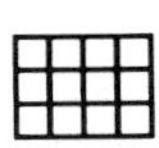	SQRT(footPrint/12)*14
2	L-shape	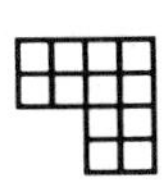	SQRT(footPrint/12)*16
3	Atrium	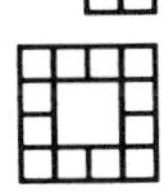	SQRT(footPrint/12)*24

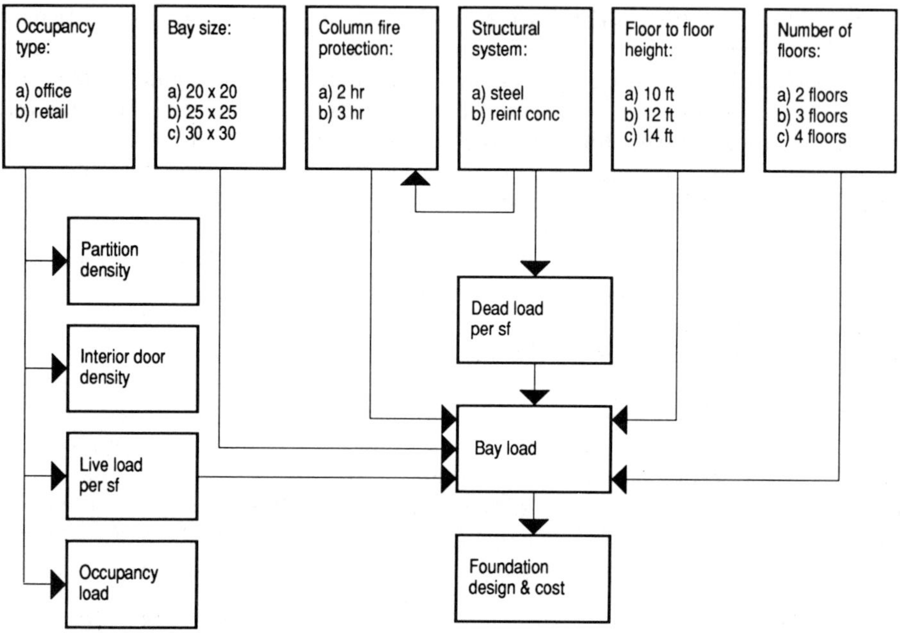

Figure 3.6. Example of dependency relations between objects and attributes.

The issues discussed in this section point to the need for a complex organizational structure for organizing design information. Choices need to be organized hierarchically to enable the designer to think about the design in either a top-down or bottom-up fashion. In addition, the definition of some choices (e.g., foundation system) can require the integration of a variety of different knowledge types, including both procedural and declarative knowledge. The definition of choices may be a simple list defined by a normative decision-making approach (as in the list of exterior wall preferences) or may require the use of a variety of different types of knowledge representation as well as substantial computational power. The need to define a range of choices increases the difficulties associated with implementing this approach.

Two prototypes have been developed to experiment with implementation issues. In one experiment, an expert system was developed to assist in the evaluation of the feasibility for reusing vacant industrial facilities. This expert system was written in PROLOG and utilized a hierarchical frame representation to organize building choices.* The second experiment organized both the database and building choices as a series of interrelated tables within the framework of an Excel spreadsheet [Johnson 1990]. The frame representation provided for a convenient mechanism to implement concepts such as inheritance and demons. The Excel spreadsheet implementation was surprisingly easy to implement and favored due to a more friendly user interface. However,

*This project was funded by the Homer Hoyt Research Institute. The author gratefully acknowledges the substantial contributions made in this research by Eduardo Sobrino, currently an assistant professor at the University of Puerto Rico.

the flat nature of a spreadsheet presented difficulties in providing a reasonable semblance of a hierarchical structure. It was also difficult for users to understand and trace the dependency relationships among cell formulas.

Decisions

Design inquiry as defined in this paper requires a close coupling of decisions, analysis of the outcome of decisions, and judgments about subsequent preferred courses of action. Decisions involve exploring a design solution by selecting from the choices that were conceptualized based on how the problem was framed. Selecting from these choices is goal oriented and may require combinations of both intuitive processes and various computational search approaches. One simple approach to this search process takes advantage of the ordering of choices from low to high cost. That is, the lowest-cost object can be retrieved by knowing that low-cost items are always located in the first row of the table that defines the object. Table 3.5 indicates how this selection scheme was implemented on a spreadsheet. The first argument in the decision formula "@VLOOKUP(1,$c__elevFloor,2)" represents a decision to select the lowest-cost elevated floor structure (first row) from the c__elevFloor choices table. This choice is defined by another @VLOOKUP() formula, which used the bay size to define its lookup index number. The bay size is a shape decision that was entered in another part of the worksheet.

Although a useful form of this model was implemented using a spreadsheet, the limited capabilities of a spreadsheet made more intelligent search mechanisms difficult to implement. Table 3.6 illustrates another form of this process, which searches for the

TABLE 3.5 Selection Process for Spreadsheet Formulas

	Unit Price Database	
p__steelJoist		
1	Steel joists, beam and slab on columns (20-ft bay)	$6.66
2	Steel joists, beam and slab on columns (25-ft bay)	$7.65
3	Steel joists, beam and slab on columns (30-ft bay)	$8.11
4	Steel joists, beam and slab on columns (35-ft bay)	$9.71
	Elevated Floor Choices Table	
c__elevFloor (Elevated floor structure)		
1	Steel joists, beam and slab on columns (25-ft bay)	@VLOOKUP($baySize/5-3,$p__steelJoist,2)
2	Composite beam and deck, lightweight slab (25-ft bay)	@VLOOKUP($baySize/5-3,$p__composite,2)
3	Steel beams, composite deck, concrete slab (25-ft bay)	@VLOOKUP($baySize/5-3,$p__steelBeam,2)
	Decision Formula	
@VLOOKUP(1,$c__elevFloor,2)		

lowest-cost improvement that can be made for conjectured design. The building hierarchy uses a frame representation. As with the spreadsheet example, design choices (in this simple example, only size and quality) are arranged from low to high cost. Design improvements are generated by incrementally increasing the choice (e.g., from minimum to average), calculating the cost of this design change, and comparing this cost to the cost of improvements proposed previously. Design improvements could be further constrained to look first at size improvements or first at quality improvements, depending on the preferences of the designer and client. Knowledge-based evaluation systems might also suggest solution types that have been especially effective in the past.

Evaluation (Analysis and Judgment)

The result of this decision process is a cost assessment by building attribute similar to that indicated in Table 3.7. This is a rather standard cost estimate report. However, the quality decision column provides an effective definition for the current state of the building design with respect to the previously defined range of building attributes. The numbers in this column are the first arguments in the @VLOOKUP() decision formula. The fact that the column consists of nearly all 1s indicates that the designer has defined a "low"-quality (and low-cost) building design. Procedures outlined in Table 3.6 together with the designer's judgment would give guidance as to where to look next to improve the quality of the building.

CONCLUSION

In this paper we have proposed a revision to existing design evaluation paradigms that are concerned with resource allocation. Traditional approaches to cost estimation are too narrowly focused on improving the predictability of costs to be of significant utility to design decision making. The primary focus of the approach that was proposed and illustrated in this paper was the integration of values and preferences (as modified by costs) within the design inquiry process. This integration is achieved through the adaptation of decision theory principles. Unlike traditional cost-estimating models, the primary purpose for computing a cost model is to help order values and preferences within the design decision process. This decision process cannot be computed because the determination of values and preferences are inherently decisions that must be made by people. However, computational approaches were suggested that play a central role in clarifying potential choices, assisting the discovery of potential trade-offs, and helping the design team to weigh the relative merits of choices they make.

There are implications for this framework that extend beyond those discussed in this paper. One such extension is the pedagogical implication of refocusing from cost estimation to value and preference assessment. A revised curriculum with this focus would provide an approach for integrating resource allocation decisions in design that goes beyond traditional courses in cost estimating. Such a curriculum may include a critical assessment of how both priced as well as unpriced values [Sinden and Worrell 1979] affect design decisions. However, the development of an appropriate computational support system would be an important component of this curriculum. Other efforts might investigate the appropriate integration of both preference and non-

TABLE 3.6 Conjecture, Refine, Analyze Model: PROLOG Example

```
/* Object-attribute-value simple generic building hierarchy */

oav(building, cost, if_needed, calc_cost).
oav(building, size, if_needed, ask_attr).
oav(building, quality, if_needed, ask_attr).

/* CHOICES (the design "Palette") */

choice(size,min,1500).    % size is number of sq ft
choice(size,avg,2000).
choice(size,max,2500).
choice(quality,min,50).   % quality is $/sq ft
choice(quality,avg,60).
choice(quality,max,70).

/* Define incremental (marginal) improvements. */

improve_attr_design(min,avg).
improve_attr_design(avg,max).
improve_attr_design(max,max). %no improvement possible.

/* Conjecture-refine-analyze */

generate_design_improvement([]).
generate_design_improvement([[Attr,Attr_choice][Tail]) :-
  improve_attr_design(Attr_choice,Attr_choice).
generate_design_improvement([[Attr,Attr_choice][Tail]) :-
  improve_attr_design(Attr_choice,Improved_attr_choice),
  test_low_cost_improvement(Attr,Improved_attr_choice),
  generate_design_improvement(Tail).

conjecture_design(Design,Cost) :-
  frame_get(building,quality,Quality),
  frame_get(building,size,Size),
  append([[quality,Quality],[size,Size]],[],Design),
  frame_get(building,cost,Cost).

conjecture_refine_analyze(Design,Cost,Rev_Design,Rev_Cost) :-
  conjecture_design(Design,Cost),
  generate_design_improvement(Design),
  frame_get(rev_building,cost,Rev_Cost).
```

preference models (including optimization approaches) of resource allocation within design inquiry, and the exploration of the critical long-term consequences of resource allocation decisions. Finally, the work presented in this paper utilizes decision theory primarily as a mechanism for ordering the decision-making process and to facilitate the identification of trade-offs. The definition of utility levels, normally key ingredient in classic decision theory, has not been included in this research because of difficulties in

TABLE 3.7 Example of Analysis Results (Partial List)

Component	Quality Decision	Unit Cost	Description of Choice	Qty.	Total Cost
Spread footings	1	$1,032.50	3 ksf soil capacity	25	$27,361
Strip footings	1	71.74	4 ft × 12 in. wall, cast in place, 3 ksf soil	400	30,418
Excavation	1	0.78	Sand and gravel, on site, 4 ft deep	10,000	8,268
Slab on grade	1	2.64	4 in. thick, nonindustrial, reinforced	10,000	27,984
Columns and beams	1	14.34	Gypsum board, 2 hours	750	11,400
Elevated floors	1	7.65	Steel joists, beam and slab on columns (25-ft bay)	20,000	162,180
Roof	—	6.86	Steel joists, beam and slab on columns (25-ft bay)	10,000	72,716
Stairs	2	4,070.00	Cement fill, metal pan with landing (16 risers)	13	56,085
Exterior walls	1	15.60	Standard brick, running bond, 8-in. CMU backup	7,875	130,221

adhering to the rigorous assumptions demanded of this theory. Solving problems associated with a more rigorous integration of utility theory and design is another topic for future research in this field.*

REFERENCES

Brandon, P. S. (1984). "Cost Versus Quality: A Zero Sum Game?" *Construction Management and Economics*, **2:** 111–126.

Cox, B., and F. Horsley (1983). *Square Foot Estimating*, R. S. Means, Kingston, Mass.

Edwards, W., and J. R. Newman (1986). "Multiattribute Evaluation," in *Judgment and Decision Making*, H. Arkes and K. R. Hammond (eds.), Cambridge University Press, London, pp. 13–37.

Fechner, G. T. (1871). "Zur Experimentellen Aesthetik," *Abhandlungen der Königliche Sachsische Gesellschaft der Wissenschaften Mathematisch-Physische Klasse*, **9:** 555–635.

Hillier, B., J. Musgrove, and P. O'Sullivan (1984). "Knowledge and Design," in *Developments in Design Methodology*, N. Cross (ed.), Wiley, New York, pp. 245–264.

Johnson, R. E. (1990). *The Economics of Building*, Wiley, New York.

Kalay, Y. and L. Swerdloff (1987). "A Partnership Approach to Computer-Aided Design," in *The Computability of Design*, Y. E. Kalay (ed.), Wiley, New York.

Lawson, B. (1980). *How Designers Think*, Architectural Press, London.

Newton, S. (1987). "Computers and Cost Modelling: What Is the Problem?" in *Building Cost Modeling and Computers*, P. S. Brandon (ed.) E.&F.N. Spon, London.

Raferty, J. (1987). "The State of Cost/Price Modeling in the UK Construction Industry: A Multicriteria Approach," in *Building Cost Modeling and Computers*, P. S. Brandon (ed.), E&F.N. Spon, London, pp. 51–53.

Rowe, P. G. (1987). *Design Thinking*, MIT Press, Cambridge, Mass.

Schön, D. A. (1988). "Designing: Rules, Types and Worlds," *Design Studies*, **9**(3): 181–190.

Simon, H. (1981). *The Sciences of the Artificial*, MIT Press, Cambridge, Mass.

Simon, H., Chairman (1986). "Report of the Research Briefing Panel on Decision Making and Problem solving," in *Research Briefings 1986*, National Academy of Sciences, Washington, D.C.

Sinden, J. A., and A. Worrell (1979). *Unpriced Values: Decisions Without Market Prices*, Wiley, New York.

Swineburne, H. (1980). *Design Cost Analysis for Architects and Engineers*, McGraw-Hill, New York.

*Research to integrate utility more fully is currently being conducted by Stephen Kirk, doctoral student at the University of Michigan and director of the Facilities Economics Group, Smith, Hinchman and Grylls, Inc, Detroit, Michigan.

4

CRITERIA FOR DESIGN EVALUATION: A PROCESS-ORIENTED PARADIGM

Gabriela Goldschmidt

Technion–Israel Institute of Technology
Haifa, Israel

Architectural research of the last two or three decades has devoted itself largely to design methodology. Systematic evaluations of design products and prescription of their desired qualities led to specifications for better designs and possible routines to achieve them. Computers have facilitated this task. The human designer, however, has largley resisted the use of innovative methods, and in this paper we claim that the reason lies in insufficient regard for innate cognitive aptitudes which are activated in the process of designing. A view of these aptitudes based on patterns of links among design moves is presented and it is proposed that process research is mandatory for further advancements in design research utility.

ARCHITECTURAL JUDGMENT

Judgment and evaluation of architecture are concepts that raise difficult questions. It is not my intention here to tackle the judicial, moral, and philosophical aspects of judgment in general, although they have serious implications for architectural judgment. Rather, adopting a pragmatic point of view, I point out what architectural evaluation concerns itself with and what it leaves untouched, for lack of clear conceptions no less than for want of evaluating tools. For the sake of argument I do not distinguish among judgment, evaluation, and assessment. To avoid broad generalities, I also restrict the discussion to judgment *in*, rather than of *architectural design*, a term I prefer to use instead of *architecture*, as it denotes quotidian and continued design efforts rather than completed accomplishments. I use two well-known examples to trace the evolution of our sense of evaluation in the past two or three decades, as a background to my own contentions in this domain.

Collins: Evaluation Criteria

In 1971, Peter Collins published *Architectural Judgement* [Collins 1971]. Although specifically dedicated to a comparison between legal and architectural judgment, Collins' scholarly and comprehensive discourse is a reliable reflection of architectural theory of its day. Perspectives such as Collins' informed diverse theoretical investigations, and I therefore use it as a representative treatise. Collins sees judgment as a rational and evolutionary process, embedded in professional contexts and using professional criteria. The criteria he elaborates on are artistic, scientific, and something that he refers to as "professional connoisseurship." The latter is a viewer-centered body of criteria "in which norms vary in complex and unpredictable ways, and depend on sensibility as well as reason" (p. 143). Collins says more about artistic criteria, and a lot about scientific ones, by which he means criteria for building performance. These, to him, are the crux of the problem of evaluation insofar as it is synonymous with "design decisions," a term that "can only refer to those aspects of architectural judgment involved in the *creation* of an architectural environment" (p. 109). Unlike viewer-centered criteria, we are here talking about designer-centered criteria, which, Collins asserts, are imbued in four types of context.

The first context is the physical and economical environment, the second is political, the third, rather complex context is historical and strongly marked by the use of precedents, and the fourth is what Collins calls the procedural context. Physical, economical, and political contexts are the subject of "scientific architectural research," which Collins sees as suffering from the following limitations:

1. It is unclear whether research methodologies can be appropriately applied to any kind of architectural judgment.
2. The establishment of precise design norms contradicts the notion of criticism and evaluation as integral parts of the design process.
3. It is unclear to what extent architectural ideals, qualitative in nature, can or should be mathematically expressible.
4. Research techniques may be restricted in terms of the number and character of building types to which they may be applicable.

Collins' global view, including the incertitudes he expresses, are typical of the period in which they were expressed. The 1960s and 1970s saw an upsurge of architectural research, largely dedicated to or based on *design methods*, which seemed to carry the promise for more advanced and rational design processes, resulting in better control over the resultant product and an overall rise in design quality. We should bear in mind that a view of design and evaluation criteria according to building performance is not something that could be taken for granted two decades ago, and it came to replace an age-old preoccupation with the contribution of different disciplines to design. A similar notion is evident in other writings of the period. For example, in a well-known article Hillier and Leaman claim that "The key idea in this conceptual model of research is the idea of building function. We propose that instead of dividing the problem of building into the contributions of different disciplines, the initial distinction should be between the different functions the building performs, and that research should be focused on these. For any building can be thought of in terms of four general functions: as a climate modifier, a resource modifier, and a cultural modifier" [Hillier and Leaman 1972, pp. 518–519]. This formulation resonates Amos Rapoport's analysis in *House Form and Culture* [Rapoport 1969], treating primitive and vernacular buildings, which suggests that climate, construction, materials, and technology have been acting as modifying factors to built form, which is determined primarily on grounds of sociocultural factors. Rapoport claims that basically, the primacy of sociocultural factors is still valid at present despite reduced physical constraints. (In this he may differ from both Collins and Hillier and Leaman. For our purpose here, these differences are of no consequence.)

But let us return to Collins' last, *procedural* context, which he defines as "the sequential influences of ideas by which the final scheme evolves from the architect's initial sketches, and the reasons which cause him to favor one form of evolution rather than another" (Collins 1971, p. 50). Collins senses that there is more to the process of designing than a response to even the most rigorous scientific criteria. He uses the term "the creative process" and states that if procedures form an integral part of the creative process, "they must necessarily be included among the criteria by which any professional art is judged" (ibid., p. 78). Collins does not take the notion of design procedures much further, but he adds that the selection of appropriate form configurations is a matter of decision making and does not entail choosing from several alternatives of completed projects; rather, it happens "at the embryonic stages in the course of the design" (ibid., p. 147). This form of decision making is really judgment, in which "the relationship between intuitively imagined forms and intellectually apprehended data is continually assessed" (ibid., p. 146).

Alexander: A Design Method

The early 1960s found many in the architectural profession hopeful that through the introduction of advanced analysis and synthesis formulas new design methods could be devised that would be less "innocent" and more effective than traditional "intuitive" form making. In essence the methods proposed by a variety of researchers called first and foremost for precise definitions of the desired properties of the designed entity, by way of decomposition of the design problem into small, manageable parts. The parts were assigned values which enabled the determination of connections among them and allowed the evaluation of products according to criteria derived from the desired qualities. Representation of connections and values was made possible through the use of interaction matrices, connection diagrams, and tree graphs [e.g., Jones 1963, Luckman 1969, Alexander 1964]. Most of these methods accepted as a basic conjecture that the design process, in a gross simplification, is divided into three phases: analysis, synthesis, and evaluation. The purpose of quantitative methods (and graph representations) was to overcome what Alexander calls "selfconsciousness," thus enabling criticism and evaluation based on logical transparent criteria, which should facilitate the achievement of "goodness of fit" between the designed form and the contexts to which it is expected to adapt [Alexander 1964].

Alexander's methodology is presented in *Notes on the Synthesis of Form* [Alexander 1964], which is so well known that it requires neither introduction nor lengthy paraphrasing. Regardless of the fact that Alexander himself later backed out of many of his own conclusions, this work, a brilliant exemplar of a "design method", has had a lasting impact on research that is still felt today, over two decades later. Alexander did not oversee the designer, his or her needs, ambitions, and creativity, but claimed that an appropriate systematic, algorithmic approach to a design problem would invariably yield better design products in today's circumstances. There are two important things to note here. First, the departure point is the designed entity. The designer, although not superfluous, has a relatively modest role to play in the production of a well-adapted design solution. This approach recalls Collins' viewer-centered attitude and is typical of methods research of that period, much in reaction to what was diagnosed as hopelessly poor design, caused by intuitive and self-conscious design procedures incapable of meeting the complex requirements posed by contemporary design tasks. Second, the particular algorithm emphasized by Alexander, the tree, is a powerful analytical tool for situations in which components of a system can be classified and categorized so as to relate them to one another in time or space (in this case, in terms of fitness among them or lack of it). The decomposition of a design task into its constituent parts was par force accompanied by classifications which made it possible to suggest the hierarchical sequences displayed by the tree. The classificatory quality of the tree is not often pointed out, possibly because it is taken for granted. But classification can lead to oversimplification when single values are attached to items that in fact fulfill several functions. If we recognized the multifacetness of many activities or spaces that served as components in Alexander's decompositions, we would be obliged to employ a highly complex classificatory system with more than one value per item. The resultant trees would be extremely messy and would, most likely, defy the purpose of their creation. Alexander himself acknowledged this fact and later suggested that other, more appropriate mathematical algorithms, such as semilattices, be used to represent complex design tasks and entities [Alexander 1965].

Taken together, Collins' articulation of criteria for evaluation and therefore also of design guidelines and Alexander's design method, hinging on decomposition of the task and its logical reconstruction according to systematic algorithms, represent the kind of thinking that informed design research for many years. At the center of attention is the designed entity and its context; research aims at devising methods for happy interactions among them, by way of facilitating balanced design decisions. Today's research, although more sophisticated and largely computerized, still centers on the same product-oriented posits. Collins' uncertainties regarding scientific evaluation criteria are by and large obsolete in view of developments in database and expert-system techniques. It is true that more attention is paid to the designer, in two capacities: first, he or she is now accepted as working in partnership with the computer on complex design issues, and it is becoming clearer what each partner can do best and in what ways it needs the other partner's help [Swerdloff and Kalay 1987, Ervin 1989]. However, this partnership is mainly relevant to decision making, which, as I argue presently, is not the proccess that I refer to when I talk about the processes that typify design behavior. Second, constructivist philosophy has succeeded in convincing us that the worlds designers live in are not uniform or homogeneous; by way of "world making" [Goodman 1978], design communities and individual architects construct their own contexts. Consequently, we now deal with differential design knowledge as a separate line of investigation, using artificial intelligence modeling techniques [e.g., Gero et al. 1988]. Interestingly,

Collins' notion of a procedural context can be seen as a forerunner of this tendency. However, his insistence that procedures involve decision making seems to indicate that he confounded process with methodology, as so many did and still do.

I propose that we take one step further in the move toward the designer and enter another variant into the "design equation," namely, the designer's thinking patterns, by which I mean not only what he or she knows, but how this knowledge is actually constructed, enhanced, and tapped while designing: the cognitive attributes of design thinking. Decision making is not synonymous with thinking patterns, it is an outcome of such patterns. If we looked at the design process only in terms of decision taking, we would miss out on the characteristics of cognitive operations that lead to specific decisions. Only fairly obvious decisions are explainable without an exploration of the trajectories that lead to them, which often stem from unexpected sources. Why this is necessary should become clear to us when we consider the fact that by and large, the design community has resisted the introduction of the so-called new design methods in the last two decades quite persistently. In fact, it is no exaggeration to claim that as of yet, "methods" have failed to make a dent in traditional design behaviors, at least in the early exploratory stages, despite wide experimentation, mostly in educational settings. I believe that the reason for this failure does not hinge on insufficient development of the methods, as usually claimed by critics [Broadbent 1973]. Design research has hitherto largely ignored inner design routines, the actual cognitive processes that take place in the course of designing. A "brave new world" approach deemed them "primitive" and worthy of eradication and replacement by new, rational, systematic methods. As it were, the primitive intuitions refused to die out and methods failed in as much as they worked against the grain of what might very well be innate cognitive aptitudes. (The notion of innateness is strongly suggested by contemporary research in various fields, including cognitive sciences; see, for example, Piattelli-Palmarini [1989].) I therefore propose to include design cognition in our research agenda as a means of understanding and evaluating design processes, and would like to open the discussion by offering a preliminary picture of what needs to be studied and how it might possibly be done.

DESIGN PRODUCTIVITY

Since I want to look at design processes not as an independent cognitive investigation that happens to be carried out in the domain of designing, but as a study of design behavior, I must bridge between design process and design product. This I propose to do by way of comparing processes that lead to successful products to processes that end with less happy results. In a full-fledged design effort, success or lack of it depends on much that is extraneous to the designer, from input by associates to commercial, political, and other pressures. Here I narrow the discussion to the front edge of designing, the "embryonic stages" in Collins' terms, in which the designer is assumed to be in full control. A design product at that stage is defined as a (usually partial) visual representation (or a set of representations) of the designed entity, which is satisficing to the designer as well as to the viewer. A process that leads to such a product is successful and is thought of as *productive designing* (after Wertheimer's *Productive Thinking*; see Wertheimer [1945]). If we succeed in determining the parameters of design productivity, we should be able to assess them. This is precisely what we do in what follows.

Decomposition of the Process: Unit of Investigation

If we drew only one lesson from the early work on design methods, it would have to be the disposition toward a fine-grained decomposition of design tasks (as well as entities) into their smallest components. Similarly, the design process as defined above must be dividable into its most basic operations in order to be accurately represented. A representation is necessary so as to assess the process and look for *correlates* between it and the products it yields. I have shown elsewhere that in reality, there does not necessarily exist a match between the success of a design product and a compliance of the designer with commonly accepted standards of design performance [Goldschmidt 1989b]. In search of appropriate correlates, my approach is therefore entirely phenomenological in that protocols of observed on-line, real-life processes are the basis for all further analyses. Our protocols are amassed from tape recordings of think-aloud design exercises, cross-referenced with sketches that are almost always produced by designers at the front-edge phase with which we are concerned. If we accept the protocol as reflecting the process, we must now answer two questions: "What are the units into which we should parse the protocol?" and "To what test should we subject those units to arrive at productivity correlates?" This is, of course, the cardinal question of the entire discourse.

There is no consensus on operational units of performance in designing. In our case the wish to adhere to the smallest coherent operation detectable in design activity leads to the *design move.* The notion of a move has been used in different investigations and has acquired somewhat different meanings [i.e., Habraken 1985]. I define a design move as an act of reasoning that presents a coherent proposition pertaining to an entity that is being designed. According to Ericcson and Simon's method of informal protocol analysis [Ericcson and Simon 1984], our protocols have been coded into individual design moves, using human expert judgment. The following is an example of moves extracted from a protocol produced and analyzed at MIT [Goldschmidt 1989a]. Figure 4.1 shows a sketch made by the designer as she spoke, cross-referenced with her verbalizations:

1. "If I look at the form again (it might also be the influence of having done entry 1, but) it seems that spatially, these are the larger directions (*w, x*).
2. I am getting one, two, three spaces here (*p*) and one, two (*q*) there.
3. They're about square, so there is a tendency to try and see them as spaces.
4. These are secondary directions within the space (*y, z*), so the entry (3) is actually moving in along the secondary directions."

Now that we have design moves, we must decide how to analyze them. Most studies of the design process proceed to establish typologies of moves or their equivalents [Akin 1978, Wang and Habraken 1982, Eckersley 1988]. For our purposes we fail to benefit from an attempt to classify the moves made by designers at the front edge of designing. Instead, we look at a noncategorical attribute of the moves: their *linkability* to other moves. Each move is inspected by a human expert investigator to determine whether or not it maintains a logical link to one or more antecedent moves in the chronological sequence of design moves as recorded in the protocol.* Let us instantiate linkaging using

*I regard the question of validity of human qualitative judgment in both determining moves and deciding on links among them as exceeding the scope of this study.

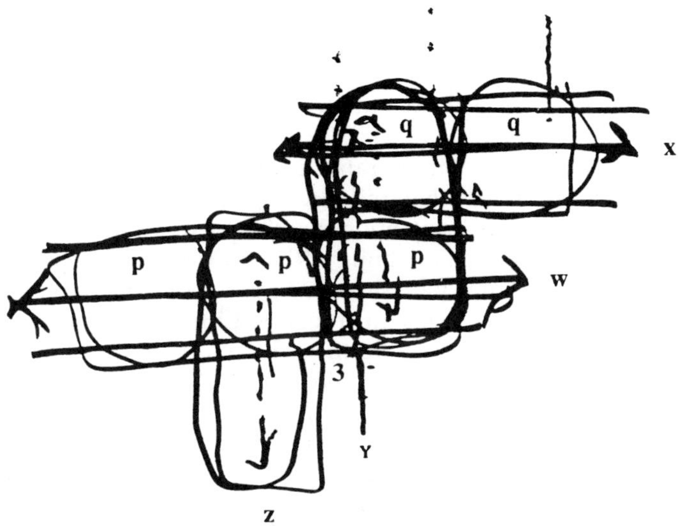

Figure 4.1. Sketch from protocol.

the four moves listed above. Move 1 cannot be linked to a preceding move as none is recorded (since the sample is taken from the middle of a protocol, in reality move 1 may very well be linked to at least one former move). Move 2, if judged only according to the verbalization given above, would be considered unlinked to move 1. A look at the sketch, however, leads to the conclusion that move 2 is indeed linked to move 1: the spaces it specifies are articulated through encircling the "large directions" of move 1 (see Figure 4.1). Move 3 is linked to move 2 as it elaborates on it but does not seem to be linked to move 1. Move 4 discontinues the spatial diagnosis of moves 2 and 3, but it returns to the question of directions which was first brought up in move 1 and is therefore linked to it. Our sample, then, includes a move that is unlinked (move 1), moves linked to immediately adjacent previous moves (2 and 3), and a move linked to a remote move that does not precede it directly (move 4). The sample is very small and does not show this, but a move may be linked to more than one other move (so far, the maximum number of links established for a move is six—a high value that occurs rarely).

I chose linkaging as the decisive test for design moves because I see the process of designing as a lot more complex than the analysis–synthesis–evaluation paradigm and its successors suggest. If design operations are not so orderly and do not proceed in fixed or predictable manners, they must be part of a wide network. Mapping links on such a network should reveal something about the nature of interconnectivity among design moves. If connections have recognizable structures, and if they are repeatedly related to the generation of successful design products, it will be possible to infer productivity correlates. Now, in representing the design process, the variables must therefore be not the moves, as in most studies, but the links among them. We need a representation that enters links among moves as input and as output provides information about the design process which we can use to assess its productivity.

Productivity Correlates

A representation system that uses links as input and displays structural design reasoning patterns as output was developed under the name *Linkograph* [Goldschmidt 1989a]. The Linkograph is a simple graphic notation in which the sequence of moves is shown on a straight line and the links are nodes at the intersections of diagonal network lines connecting two related moves. A Linkograph for the four moves cited above is shown in Figure 4.2.

Figure 4.3 shows a longer Linkograph, generated for the study quoted above [Goldschmidt 1989a]. We notice that links are unevenly distributed and this is a most important observation. It is my contention that the distribution of links among design moves reflects the status of design reasoning. When the reasoning is productive, the link distribution acquires particular patterns, which are geometrically discernible in the Linkograph. Low productivity yields unstructured Linkographs in which no patterns are discernible. I shall briefly describe the three patterns that have thus far been identified.

1. **Chunk.** A *chunk* is a block of links among successive moves which form links almost exclusively among themselves and are barely or not at all interconnected with other moves. In Figure 4.3 we discern two chunks, comprising moves 2–17 and 21–32. A semi-independent chunk (with more interconnections with moves beyond its boundaries) can be observed between moves 31 and 37. In the geometry of the Linkograph, the chunk takes the form of a triangle enclosed by the move line and link lines of the two extreme moves (first and last) of the chunk. Five of eight Linkographs that were prepared displayed at least one chunk each. Typically, chunks are generated by one to two dozen moves, regardless of the number of links enclosed in them.

2. **Web.** A *web* is formed when a large number of links is formed among a relatively small number of moves. As in the case of the chunk, the geometry of the Linkograph dictates a triangular enclosure of the web. We might say that the web is a piece of network in which the density of links is especially high. Webs are smaller than chunks and comprise, in our samples, up to nine moves (the minimum number of moves was set at five). They are also quite rare: only three of eight Linkographs display webs, and only in one case do we find more than one web in a single notation. The Linkograph in Figure 4.3 has a web between moves 26 and 32: 14 links per seven moves.

3. **Sawtooth track.** In some instances a sequence of several moves link each to the one preceding it. In such cases the link lines along the move line describe a pattern similar to a sawtooth. When a designer builds one proposition upon another in a

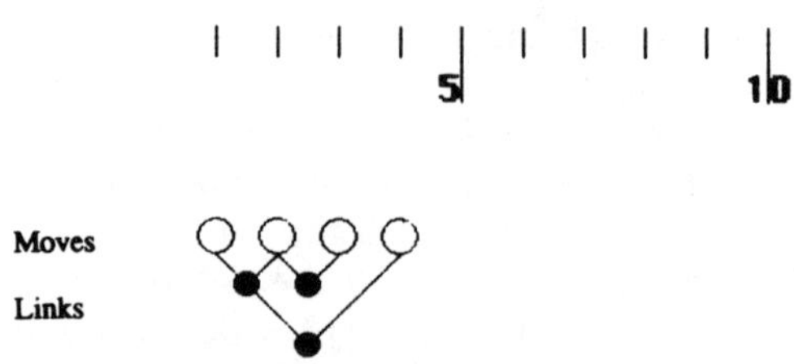

Figure 4.2. Linkograph for four moves.

sequential order, he or she is said to be tracking and the resulting track is called a *saw-tooth track.* In Figure 4.3 tracks can be seen in the following strings of moves: 7–11, 19–22, and 28–32 (tracks with fewer than three links are not counted).

Before I comment on these patterns I shall introduce two more indicators related to links, inferable directly from the analysis of protocols, but which are easy to recognize in the Linkograph.

4. **Link index.** The *link index* is the ratio between the number of links and the number of moves that generate them in a given design cycle. Webs, which by definition have a high link density, are marked by a high link index. In the cases studied, the overall link indexes for long cycles (up to 64 moves) range between a low 0.83 and a high 1.73 (with higher and lower ratios for selective sections). The cutoff value seems to be 1.0; any smaller value is very low. In the Linkograph of Figure 4.3 there are 37 moves and 52 links, resulting in a link index of 1.4.

5. **Critical moves (critical path).** Each move can have two sets of links. Links that are inferred from the protocol connect a move to previous moves as described above and are called backlinks. At the time the move is made, only its backlinks can be determined. When the cycle is completed and all backlinks are notated, the Linkograph also shows the backlinks of later moves to a particular posterior move. These links are that move's forelinks. Since each link connects two moves, it is one move's backlink and the other's forelink. The uneven distribution of links is reflected in the fact that moves differ greatly in the number of their links, in both the backward and forward directions. Moves that are particularly rich in links (in one or both directions) are called *critical moves.* When the richness does not result from repetitious propositions (which is sometimes the case), their ensemble is thought of as the cycle's *critical path.* For our purposes here it is significant to see where in relation to the structural patterns listed above we identify critical moves, and what that means in terms of the design process and its evaluation.

We can now proceed to check whether or not there is a correlation between a qualitative evaluation of the productivity of processes and the values reached for the patterns and indicators presented above. Much in line with our expectations, we find that Linkographs of the more productive processes display higher link indexes, more chunks, and

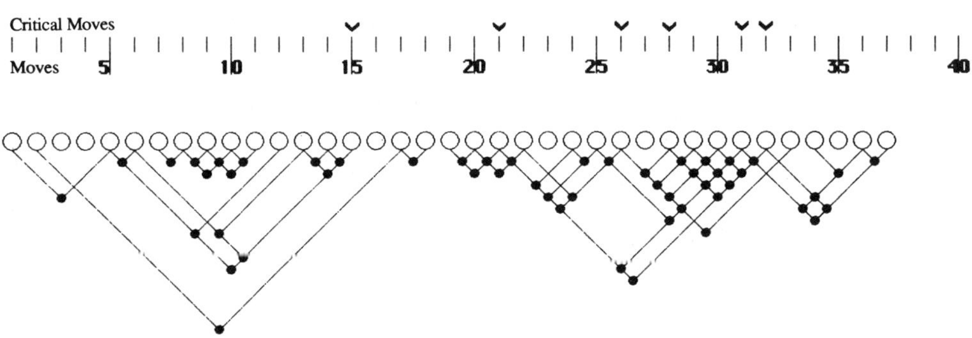

Figure 4.3. Linkograph for front-edge design cycle.

more webs. They do not necessarily have more or longer sawtooth tracks or critical moves, but their critical moves are consistently placed at strategic points, as I shall show presently. Because of the small number of cases that were checked I do not present statistics, which might not be significant enough. However, the tendency is clear-cut: productive design processes that end in some concrete visual representation(s) of a design entity tend to have structure: design propositions are made neither at random nor as a nondifferentiated continuous progression. Moves in productive design processes tend to be aggregated in groups; the productive designer thinks through a group of interrelated issues before he or she moves on to another group of potentially interconnected issues. The most significant design concepts are constructed from a tightly knit web of arguments that support one another. Generally speaking, the productive designer only elicits moves that have a high potential for connectivity to other moves, and therefore his or her link index is high. The less productive designer makes more random trials; more moves are produced which do not have a high potential for contribution to the design concept because they do not interlink much with other moves. The thinking in those cases is less structured and issues are not systematically investigated before the designer goes on to look at other issues. There are no "strong" concepts, which require the backing of interconnected arguments. And critical moves, when they exist, are largely repetitious and are randomly placed relative to the groupings of moves and links in the process (which are themselves random, so we get double randomness).

It is particularly interesting to look at sawtooth tracks, which do not correlate with productivity. A careful reading of the protocols, however, elicits the hypothesis that they do correspond to *design thinking.* In some of the cases in our sample, designers interpreted their task as commenting on the design exercise presented to them, without actually trying to reach much of a design product. This was especially true for design instructors, who used the opportunity to lecture on design, sometimes at great length. When not attempting to make something new, as when designing, but talking about things that these designers were already familiar and comfortable with, the reasoning took a somewhat different shape and arguments followed one another and built on one another in a straight, linear manner, which produced the sawtooth tracks. In the Linkograph of a senior design teacher 68 percent of the moves participate in sawtooth tracks, of which the longest spans seven moves. I conclude from this finding that general expert reasoning differs from expert design reasoning in the amount of networking involved, which is higher in a design search, while linearly continuous reasoning is dominant otherwise.

Let us return to Figure 4.3 and inspect the position of critical moves in the Linkograph (more than three links in one direction is considered "rich" and qualifies a move as critical). The moves in question are <15, 21>, 26>, 28>, <31, <32 (< stands for backlinks, > for forelinks). Five of the six critical moves occur in the second half of the design cycle, after the noticeable break between moves 18 and 19. The first subcycle seems preparatory: it has a low link index and no webs. Move 15 concludes a subchunk between moves 5 and 15. In the second phase of the process, move 26 inaugurates a new chunk and the last critical move, 32, ends it. Move 26 is also the first move of the single web in this Linkograph, and critical moves 31 and 32 frame it at the other end. Moves 31 and 32 are also the point of overlap between the chunk in this phase and the semi-independent chunk that follows it. Move 28 can be seen as the beginning of a web within a web (moves

28 to 32). Critical moves, then, have a tendency to frame webs and chunks, while they do not relate to sawtooth tracks. In an experiment I made, I compared the critical path that I inferred from the Linkograph of a protocol prepared by Wang and Habraken [Wang and Habraken 1982] to the critical path they inferred, based on their own qualitative judgment. The comparison yielded an almost identical critical path [Goldschmidt 1989a], which means that significant events in that design episode are link laden. This result supports a hypothesis that important events in the process of designing hinge on links among design moves. The Linkograph shows the nature of occurrences in the process of designing and suggests that those which are significant and lead to results (as specified by productivity) take on a distinct form that results from the topological relationship among links. Since no attempt was made to classify the moves (or the links), I would like to propose the primacy of structure over content in the cognitive operations that govern designing. The next and much heavier task is the tieing together of structure and content: if we did introduce a typology of moves or links into the investigation, what would it teach us about designing? This question is as yet unanswered.

IN CONCLUSION

The purpose of this exposition was to argue in favor of a process-oriented view of design. We evaluate designs to establish how well suited they are to their purposes, in the hope that what we learn from the evaluation will help produce better designs. Should we not also look at what makes a design process productive, in the hope that if we knew, we might be able to enhance the productivity of processes? I shall attempt but a brief and careful venture into this query, as research has only just exposed the tip of the iceberg and most of the work still lies ahead.

The major conclusion to which I can point is that the most effective design concepts are well-structured constructs built up from many interrelated individual design propositions. This is to say that strong design ideas are not conceived "in toto" when first elicited. Rather, many partial complementary and mutually reinforcing propositions must support one another for a useful design concept to emerge. A model of designing that rests on the building up of concepts is different from the common CAD "generate and test" model. In the new combinatory model sufficiency depends on a critical mass of arguments that acquires its viability and relevancy through the links among its constituent parts (but whose value may exceed that of the total sum of its parts). The great advantage of this structural model is its open-endedness: each element can potentially create any number of new links to propositions that participate in the formation of additional concepts. The most productive processes display a pattern which suggests that key propositions (critical moves) are likely to control the transformations from one (partial) candidate concept to another.

Inevitably, we now ask ourselves what the implications of this model are for designing and for design evaluation. It is not easy to predict how any of the results presented in this paper could be implemented in designing. The question merits responses on two different levels: that of basic research and that of design aids. First there are the theoretical problems pertaining to the understanding of designing and its computerized modeling. The biggest limitation of the structural process analysis I have conducted is the fact that, as yet, it does not interface with move typologies. If a typology or typologies of design moves could be devised that would prove to correlate with linkage patterns, we would have made a large step forward. Having said that much, I must add that in face of my own failure, so far, to make any progress in this direction, it is impossible for me to predict what findings could be expected, if any. Second, in a more pragmatic vein perhaps stands the challenge of enhancing productivity in processes. In educational settings in particular, design processes could be submitted to interim analyses of moves and links, in order to answer questions about the prospective viability of partial concepts. The results, possibly in the form of Linkographs, would serve as evaluations of the process. Such evaluations are invaluable to designers who wish to increase their awareness for strengths and weaknesses of their own processes. Finding out what links were formed in what patterns and what the critical path was reveals what combinations of moves proved especially beneficiary.

Finally, it may be of no lesser consequence to learn which moves proved futile: every Linkograph includes unlinked moves and their number is only somewhat smaller in the most productive processes. Trial and error, it seems, will continue to be part of human design performance. Our aim should not be to eliminate it but to control it through awareness.

REFERENCES

Akin, Ö. (1978). "How Do Architects Design?" in *Artificial Intelligence and Pattern Recognition in Computer Aided Design*, J.-C. Latombe (ed.), IFIP, North-Holland, New York, pp. 65–104.

Alexander, C. (1964). *Notes on the Synthesis of Form*, Harvard University Press, Cambridge, Mass.

Alexander, C. (1965). "A City Is Not a Tree," *Architectural Forum*, **4–5**.

Broadbent, G. (1973). *Design in Architecture*, Wiley, London.

Collins, P. (1971). *Architectural Judgment*, McGill–Queen's University Press, Montreal.

Eckersley, M. (1988). "The Form of Design Processes: A Protocol Study," *Design Studies* **9**(2): 86–94.

Ericcson, K. A., and H. A. Simon (1984). *Protocol Analysis: Verbal Reports as Data*, MIT Press, Cambridge, Mass.

Ervin, S. M. (1989). *Computer-Aided Diagramming and the "Generator-Test" Cycle*, Report to the National Science Foundation, Grant 8611357-DMC.

Gero, J. S., et al. (1988). *Chunking Structural Design Knowledge as Prototypes*, Technical Report EDRC-12-25-88, Carnegie Mellon University, Pittsburgh, Pa.

Goldschmidt, G. (1989a). *Graphic Notation for the Process of Designing*, Report to the National Endowment for the Arts, Grant 87-4251-0169.

Goldschmidt, G. (1989b). "Problem Representation Versus Domain of Solution in Architectural Design Teaching," *Journal of Architectural and Planning Research*, **6**(3): 204–215.

Goodman, N. (1978). *Ways of Worldmaking*, Hackett, Indianapolis, Ind.

Habraken, N. J. (1985). *The Appearance of the Form*, Awater Press, Cambridge, Mass.

Hillier, B., and A. Leaman (1972). "A New Approach to Architectural Research," *Royal Institute of British Architects Journal*, December.

Jones, J. C. (1963). "A Method of Systematic Design," in *Conference on Design Methods 1962*, J. C. Jones and D. G. Thornley (eds.), Pergamon Press, Oxford.

Luckman, J. (1969). "An Approach to the Management of Design," in *Design Methods on Architecture*, G. Broadbent and A. Ward (eds.), Lund Humphries, London.

Piattelli-Palmarini, M. (1989). "Evolution, Selection and Cognition: From 'Learning' to Parameter Setting in Biology and in the Study of Language," *Cognition*, **31**: 1–44.

Rapoport, A. (1969). *House Form and Culture*, Prentice Hall, Englewood Cliffs, N.J.

Swerdloff, L. M., and Kalay, Y. E. (1987). "A Partnership Approach to Computer Aided Design," in *Computability of Design*, Y. E. Kalay (ed.), Wiley-Interscience, New York, pp. 315–336.

Wang, M.-H., and N. J. Habraken (1982). *Notation of the Design Process: The Six Operations*, manuscript, MIT, Cambridge, Mass.

Wertheimer, M. (1945/1959). *Productive Thinking*, Harper Torchbooks, New York.

PART TWO

FRAMEWORKS FOR INTEGRATION

5

DESIGN FOR PERFORMANCE

Gerhard Schmitt

Department of Architecture
Swiss Federal Institute of Technology
Zurich, Switzerland

Design for performance describes a generative approach toward fulfilling qualitative and quantitative design requirements based on specification and existing cases. The term *design* applies to the architectural domain; the term *performance* includes the aesthetic, quantitative, and qualitative behavior of an artifact. In achieving architectural quality while adhering to measurable criteria, design for performance has representational, computational, and practical advantages over traditional methods, in particular over post-facto single- and multicriteria analysis and evaluation. In this paper a proposal for a working model and a partial implementation of this model is described.

INTRODUCTION

> "Are you implying that the very same object can be beautiful and ugly at the same time?" "Yes, by Zeus," said Socrates "and also simultaneously good and bad... because everything is as beautiful and good as it fulfills its purpose and as ugly and bad as it fails to fulfill its purpose."
>
> —XENOPHON, *Memories of Socrates*, fourth century B.C.

To arrive at a design solution that is "good" and "beautiful," no single computational method is sufficient. A building design program traditionally contains a problem description as a list of functions and associated areas that implicitly contains high-level specifications. There may also be a set of explicit specifications and requirements to be met in the form of building regulations. Design then develops according to the program, and the results appear directly and graphically on external media. However, performance and later behavior of a building are rarely deducted from traditional methods of presentation, and the display of numbers and charts may be an inappropriate means to explain why and how a building will or will not perform as a whole.

The thrust of this paper is therefore: (1) to present performance criteria as direct design specifications so that the resulting artifact is more likely to behave according to specifications, and (2) to redefine the role of evaluation techniques as important feedback components in an expanded design system. In the paper we first describe the overall concept, then concentrate on the components of a system to support the concept, subsequently present enabling methods used by the components, and finally introduce an example that realizes some of these ideas.

Components of the proposed model that will enable a computer system to produce design that has acceptable architectural quality and will fulfill performance requirements agreed upon at the beginning are (see Figure 5.1):

1. Building program
2. Knowledge bases
3. Performance specification
4. Database of complete and indexed cases
5. Single-criterion evaluation
6. Multicriteria evaluation

Enabling methods, techniques, and operators that support high-level interaction with the components of the overall model are needed. Of relevance for the proposed model—besides traditional techniques such as object-oriented modeling—are the following methods (see Figure 5.1):

1. Constraint-based reasoning
2. Specification-driven design
3. Case-based planning

Design by performance specification and learning from previous cases both take into account qualitative aspects and offer active design support. As an example the paper presents Archplan, a program for architectural prototype refinement.

CONCEPT: DESIGN FOR PERFORMANCE

Design performance cannot be enhanced with cosmetic operations after the design has been completed but must be a guiding force from the beginning of the design process. A stream of activities reaches from attempts in the late 1960s to formulate one design strategy as performance specification to an ongoing discussion at Carnegie Mellon University to base design processes on accountability and performance as fundamental principles. The research group around the architect Fritz Haller has developed a practice-oriented approach to the problem [Haller 1988].

Traditional single- and multicriteria evaluation deals with a design that has already been created. Feedback from the performance evaluation of a design may or may not influence the architectural result. In contrast, including performance specification in the design process as a driving force is a more promising means to create design in accordance with criteria that client and architect define and agree on from the outset.

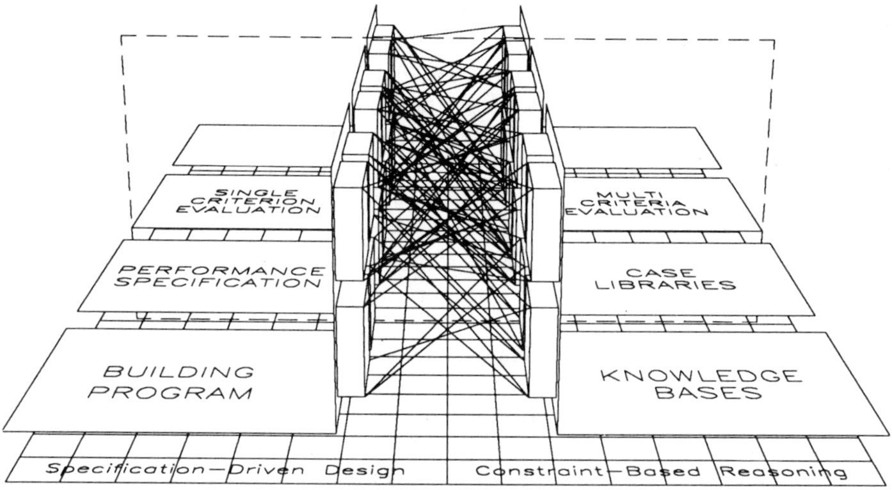

Figure 5.1. Components of a computer-aided design system: building program, knowledge bases, performance specifications, case library, evaluators. Lines connecting components depict necessary interactions. Underlying methods are, among others, specification-driven design and constraint-based reasoning.

Furthermore, design generation is simpler than single- or multicriteria evaluation, as the generating parameters are under the control of the designer.

As a concept that utilizes this history as well as available computational methods, we formed the idea of combining specification-based design with case-based planning. To lead to realistic and nontrivial results, other components are needed in a computer-aided design system. We propose to base a design model on the following components (see Figure 5.1):

1. **Building program.** The building program is the best known form of goal-oriented specification. It lists all essential functions, spaces, and requirements. It contains various levels of specification but is not always complete and may contain contradictions. This observation is particularly evident in the design of private residences.

2. **Knowledge bases.** These repositories give efficient access to algorithmic and factual knowledge needed to supplement the building program. Examples are local building codes and calculation methods, but also causal or heuristic relations between design features and effects derived from comparing or generalizing cases (see component 4).

3. **Performance specifications.** They are often separated from the building program and include requirements and constraints. By loosely coupling the design activity of generation with performance specification and graphical feedback, the designer has direct control of steering the evolving design utilizing knowledge from previous generations and existing cases.

4. **A database of complete and indexed cases.** The repository and memory of complete and indexed cases is perhaps the most important but computationally least explored

part of the driving forces for design. The case-based reasoning approach attracts growing attention for other tasks, such as planning [Hammond 1988].

5. **Single-criterion evaluation.** This is understood as judgment applied to analysis, where analysis is the judgment-free computation of performance values. Performance judgment is the process of applying past experience to the results of a computation and deciding whether or not the proposed solution is satisfactory. While computationally straightforward, the method cannot be used as a generator for design.

6. **Multi-criteria evaluation.** This is a more realistic form of judgment, as it considers multiple criteria and their interactions. It provides a view of a design project in terms of its performance, limited only by the number and complexity of the parameters.

Real judgment of performance is possible only after the design has been built and measured. An acceptable compromise is the simulation and evaluation of the object and the translation of feedback into new design specifications. Single- and multicriteria evaluations are interesting tools for judging traditionally completed design as well as evaluators and providers of feedback in the proposed computer aided design system. Single- and multicriteria evaluations concentrate on isolated, measurable, and quantitative aspects of performance prediction which are scientifically sound and result in a more comprehensive understanding of design and in the development of valid simulation tools [Schmitt 1982].

COMPONENTS OF THE SYSTEM

As meaningful analysis is more difficult than simulation and simulation is more difficult than generation, generation of design based on performance specification and existing cases seems a promising technique for building practical design systems. Design by performance specification its not meant to decrease the importance of the architect who has the ultimate responsibility for the building, but to ensure that the design solution is, from the beginning, in accordance with specifications. The necessary artifact representation is challenging, as it requires commitment to define mappings between performance syntax and architectural semantics. Design by performance specification is also different from single- or multicriteria optimization [Radford and Gero 1988], as it does not claim to be within an optimum but within a specified range that might and will include nonoptimal values. The following sections describe four modules of the proposed computer-aided design system.

Performance Specification

Performance specification we define as goal-oriented requirements concerning the behavior of an artifact after construction. Geometric and aesthetic appearance is one of the few performance factors for which a direct mapping between representation and artifact exists. Plans, sections, perspectives, and geometric models in general fulfill this purpose. Other performance factors, such as temperature, light, or air quality in a space, are much more difficult to display at the same level. In these cases we must assume that the design does not prevent the achievement of a required performance or that additional equipment is needed to guarantee required performance levels, often alleviating the devastating effects of poor design decisions.

Contrary to widespread opinion, design according to performance specifications does not prevent good architecture. Impressive in this respect is the work of the Viennese architect Anton Schweighofer, whose residential buildings fulfill such difficult and potentially conflicting specifications as low cost, good spatial quality, daylighting of every space (including basement and staircases), and decent access to and egress from underground parking. In the following sections we describe some methods to specify performance requirements.

Numerical Specification. Design by numerical specification utilizes constraint-based reasoning methods. However, no set of constraints is complete and meaningful enough to guide the design process intelligently, and the same holds true for rule-based design systems. The additional components shown in Figure 5.1 are necessary to create meaningful design. Numerical specification of design performance is acceptable in quantifiable domains for which appropriate numerical models exist and for the refinement of prototypes. Three types of numerical specification are those of threshold, range, or discrete values:

1. **Threshold.** Specification of minimum or maximum values. Examples are building code (fire rating, distances to closest stair), cost, *U* value, or time of completion.
2. **Range.** Specification of a range of acceptable values. Examples are cost, quality of material, and diurnal temperature swings.
3. **Discrete values.** Possibly arbitrary numbers derived from previous specifications. Examples are the number of light fixtures as a result of architectural requirements.

For known or routine design tasks, appropriate values are stored in databases or attached directly to the objects. Figure 5.2 demonstrates design by specifying numerical values in fractal and shape grammar algorithms. Success is measured by the degree to which the generated ensemble fulfills spatial and aesthetic requirements, its structural and organizational performance is guaranteed by specification. Figure 5.3 displays a design object created by specifying parameters of a prototype. Performance is measured by the degree to which the result recreates the Roman original.

Graphical Specification. The difference between numerical and graphical specification lies mainly in the user interface. A possibly confusing issue is the level of abstraction on the display. As design objects are necessarily abstracted as well, differences between design objects and specification objects may be obscured. We propose a three-dimensional graphical feedback and a two-dimensional specification objects to avoid this problem. In Figure 5.5, for example, the designer selects values in a range between low and high for the parameters by sliding bars. The resulting profile of the circulation scheme is a form of graphical specification. As a general rule we suggest design objects to be at least one order of realism higher than specification objects. Implementation of this idea makes use of the following tools:

1. **Buttons.** Selection of one or more choices from a set of options. Commercial examples are buttons in Hypercard and Suntools, or the graphical UNIX interfaces of Next or Silicon Graphics.
2. **Bars and other two-dimensional specification objects.** Definition of allowable ranges and for specification of values within a given range.

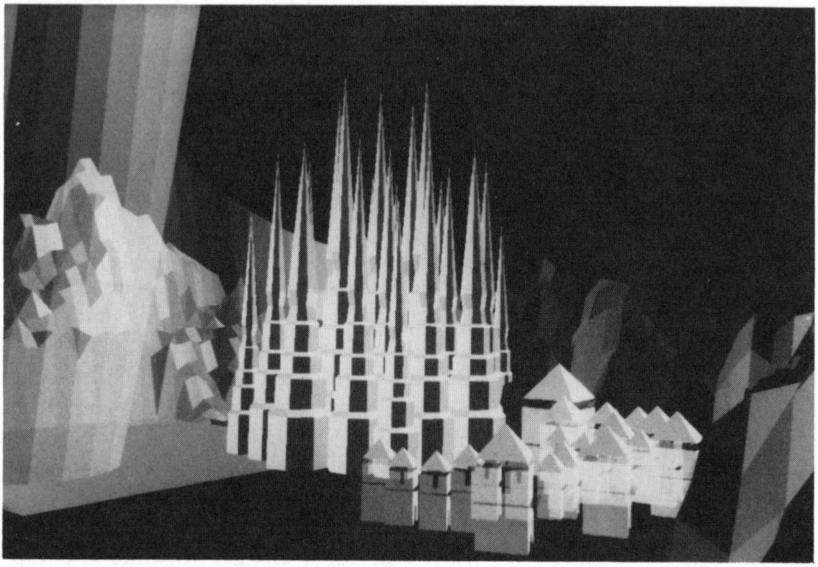

Figure 5.2. Grammar-generated buildings in fractal landscapes. Discrete numbers uniquely specify the result. Structural and organizational performance are guaranteed by not exceeding defined ranges.

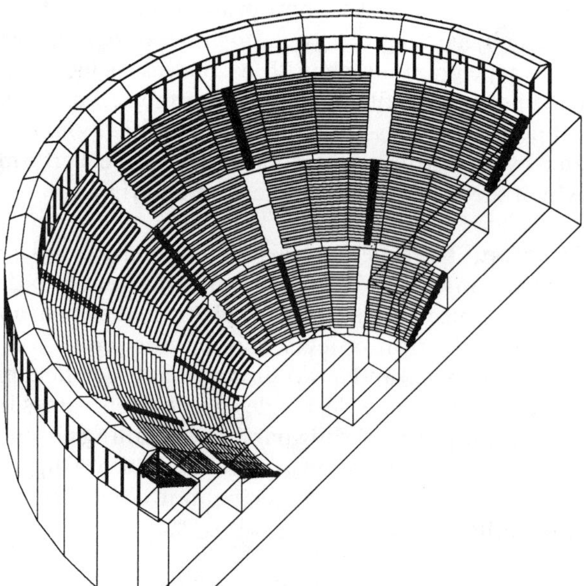

Figure 5.3. Reconstruction of the Roman theater at Aventicum in Switzerland, derived by specifying parameters of a building prototype.

3. **Three-dimensional volumes.** Cuboids or spherical objects facilitate specification of values from a three-dimensional selection space. Color, sound, and thermal comfort are examples.

4. **Direct object specification.** In the case of causal relations between object and evaluation criterion, manipulation of the object will indirectly specify the criterion: It is sometimes easier to change the object directly in order to define a specific criterion. Color, for example, may appear differently when applied to a building model and in context than when selected or specified from a neutral color space in isolation. The specification value is then stored for further reference.

Case Libraries

By observing designers at work and from common knowledge researchers came to the conclusion that the use of memory and cases is an effective approach toward design. Akin claims that with growing experience, designers reuse chunks of complex geometric and design knowledge to offer solutions to new problems, thus reducing the search space significantly [Akin 1986]. His experiments also show that unexperienced designers tend to rely less on cases, but use various search methods more extensively, and arrive at solutions later. Schank indirectly [Schank 1982, 1986] and Hammond directly [Hammond 1988] describe architectural activities as inherently case based.

A case in computer-aided design is an ideally complete description of a building with all its properties. While collecting cases, known properties are indexed. This includes those that are directly related to the investigation domain but also any recognizable and describable other properties. Particularly those not of direct interest may evolve to be of importance later. Description can range from geometry and indexed properties to analogical representation. Prototypes are a special type of case, as they are more abstracted and contain compiled knowledge in form of established relations. Cases, on the other hand, also contain unstructured and uncompiled, yet indexed data. Examination and comparison of cases was one important step in the development of ARCHPLAN (see Figure 5.4).

Our experience shows that achieving guaranteed performance by improving on previous cases is easier and more stable than achieving good overall architecture: attempting to be architecturally successful based on cases only (e.g., Palladio, Richard Meier, Daniel Libeskind) can and will quickly lead to repetition and neostyles because every new instance of design changes the overall perception of architecture. It appears that application of case-based reasoning together with performance specification could be much more useful because good performance will never become boring and indeed increases our understanding of quality.

A striking example for the application of a private case library by a famous architect is Mies van der Rohe's Berlin National Gallery. It shows both the potential and the dangers of the approach. An initial project for the Bacardi office building in Santiago, Cuba (1957–1958) has strong similarities with the Schweinfurt Museum Project (1960–1963) and the executed National Gallery in Berlin (1962–1967). Franz Schulze comments on the surprising similarity with "here, he recognized, conditions were ideal for a Bacardian structure to be transformed into a Miesian museum space" [Schulze 1985]. Obviously, the Bacardi office building fulfilled all architectural performance criteria that Mies had set for the museum.

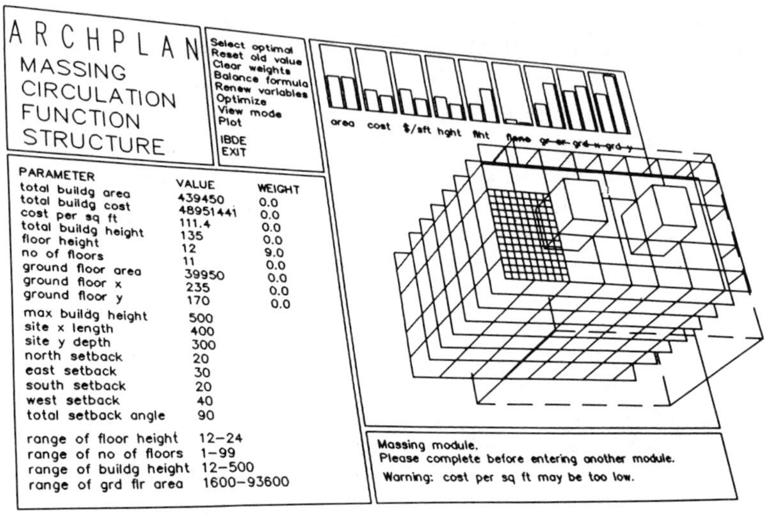

Figure 5.4. Site, cost, and massing module of ARCHPLAN.

Comparison of present design to a library of existing cases and the derivation of performance prediction by interpolating between design artifact and case library entries is possible. Matching specifications against cases in a library and modifying the closest match could save much time and guarantee a continuum of development while delivering a desired performance.

Single-Criterion Evaluation

Definition. As single-criterion evaluation we define judgment applied to the computed performance of one aspect of an existing artifact (in the following, referred to as analysis) or of a simulated design (in the following, referred to as simulation). Single-criterion evaluation requires the existence of models that sufficiently represent the entire object and the relations between the object, the criterion, and the world with its physical phenomena.

Examples. Thermal analysis and lighting simulations are good examples. For thermal analysis, an object is represented as simple as a geometric model with various attributes (first model). Causal relations between the object and its thermal behavior are simulated by applying first physical principles (second model), resulting in a set of statements that describe the thermal performance of the object. Incompleteness of and incompatibility between models cause well-known discrepancies between simulation and validation, which can reach 200 percent even within well-established and respected simulation programs [Maver 1989]. Lighting simulations attempt the visualization of light in space. Interaction between the object model, including geometry and material properties, and the model for the physical nature of light is approaching satisfactory levels of completeness. Validation is easier as lighting simulation provides visual feedback that can be checked against known conditions.

Traditional Applications. Single-criterion evaluation is a well-developed technique in architecture. Technical examples have their counterparts in the more difficult to describe, yet equally exclusive evaluation techniques concerning appearance, style, or taste. In fact, architects for most architectural competitions must concentrate on criteria that may be transmitted visually and instantly.

Computer Applications. Computer-based analysis and simulation require representation of models in appropriate data structures and enabling interaction between them [Clarke 85]. Object-oriented programming facilitates this request as objects contain both data and operations. The support of classes, inheritance, and encapsulation further aids development and testing of simple models. Single-criterion evaluation has a major advantage, as it allows the divide-and-conquer strategy:

1. Knowledge about first principles is added incrementally, thus refining the model.
2. Concentration on one criterion enables the exclusion of possible other interactions and disturbances of results.
3. A designer or client who is interested in one particular aspect of design will receive results in the fastest possible way.
4. Single-criterion evaluation is well suited for teaching, as causal relations between models can be well demonstrated.

While single-criterion simulation is representationally and computationally straightforward, the designer is often left with results that are difficult to implement without violating other performance requirements. Shortcomings are:

1. As models of an object or of physical, financial, and other phenomena are never complete, results of any simulation or analysis based on these models will not reflect reality.
2. The accuracy of a single-criterion analysis is proportional to the quality of the input.
3. It is difficult to validate the accuracy of a single-criterion stimulation.

Our strongest experience with single-criterion evaluation lies in energy simulation and analysis. From this experience, the need for multicriteria evaluation and lately for case-based reasoning arose, as it turned out that for a realistic single-criterion evaluation necessary knowledge about the object and physical phenomena is so massive that traditional representation is insufficient. To achieve a close fit between analysis and validated results, we found ourselves supplying knowledge (rules or references to previous cases; see the need for knowledge base and case library in Figure 5.1) for which even the largest program made no representational provisions. An example is the effect of air infiltration on energy consumption. Case-based knowledge from previous validations must be transformed by the designer to achieve correct results. Once meaningful input is available, hourly simulation programs such as DOE-2 and BLAST will model the buildings correctly.

Multicriteria Evaluation

Definition. Multicriteria evaluation includes the recognition and reconciliation of trade-offs [Schmitt 1988] after the application of judgment to the computed performance of more than one aspect of an existing object or of a simulated design. In addition to the models that represent the object, the relations among the object, the criteria, and the world with its physical phenomena, a model to represent trade-offs is necessary.

Examples. Trade-offs between energy and daylighting performance may serve as an example. For the passive solar heat gain of an office building large south- and east-facing windows are a possibility to reduce heating loads in winter. From the daylighting aspect, tall windows that allow deep daylight penetration into the space are advantageous. Fulfilling both requirements while guaranteeing satisfactory thermal and lighting conditions over the course of a day or a season may cause severe thermal and visual comfort problems—large heat loss in the morning and possible glare from low sun angles—that completely offset advantages of the theoretically beneficial architectural solution. In this case a trade-off mechanism that is able to recognize conflicts and to suggest possible solutions is necessary.

Traditional Application. In the past, the architect "just knew" about possible trade-offs between solutions. On the other hand, some trade-offs were knowingly ignored in favor of an overriding goal. Stylistic, aesthetic, and image-driven decisions occur continuously on the client's and the architect's side, often refusing solutions that would otherwise be preferable. This fact only proves that no set of criteria is complete and that our understanding of trade-offs must improve.

Computer Applications. Computer models for multicriteria evaluation are similar to those needed for single-criterion evaluation. Within the object-oriented programming paradigm, models are scalable and new representations may be added. Linkage to databases of various kinds becomes essential (see Figure 5.1). Computer-based multicriteria evaluation can produce a realistic view of the performance and behavior of a designed object. The danger of one-dimensional building design that is acute when a simulation considers only one criterion is reduced. As a negative aspect, the complexity of models and representations needed for multicriteria evaluation its considerable, easily allowing for input errors or loss of control. Whereas single-criterion evaluation can rely mainly on compiled knowledge in the form of algorithms, trade-offs are inherently less well defined. Simulating the negotiation of compromises is not the strength of current programming techniques.

Our experience with muiticriteria evaluation has developed in two stages. In the first stage we placed the results of a simulation of up to eight criteria next to each other in a performance chart, defined a threshold of acceptability for each criterion, and showed the normalized result of each criterion relative to the threshold. Major problems were the continuous normalization of often discrete results and the fact that any trade-offs occurred on the designer's side [Flemming and Schmitt 1986]. In the second stage, more knowledge on how to resolve trade-offs and a better understanding of computational tools suggest that the trade-off mechanism could be automated as well. The computational treatment of trade-offs has much in common with the handling of critiques as it occurs in the IBDE project [Terk 1989].

ENABLING METHODS

Three methods have particular relevance in this project: constraint-based reasoning, specification-driven design, and case-based planning. There are some similarities between the first two methods in that both allow design to develop only within defined limits. Both of them belong to one set, but the employed reasoning is different. Constraint-based reasoning utilizes constraint satisfaction with forward search as a weak method to find a solution. Specification-driven design is goal driven but also uses constraints. If the problem at hand is of hierarchical nature, constraint propagation is a feasible approach. If it is not hierarchically decomposable, optimization as a strong method and the creation of a Pareto set of feasible solutions are chosen. Having access to complete and indexed cases containing solutions to similar problems should help to solve ill-defined design problems and significantly reduce search time while allowing innovative design solutions.

Relations between these methods we describe as follows: a main design idea and a design concept are at the highest level of requirements in design. The idea is part of an existing case (case-based planning) or newly invented. It may sometimes be in conflict with specific requirements, which are introduced at the design specification level (specification-driven design). Design specifications are finally mapped into constraints (constraint-based reasoning).

Two practical examples may clarify the idea. Applied to a *renovation project* whose goal is the reduction of energy consumption of an existing building, the case base will ideally contain past renovations with before and after consumption values, renovation costs, materials, details, and images of the result. The specification is: reduce energy consumption to the lowest level at reasonable cost while maintaining air quality. As the problem may be of hierarchical nature—find and identify locations of energy loss, suggest improvements, check if sufficient air changes remain—constraint propagation is chosen. Applied to a *new design* where a client has only vague overall ideas of the building but knows exactly some specifics, such as the site, the cost, and the desire for a fireplace and a large kitchen, the case base would be scanned for examples that fulfill the requirements and for additional specifications the architect fills in. The result could be a prototype that fulfills the client's wishes and whose parameters would then be determined using constraint satisfaction. Search would appear only in cases where the architectural result is not satisfactory or the case base does not yield appropriate information.

We see cases as indexed depositories of structured or unstructured information with an attached recording mechanism for dependencies between features and effects that were observed over time. Examples for structured information are dimensions or actual yearly energy consumption. Examples for unstructured information are bit-mapped images of the building, reports from its inhabitants, or a history of repairs. It is important that both types of information be available; otherwise, the outcome of the cased-based process is trivial because it would lie in a possibly vast, yet finite and causal space. Although they do exist, causal relations between important features of a building and its true (architectural) meaning are rare. A depository of cases that grows over time will in fact lead to the discovery of more causal relations between feature and effect in a building. It is important that the case base remains accessible throughout the entire circle of design, particularly in the event of an impasse generated in utilizing specifications and constraints. Cases can be stored in frames or binary large objects (BLOBs), supported by a tool such as the Informix-OnLine database engine.

Specification-driven design formulates goals utilizing existing knowledge about causal relations to achieve design effects. It is most useful to generate parts of an artifact or a function once a general idea of the building exists, but meaningless for generating an entire design. Application examples are building subsystems such as HVAC, elevators, well-defined room layouts, or production streets. To specify a conference room for 12 people will produce the necessary components, satisfy the functional relations, and map them to a geometric description. To specify a design as a zero-energy building will only result in the known problems of underdefined specifications.

Constraint-based reasoning has produced results in certain areas of the design problem space, such as geometric reasoning. Constraints are defined as information that limits design variables to a specific set of permissible values that may be either qualitative or quantitative, continuous or discrete [Stauffer and Slaughterbeck-Hyde 1989]. Applied to the example of the conference room, necessities such as the existence of a projector and a projection surface could be mapped to constraints that govern the generation of the architectural space. In general, specification-based design and constraint-based reasoning stay closely related, with the specifications generating solutions and the constraints guaranteeing their compliance.

Constraints only are not sufficient to generate satisfactory design, neither are specifications. The design concept or the design idea is the governing and driving force behind design. Specification and constraints help solve the more low-level concepts. However, these two techniques may be applied to the evaluation and establishment of design concepts as well.

EXAMPLE: ARCHPLAN

A handcrafted design system for a small domain from which the proposed system evolved does exist and demonstrates one possible use of performance specification and case library. The Integrated Building Design Environment (IBDE) is a set of six engineering expert systems which deal with specification and design of speculative high-rise office buildings. The seven systems ARCHPLAN, CORE, STRYPES, STANLAY, SPEX, FOOTER, and PLANEX together form IBDE [Fenves et al. 1989].

ARCHPLAN is the architectural front end to IBDE and suggests a design based on interactive specification and definition. CORE accepts input from ARCHPLAN, generates a layout for elevators and lobbies, then produces layout solutions for the entire core, and finally displays the best alternative [Flemming et al. 1988]. STRYPES and STANLAY use constraint-directed search to accomplish the preliminary configuration and layout of the structural system. They are both implemented in EDESYN, an expert-system shell for engineering design [Maher and Longinos 1987]. SPEX performs the preliminary design of structural components selected by STANLAY. Design standards requirements are represented as decision tables [Garrett and Fenves 1986]. FOOTER decomposes the design of foundations into the subproblems of selection of foundation type, material type, casting type, excavation type, and parametric foundation design. The purpose of PLANEX is construction planning. It relies on a number of distinct knowledge sources in the form of decision tables which are rule sets. Decision tables are also used for knowledge acquisition to elicit rules [Hendrickson et al. 1987].

ARCHPLAN: The Program

ARCHPLAN represents a building and its functions as a prototype, expressed as a set of objects. The present design process approach is that of prototype refinement. It employs LISP functions and objects for procedural and declarative knowledge representation. Forward and backward chaining and hillclimbing help to find solutions to specific domain problems [Schmitt 1989]. The program consists of four individual modules that manipulate a common prototype, the building object. Design in each of the four modules, also represented as prototype objects, occurs in a cycle of specification, local prototype refinement, feedback, and eventual global prototype object update.

1. **Site, cost, and massing.** Design begins with the presentation of a set of default parameters that the designer may change at will to adjust to the problem at hand (see Figure 5.4). ARCHPLAN gives immediate graphical feedback on design decisions. Typical factors are the building site, defined by length and width, and a set of constraints given by local zoning laws. These constraints include setback distances, building height, and building setback angle. As these factors are usually nonnegotiable, their value is treated as fixed once defined. The values of a second set of constraints are more flexible. Examples are specifications of total budget, cost per square foot, number of floors, or floor-to-floor height. The designer expresses commitment to these values in the weighting factors assigned to each parameter.
2. **Function.** After the massing module is complete, a designer may continue with any module. The Function module defines distribution and three-dimensional placement of office, retail, atrium, parking, and mechanical spaces in the building, utilizing heuristics derived from existing cases.
3. **Circulation.** Vertical distribution of people and goods is the purpose of this module. The system contains knowledge about maximum distances between offices and elevators. Depending on preferences set by the designer, different solutions are possible for the placement of elevators (see Figure 5.5).
4. **Structure.** The structural grid of the building is already proposed in the massing module, but the designer may modify it at this point. Also, a set of structural types is available for selection. If a type is not feasible due to previous design decisions, ARCHPLAN will not allow its selection.

Evaluation does not take place except for visual judgment of the result. Specifying the input variables guarantees that the design is consistent with client requirements. ARCHPLAN serves as a test bed for user interfaces and specification-driven design. It has gone through several versions and is under development since 1987.

ARCHPLAN Component: The Case Library

For the development of ARCHPLAN we selected three cases as depositories of complete building solutions. The learning process was important in understanding, indexing, and

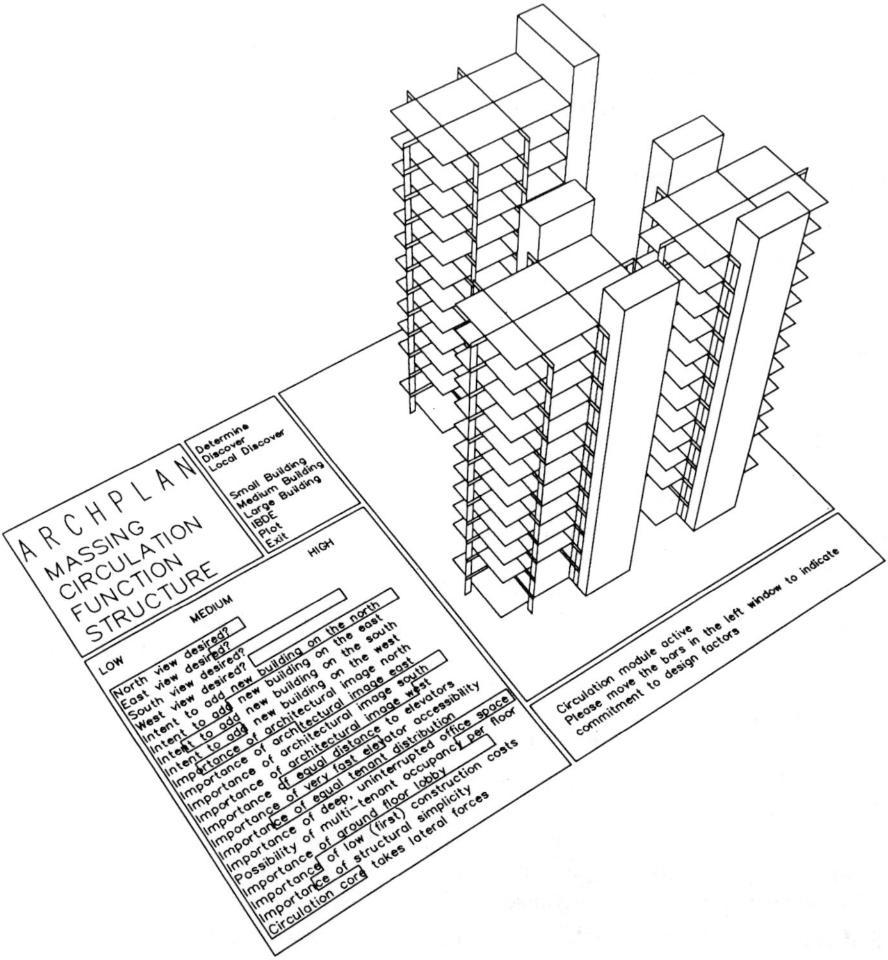

Figure 5.5. Circulation module of ARCHPLAN.

generalizing from these cases. The most interesting observation was that after the generalization of principles extracted from the cases, the reverse process of instantiation functioned well. The cases selected were:

1. **The Hongkong bank by Norman Forster, 1979–1986.** It is a high-rise building consisting of two vertical parallel slabs and a large atrium space in between. From this building we deduced principles of placing a project on a site and of general massing [Murotani 1986].

2. **The Lloyd's of London building by Richard Rogers + Architects, 1978–1986.** It is a complex building with significant daylighted office area and a large atrium. Interesting features are the elevators and their access. From this building we deduced principles concerning circulation systems [Papadakis 1985].

3. **The Pittsburgh Fifth Avenue Place by Hugh Jacobson (1985–1987).** A point symmetric mixed-use high-rise tower and a challenging structural system in the upper floors. From this building we learned about the three-dimensional distribution of functions.

At a later stage of the project we added an extended library of cases in the form of prototypes as a starting point for design. However, we did not use more than one case during program execution, as we lacked efficient storage, access, and control mechanisms. The extensive use of case-based information during the design process shows the necessity for dynamic access to this form of knowledge.

In addition to cases, ARCHPLAN makes use of a set of databases. The most prominent is the relationally organized global data store. A data transfer manager and a data translator transform data from and for each IBDE expert system in the form needed, most likely frames. After or during execution ARCHPLAN deposits data in the global data store, which may then be accessed by the other programs.

The second set of databases is external, non-project-related data. Implemented is a subset of the Means data on building costs which is derived directly from actual cases. The Means catalog presents overall and cost per square foot. Empirical cost parameters are building perimeter, building height, and building volume, among others. ARCHPLAN uses these small databases for some simple search for the best solution, given the conflicting requirements of daylighting and cost, using hillclimbing.

ARCHPLAN Component: Performance Specification

The site, cost and massing module of ARCHPLAN utilizes numerical specification. After the selection of a prototype from the case base, the designer enters performance numbers in dedicated fields such as total building area or building cost (see Figure 5.4). Those numbers will be inconsistent with regard to the prototype building model. They are guaranteed to be consistent only after the balancing function (Balance formula, top center window in Figure 5.4) has been executed. This function parses the input values and related waiting factors. Parameters represent nodes and weighting factors the relative strength of arcs in a semantic net. Building area, cost, cost per square foot, height, floor height, number of floors, ground floor area, and the ground floor dimensions need additional weighting factor input. If not specified, the default weighting factor is zero, signifying that this particular parameter is most flexible. If any of the numerical values is out of a range specified previously, ARCHPLAN will notify the designer in the lower right window. Ranges for optimization at the bottom left of the left window are also described numerically. Choosing the Balance formula option adjusts the bar graphs in the top right window to be in accordance with all numerically specified values. The bar graphs are normalized between minimum and maximum allowable values. The right graph shows values derived from optimization for cost, daylighting, or both; the left graph depicts the profile resulting from manually specifying one or more parameters.

The circulation module of ARCHPLAN employs *graphical specification*. None of the factors that contribute to define the position of the vertical circulation elements in high-rise buildings are of Boolean value; rather, they vary seamlessly in importance from low to high. The intent is the creation of a profile that, applied to the object created by previous design decisions, is able to suggest appropriate positions for the elevators. Figure 5.5 explains the profile, which consists of the following parameters:

1. Desired views in each direction (north, east, south, west).
2. Intent to add new buildings on each direction.
3. Intent to project a strong architectural image on each direction. This factor is very subjective and a certainly incomplete attempt to equate external elevators with interesting image quality.
4. Importance of equal distance from elevators. A high value will cause the elevator to be located toward the center of the building.
5. Importance of quick elevator accessibility. A high value will increase the number of elevators.
6. Importance of equal tenant distribution. A high value will suggest a symmetrical placement.
7. Importance of deep, uninterrupted office space. A high value will suggest external elevators.
8. Possibility of multitenant occupancy per floor. A high value will encourage nonsymmetrical solutions.
9. Importance of ground-floor lobby. A high value will suggest a large, uninterrupted lobby space.
10. Importance of construction costs and structural considerations. Archplan considers the structural efficiency of free-standing central cores versus symmetrically placed external elevators.

Specification occurs by sliding a movable bar over the parameter description. Short bars signify low values, long bars signify importance placed on a particular value. For each profile, the Determine and Discover options in the top center window produce the most likely solution, displayed graphically in the top right window. Choosing Local Discover suggests the next most likely solutions. It would be very difficult to enter, control, and, remember such complex preferences in a nongraphical way, for example in a spreadsheet. This type of input is attractive, as any change in the specification profile triggers almost immediate changes of the design object. Direct object manipulation would be achieved if, in addition, the graphical manipulation of the three-dimensional model would cause the specification profile to change.

CONCLUSION

With the move from isolated CAD packages toward integrated solutions, design for performance becomes a feasible alternative to post-facto evaluation. The traditional approach of generating design followed by evaluation could polemically be compared to generate and test. The approach proposed in this paper generates design according to evaluation criteria agreed upon in advance. Single- and multicriteria evaluation thus become program internal checking devices or maintain their previous role for traditionally created design and renovation projects.

Design for performance must differentiate between the architectural and performance aspect: while performance increases with reliance on tested good cases, architecture generally will not. Until we understand design more comprehensively, computer representation and process of design performance simulation are dependent on several modules:

1. The *building program description* as the most direct, yet not necessarily complete form of specification.
2. The *building performance specification*, including goals and constraints. Performance specification can become more high level, as better knowledge and case bases are available.
3. *Data and knowledge bases* to give efficient access to algorithmic and factual knowledge. Knowledge and databases should contain only proven or highly likely relations between design features and performance effects.
4. A *case library* as a repository for complete examples of good (and poor) performance with a recording mechanism for observed causal relations and effective search methods. Performance specification and case base are closely linked.
5. *Simulation, evaluation, and feedback* as mechanisms to evaluate the usefulness and quality of design. They come into play at any time during the design process to provide feedback and at the end to verify compliance of specification and actual performance.

ARCHPLAN as a partially implemented example of this approach reveals some of the strengths and weaknesses: the system guarantees consistency with performance specifications and the integrity of any derived solution. On the weaker side, the program is at the moment open only to the programmer and each addition of a new object requires careful integration. Given the example above and the improving representational capabilities of development software, it may be that computer-aided design will improve design performance much faster than it could improve traditional architectural design quality. So far, we seem to understand the mechanism of specified growth best.

ACKNOWLEDGMENTS

The author wants to thank his colleagues who helped to implement ideas of specification-driven design, in particular Maia Engeli and Annelies Zeidler. Chen-Cheng Chen, Shen–Guan Shih, and Chia Ming Chen programmed ARCHPLAN. The IBDE project is supported by the National Science Foundation of the United States at the Engineering Design Research Center at Carnegie Mellon University. Research on case-based reasoning in architecture is supported by the Swiss Nationalfond at the Swiss Federal Institutes of Technology in Zurich and Lausanne.

REFERENCES

Akin, O. (1986). *Psychology of Architectural Design*, Pion, London.

Clarke, J. A. (1985). *Energy Simulation in Building Design*, Adam Hilger.

Fenves, S., U. Flemming, C. Hendrickson, M. Maher, and G. Schmitt (1989). "An Integrated Software Environment for Building Design and Construction," in *Proceedings of the 2nd International Symposium on Computer Aided Design in Architecture and Civil Engineering, ARECDAO'89*, Barcelona, Spain.

Flemming, U., and G. Schmitt (1986). "The Computer in the Design Studio. Ideas and Exercises That Go Beyond Automated Drafting," in *Proceedings of ACADIA'86*, Houston, Tex., October.

Flemming, U., R. F. Coyne, T. J. Glavin, and M. D. Rychner (1988). *A Generative Expert System for the Design of Building Layouts, Version 2*, Technical Report EDRC-48-08-88, Engineering Design Research Center, Carnegie Mellon University, Pittsburgh, Pa.

Garrett, J. H., and S. J. Fenves (1986). *A Knowledge-Based Standards Processor for Structural Component Design*, Technical Report R-85-157, Department of Civil Engineering, Carnegie Mellon University, Pittsburgh, Pa., September.

Haller, F. (1988). *Bauen und Forschen*, Exhibition: fritz haller bauen und forschen, Kunstmuseum Solothurn, Switzerland, August.

Hammond, K. J. (1988). *Case-Based Planning: Viewing Planning as a Memory Task*, Academic Press, Boston.

Hendrickson, C. T., C. A. Zozaya-Gorostiza, D. Rehak, E. G. Baracco-Miller, and P. S. Lim (1987). "An Expert System for Construction Planning," *ASCE Journal of Computing in Civil Engineering*, **113**(5).

Maher, M. L., and P. Longinos (1987). "Development of an Expert System Shell for Engineering Design," *International Journal of Applied Engineering Education*, Vol. 3, pp. 279–286.

Maver, T. (1989). "The Computability of Technical Performance of Buildings," in *Proceedings of the First Convègno Nazionale, Consìglio Nazionale delle Ricerche, Progètto Finalizzato Edilìzia*. Sorrento, Italy, May.

Murotani, B. (1986). *Process: Architecture*, No. 70, Bunji Murotani, Tokyo.

Papadakis, A. (1985). *Architectural Monographs*, Richard Rogers + Architects, London.

Radford, A. D., and J. S. Gero (1988). *Design by Optimization in Architecture, Building, and Construction*, Van Nostrand Reinhold, New York.

Schank, R. C. (1982). *Dynamic Memory: A Theory of Reminding and Learning in Computers and People*, Cambridge University Press, New York.

Schank, R. C. (1986). *Explanation Patterns: Understanding Mechanically and Creatively*, Lawrence Erlbaum, Hillsdale, N.J.

Schmitt, G. (1982). "Auswirkungen des Energieproblems auf die Architektur unter besonderer Berücksichtigung von Computer Aided Design und Computersimulationen als Entscheidungshilfen im Entwurfsprozess," Ph.D. dissertation, Technical University of Munich.

Schmitt, G. (1988). *Microcomputer Aided Design*, Wiley, New York.

Schmitt, G. (1989). "Architectural Pre-processor to Engineering Expert Systems," in *Proceedings of the IABSE Colloquium 1989 on Expert Systems in Civil Engineering*, Bergamo, Italy, October.

Schulze, F. (1985). *Mies van Der Rohe: A Critical Biography*, University of Chicago Press, Chicago.

Stauffer, L. A., and R. A. Slaughterbeck-Hyde (1989). "The Nature of Constraints and Their Effect on Quality and Satisficing," in *Design Theory and Methodology: DTM '89*, W. H. Elmaraghy (ed.), DE-Vol. 17, American Society of Mechanical Engineers, New York.

Terk, M. (1989). "Integrated Environment for Computer-Aided Facility Development," Ph.D. thesis, Department of Civil Engineering, Carnegie Mellon University, Pittsburgh, Pa.

6

DESIGNER-AUTOMATED ALGORITHM PARTNERSHIP: AN INTERACTIVE GRAPHIC APPROACH TO FACILITY LAYOUT

Robin S. Liggett

Graduate School of Architecture and Urban Planning
University of California, Los Angeles
Los Angeles, California

In this paper research focusing on the use of interactive graphic interfaces for the solution of facility layout problems that have quantifiable but sometimes competing criteria is described. A system that implements a designer-automated algorithm partnership is presented.

INTRODUCTION

Automated solution techniques for spatial allocation problems have long been an interest of researchers in the computer-aided design field. The first algorithms for solving layout problems were developed over 25 years ago. There is currently a renewed interest in these types of problems, particularly at the commercial level. This is due largely to the increasing use of computer-aided design and drafting systems in the building design and management industry, and the development of facilities planning and management information systems, which provide easy access to inventories of data on space use and needs. New approaches to the solution of spatial allocation problems have also become possible with the availability of inexpensive, fast personal computer platforms and the introduction of user-friendly graphic interfaces that have become the industry standard.

Although the computer-aided design environment has changed dramatically, the issues of automated layout have largely remained the same. The problem remains one of allocating a set of activities within a facility such that space needs are met and a number of competing criteria optimized. There is a need to maximize operating efficiency, which is usually defined in terms of contiguity of working groups or ease of communication between employees while minimizing fixed operating expenses such as lease costs or the amount of wasted or unused space.

Early approaches to computer-aided facility layout attempted to capture competing criteria in the form of a single objective function which was then used to drive an automated algorithm. At the other extreme, the so-called "layout" features of some CAD systems simply provide the graphic interface for the user to lay out a plan in manual mode with little or no access to information concerning the layout criteria.

Both methods have proved inadequate. The manual method would certainly benefit from feedback of information that might aid in the solution generation. Although the solution generated by an automated algorithm is a helpful starting point for future interactive manipulation, a single cost function captures only one aspect of a designer's concerns in any realistic context.

In response to these concerns the newest systems being developed today are attempting to implement a designer-automated algorithm partnership. In this paper we describe research focusing on the use of interactive graphic interfaces for the solution of facility layout problems which have quantifiable but sometimes competing criteria. The ideas presented in the paper have been implemented in a personal computer system for facility layout which will soon be commercially available. Although examples from this system are used to demonstrate the feasibility of the concepts, they should be applicable to other types of optimization problems in the computer-aided design area.

THE FACILITY LAYOUT PROBLEM

Facility layout problems range in scale from assigning activities to cities, sites, campuses, or buildings, to the location of equipment and personnel groups on a single floor of a building. A problem can be large and complex, ranging across many scales, or can be narrow and focused at one scale. Consider, for example, a large corporation with a complex facilities planning problem of reassigning its 1400 employees to seven different buildings around the city. Working groups of employees must first be assigned to buildings. Then within each building, groups must be assigned to floors (i.e., individual stacking problems). Finally, at a more detailed level, individual employees must be assigned to locations on the plan of each floor. In this problem, key criteria are communication requirements between corporate groups as well as management location preference considerations.

In contrast, consider a more narrowly focused plant layout problem studied at IBM Rochester that looked at minimizing the movement of automatic guided vehicles (AGVs) used in a manufacturing plant [Gaskell and Martin 1989]. The goal was to determine a plant layout that would satisfy requirements of the manufacturing areas and optimize the efficiency of material handling with AGVs. In this application, critical information to be modeled was the travel network of the AGVs and the required flow of goods between areas.

A layout problem can surface in the design and allocation of space in a new building or the reassignment of space in an existing building. During the conceptual design phase, allocation of space within a new building can be a powerful tool to test alternative options for building configuration. Alternative plans can be evaluated with respect to best use of space in order to determine such things as the optimal number of floors, perimeter of the plan, and so on.

In an existing building, there is the ongoing problem of space management. For example, as project groups increase or decrease in size, how should employees be located within an office so that group contiguity is maintained with a minimum number of workspace moves? How can unused space be consolidated effectively to minimize lease costs?

Whether the problem is multistage or single-stage, whether it centers around planning for a new building or an existing one, a number of common goals have been identified. These goals include the following: satisfaction of the space needs of all activities; maximizing ease of communication, material, and information flow; achieving cohesion of groups that must work together and contiguity of closely interrelated groups; minimizing fixed operating expenses, particularly rent; minimizing wasted and/or unusable space; maximizing location preferences of management; and projecting growth, reduction and rearrangement in spatial arrangements over time [Hamer 1988].

Goals can be classified into two types of criteria: those that measure the relationship between the activity and the space itself (i.e., size, cost of space, management preference, etc.) and those that measure relationships among the activities (i.e., maximizing ease of communication, material and information flow, achieving cohesion and contiguity of groups, etc.). Automated (computer-aided) approaches to facility layout must be concerned with both the *representation of space* (i.e., the set of activities to be located and the space in which to locate them) and the *representation of the criteria* used for generating and evaluating a particular arrangement of the activities in the space. The selected representation of both the physical features of the problem (i.e., space) and the criteria will influence the type of problem that can be addressed, the potential automated solution methods that can be used, and the level of satisfaction with the solutions generated.

SIZE AND SHAPE OF SPACE

Space can be represented in different ways, thus providing a method of classifying alternative types of layout problems:

1. Space as discrete objects (one-to-one assignment problem)
2. Space as area (stacking problem)
3. Space as area and shape (blocking problem)

The simplest layout problem is the assignment of a set of discrete activities to a set of discrete locations in such a way that each activity is assigned to a single location. This is called a one-to-one assignment problem. Although not widely used, it has some applications. For example, the assignment of buildings to sites or the assignment of employees to preexisting offices or workstations can be a one-to-one assignment. The issue of size does not enter into the layout process.

Generally, however, space planning applications are not as straightforward as one-to-one assignment. The areas required by activities are not necessarily equal, so it is not feasible to match activities and locations on a one-to-one basis. In stacking plan problems, more than one activity can be assigned to a single floor or a single activity can occupy multiple floors (many-to-one or one-to-many assignment). How the area of an activity is apportioned among floors can be an important consideration in generating and evaluating a plan. A typical graphic representation of a stacking plan is shown in Figure 6.1*a*. In this case the size of the bar associated with each activity represents the amount of floor area it will occupy. Size is still a relatively simple issue, as actual activity shapes are not considered.

The most difficult problems to represent are those at the block-plan level. An activity is represented as a polygon on the plan. This polygon should be able to take on any shape and location while maintaining the required activity area. While such a representation is feasible in a manual layout process using a CAD interface, it is not generally possible when using an automated layout solution technique.

In the typical representation used by an automated system, each activity is comprised of some number of equal-sized modules according to required floor area. Correspondingly, the plan is partitioned into location modules equal in size to the activity modules. The problem is then one of assigning activity modules to location modules. The effect of such modularization can be seen in the block-plan graphic representation shown in Figure 6.1*b*.

EVALUATION CRITERIA

If the layout of a plan is created manually by a user with a CAD system, there may or may not be information to aid the user in evaluating a particular alternative either during or after the design process. Automated algorithms, however, require some method of evaluation in order to guide the layout process. Therefore, it is necessary to capture the goals of the layout process in some quantitative form. Depending on how the criteria are formulated, different solution algorithms are possible. Standard optimization models have been chosen to illustrate the effects of alternative *representations* on the types of problems that can be addressed. While they represent only one type of approach taken to automated layout, they illustrate many of the limitations seen in purely automated methods.

Relationships Between Activities and Space

The simplest formulation is one in which activity space and location space is represented numerically in terms of area required (e.g., total square feet) and area available

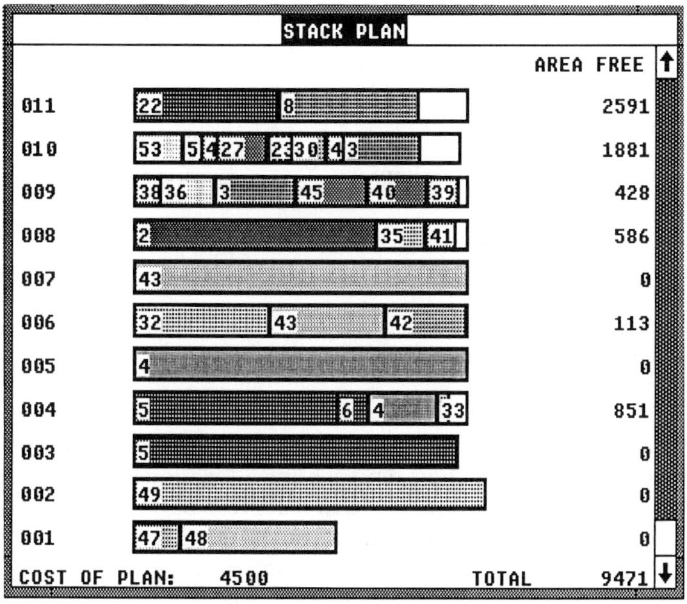

(a)

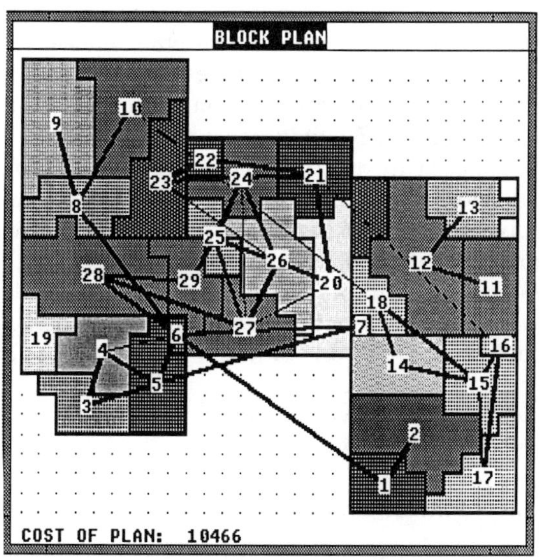

(b)

Figure 6.1. Typical graphic representations for layout problems: (a) stack plan; (b) block plan.

for assignment (e.g., again total square feet). Using a classical linear optimization model, the problem can be stated:

Minimize $\sum_i \sum_j F_{ij} x_{ij}$

subject to $\sum_j x_{ij} = A_i$ for each activity i; $\sum_i x_{ij} \leq L_j$ for each location j

where x_{ij} is the area of activity i assigned to location j
A_i is the required area for each of M activities
L_j is available area at each of N locations
F_{ij} is the cost of assigning activity i to location j

This formulation has some advantages. First it has a simple and straightforward representation. Although it may be desirable to display the results of this representation in some graphic form (such as a bar chart representing allocation of space), results can be portrayed simply as a table of numbers. Such a representation may be adequate for problems of allocating space to a site, or to floors in a multistory building, as in a stacking problem. The second and most attractive advantage is that powerful linear programming algorithms exist for the solution of such problems. Although limited, this formulation does have some potential applications; for example, the assignment of activities to buildings where lease costs plus any special startup costs are the overriding concern.

The simplicity of the formulation also leads to major disadvantages. It obviously considers only one type of criteria, those concerned with the relationship between activities and the space. Any relationships between activities are ignored. The representation of space as area prohibits the use of this formulation for solving block plan problems in which the location and shape of an activity must be considered.

Of more immediate concern are the limits it places on the type of problems that seem to lend themselves to this formulation. There is no way to guarantee that the entire area of a particular activity is assigned to a single location. Maintaining contiguity of a group may be of higher priority than the cost associated with assignment. The issue of activity contiguity can be handled by treating an activity as an indivisible object. If locations are also considered indivisible objects such that one and only one activity is assigned to each location, the problem becomes one of linear assignment where

$$x_{ij} = \begin{cases} 1 & \text{if activity } i \text{ is assigned to location } j \\ 0, & \text{otherwise} \end{cases}$$

and the area constraints of the original formulation are replaced by the following:

$$\sum_i x_{ij} = 1; \qquad \sum_j x_{ij} = 1$$

While activity contiguity is no longer an issue, actual areas of groups and activities are now ignored. In most stacking problems, multiple groups must be assigned to a single location (i.e., a floor in a building), or one group may need to occupy more than one floor. By modifying the representation to solve one problem, another emerges.

By adding the 0–1 requirement to the decision variables of the original linear model the problem becomes one of many-to-one assignment with area constraints. However, the major advantage of the original formulation, the fact that it could be solved by powerful LP algorithms, is lost. The new problem belongs to the class of integer programming problems (i.e., combinatorial problems) for which no sure solution technique exists. Although this formulation will ensure an entire activity is assigned to a particular location, it does not account for the fact that large activities may have to be split between locations. In addition, the set of criteria that represents the relationships among activities is not included in the model. Traditional facility layout has focused on generating arrangements that meet this set of requirements.

Relationships Among Activities

An optimization model that considers both types of criteria (i.e., activity-to-space relationships and activity-to-activity relationships) in a single function is known as the *quadratic assignment problem.* Formulated as a 0–1 programming problem, it looks like this:

$$\text{Minimize} \quad \sum_i \sum_j F_{ij} x_{ij} + \sum_{i1} \sum_{i2} \sum_{j1} \sum_{j2} Q_{i1i2} D_{j1j2} x_{i1j1} x_{i2j2}$$

$$\text{subject to} \quad \sum_j x_{ij} = 1 \quad \text{for each activity } i; \quad \sum_i x_{ij} = 1 \quad \text{for each location } j$$

where x_{ij} = 1 if activity i is assigned to location j
= 0, otherwise

Q_{i1i2} is some measure of interaction between activities $i1$ and $i2$

D_{j1j2} is some measure of distance or travel time between locations $j1$ and $j2$

F_{ij} is some measure of fixed cost of assigning activity i to location j

The two parts of the objective function are known as the fixed cost term and the interactive cost term. The fixed cost term is equivalent to the objective function of a linear assignment problem. The second term, which is the quadratic portion of the objective function, introduces costs caused by the interdependence of assignments. This criterion function is given in the specialized form first stated by Koopmans and Beckmann [1957] for problems concerned with the assignment of indivisible plants to locations. In the Koopmans–Beckmann formulation, Q values represent the flow of goods between plants and D values represent the cost of transportation between the locations to which plants are assigned.

This is a one-to-one assignment problem and is most easily visualized in terms of mapping a set of activities onto a set of locations. The problem consists of finding, from among the set of possible alternative solutions, a map that minimizes the value of the objective function. Although there are some applications of the pure one-to-one assignment problem, generally both activity and location space will have to be considered in terms of a set of equal-sized modules. Activity contiguity can be maintained by creating artificially high interaction values between modules belonging to the same activity. In this way trade-offs between fixed costs of assigning activities to a particular location and the objective of ensuring activity contiguity can be considered.

The quadratic assignment problem belongs to a class of combinatorial problems that are termed *NP-complete.* It is generally accepted that the efficient solution of NP-complete problems is impossible in principle. However, a number of good approximate solution strategies do exist. (See Liggett [1980, 1985] for a review of such techniques.)

AN INTERACTIVE GRAPHIC APPROACH

There are limitations associated with all the representations discussed above. To create a stack plan in manual mode, a designer would probably just assign activity areas to floors, splitting activities as necessary while maintaining contiguity if possible. In a one-to-one assignment problem, activities would be assigned to offices by matching area requirements. In block planning, the perimeter of the activity would be drawn on the plan without forcing it to conform to some artificial modular grid.

While the underlying representation of the system always limits the mode of operation, most CAD systems would be able to perform the simple operations discussed above. If an automated layout system is used to generate a solution, however, some sort of limiting representation is inherent in the problem formulation. Is, then, an automated layout approach either necessary or desirable? For small layout problems (in the range of 20 activities), the answer is "probably not." It should be relatively easy to solve problems of this size manually given appropriate visual aids. For larger problems, however, it is time consuming and difficult to create an initial layout by hand. A complete solution generated by an algorithm would be a helpful starting point for future interactive manipulation.

Therefore, rather than disregard the potentially powerful formulation and solution techniques of automated algorithms, the system described in the remainder of the paper combines an automated algorithm with an interactive graphic interface. User interaction is used to overcome many of the limitations of the automated algorithm. The process can be viewed in two ways. First, results generated by the automated algorithm can be improved by including the user in the solution process. Graphic information is used to help the user guide an algorithm to a good solution. Second and conversely, the process can be viewed as a manual one in which information from the automated algorithm is used to help guide the user to a good solution.

The interactive system was developed to run on an IBM AT or PS/2 personal computer in a Microsoft Windows environment. A color graphics monitor is required and a math coprocessor is desirable. It was designed to interface with two important components of an overall facilities management system: a facilities management information system, which provides information on space needs and resources, and a computer-aided drafting system, which can provide input on building perimeters. Layouts created using the system can in turn be sent back to a CAD system to use for more detailed design development.

The system itself is designed to solve any of the three types of layout problems mentioned earlier: the stack or zone assignment problem, the one-to-one assignment problem, and the block plan problem. Although each problem has a different graphic representation, much of the input data, the automated solution strategy, and the user interaction remains the same. The system incorporates both automated and manual methods for generating a layout. In both cases graphic output is used to guide the layout process and to aid in evaluating a particular solution.

THE AUTOMATED ALGORITHM

The automated algorithm selected as the basis for the interactive system is one developed by the author. The algorithm [which is known as SABA (stacking and blocking algorithm)] has been marketed commercially for a number of years by the Computer Aided Design Group and is currently licensed and utilized by several major CAD companies [Liggett and Mitchell 1981]. Although it has been considered one of the best of its type, it suffers from the problems of representation discussed earlier.

SABA has been formulated as a quadratic assignment problem, and the solution method is closely related to the implicit enumeration method for solution of quadratic assignment problems developed by Graves and Whinston [1970]. The layout task is viewed as an n-stage decision process where, at each stage, an activity is selected for assignment to a particular location. Since it is impossible to tell at any point the exact cost of making a particular assignment (this will depend on the subsequent location of other activities), the expected value of the objective function that results from the assignment is calculated instead. In the fully automated approach, the expected value of the cost function is calculated at each step for each possible activity–location combination. That combination which yields the minimum value is selected for assignment at this stage. This cost information is then fixed for later-stage evaluations. Execution of this algorithm traces a path through the decision tree to a very good layout, although the result cannot be guaranteed to be a global optimum.

It should be noted that existing approximate solution strategies are of two basic types: constructive initial placement strategies and iterative improvement strategies. A constructive initial placement strategy locates activities one by one, building a solution from scratch in a step-by-step fashion. The SABA algorithm falls in this category. An iterative improvement strategy begins with some initial arrangement and attempts to improve it incrementally. The most widely used improvement procedure is the simple "pairwise" exchange. Starting from an initial solution, the procedure consists of systematically evaluating possible exchanges between pairs of activities and making an exchange if it improves the value of the criterion. There has been considerable debate in the past over the relative performance of the two strategies for automated layout. Although there may be some possibilities for including user interaction in an improvement strategy, a constructive approach is needed for the type of user interaction presented in this paper. However, a pairwise exchange procedure has been included in the system for automated improvement of a constructed solution if desired.

DATA REQUIREMENTS

Information describing activities (termed *groups* in this system) is provided by the facilities management database. These data include group IDs, names and area requirements. At the block level additional information may be specified, such as proportion constraints, or actual group perimeters if a group has been preassigned. Location data (termed *zones* in this system) are provided by the facilities management database (i.e., area available for assignment) or can be input from a CAD system in terms of actual plan perimeters for a one-to-one or block problem. Information related to the evaluation criteria is divided into the two types discussed earlier: fixed costs and interactive costs.

Interactive Costs

Interactive costs involve some measure of interaction or adjacency requirement between groups to be located and the distances (or some measure of cost of travel) between locations on the plan. Interaction scores between pairs of groups are multiplied by the distance between locations to which they have been assigned. In the original Koopmans–Beckmann formulation, interactive costs had a physical (i.e., dollar) interpretation—the actual dollar cost of shipping goods. In a typical layout problem this measure is much more subjective and difficult to interpret.

The selection of a meaningful measure for interactions is a crucial consideration in a floor-plan layout application. Interaction values can be obtained from a subjective scale representing judgments about grouping requirements, or they may represent an actual measure of flow of goods or employees. This system allows data to be entered in terms of actual values (e.g., number of trips) or on some arbitrary scale (e.g., 3, most important; 2, less important; 1, least important). Facilities are provided for easily translating interaction data to another scale in order to experiment with different weighting options (e.g., 8, most important; 4, less important; 2, least important). Negative interactions may also be specified.

Travel Time Data

Getting meaningful interaction data is only one part of the equation. Actually measuring distances or travel times between locations can be difficult. When allocating space in an existing building, actual circulation distances or travel times between locations can be empirically determined. However, if the layout is for a new space where circulation has not been predetermined, it will have to be approximated as either straight-line distances or city-block distances between locations. Straight-line distances are suitable in situations such as open office floors, where it is usually possible to walk fairly directly from one location to another. City-block distances are suitable in situations such as partitioned office floors with orthogonal directions. The program allows the input of actual travel times or a circulation network when data are available as well as options for approximate calculations.

Fixed Costs

Fixed costs are those costs associated with assigning a particular group to a particular location irrespective of communication costs between groups. Fixed costs might represent rent or special facilities construction requirements. Fixed costs may also be used to represent preference information. For example, a first-floor location might be preferred for a computer room because infrastructure services (power, air conditioning, delivery access, and so on) are more readily or economically available on that particular floor. On the other hand, a first-floor location might not be considered appropriate for a computer room because of security issues.

In this system fixed costs may be entered either as actual detailed cost values for each group–zone combination or they may be entered in terms of a special simplified code: (P) *preferred* for assignment, (D) *discouraged* for assignment, or (X) assignment *prohibited. Preferred* for assignment implies a zero fixed cost. If a location is *discouraged* for an assignment a dollar/unit cost is generated. Although it is possible for a group to be assigned to such a location, a cost is incurred if the assignment is made. If an assignment is *prohibited*, it will not be made. Alternatively, to ensure a particular assignment, groups can be preassigned to locations.

When fixed and interactive costs are not expressed in terms of actual dollar costs (or on some common scale), as in the classic warehouse assignment problem, interpreting the meaning of the combined objective and weighing the two terms appropriately is a major issue. The system uses a graphic representation to help the user make this trade-off between the two types of costs.

THE INTERACTIVE INTERFACE

The system's main window is divided into two permanently displayed windows: the *plan* window and the *groups* window. The *plan* window displays graphically the type of plan appropriate for the problem. For example, Figure 6.2 shows a typical plan for *stack*. This graphic representation is used for all problems concerned only with allocating group areas to zones. A different type of graphic representation is used for *1to1* and *block* plans. In these cases the actual perimeters of the zones on the plan are shown.

The plan window displays the current assignment of groups to zones. The user can interact on the plan window, using the mouse to place groups, move groups, or to select groups or zones for a more detailed description. Interaction specific to each type of plan window is described in a later section; however, the nature of user interaction is fairly consistent across plan types.

The *groups* window describes the groups that are to be located on the plan. They are identified by their ID, name, and area requirements. Specific codes referring to their current assignment status are shown as a single letter preceding the group sequence number (see Figure 6.2). These codes, which give the user a quick visual reference to the status of the plan, are:

blank	Unassigned
p	Preassigned
*	All group assigned to a single zone
s	The group is split between zones (for block plans the "s" means that the group has noncontinguous parts)
+	Part, but not all of the group has been assigned.

If a dollar sign ($) precedes the assignment status symbol, there are preference costs (fixed costs) associated with the group–zone assignment (see e.g., the listing for group 5 in Figure 6.2). This position is blank if there are no preference costs.

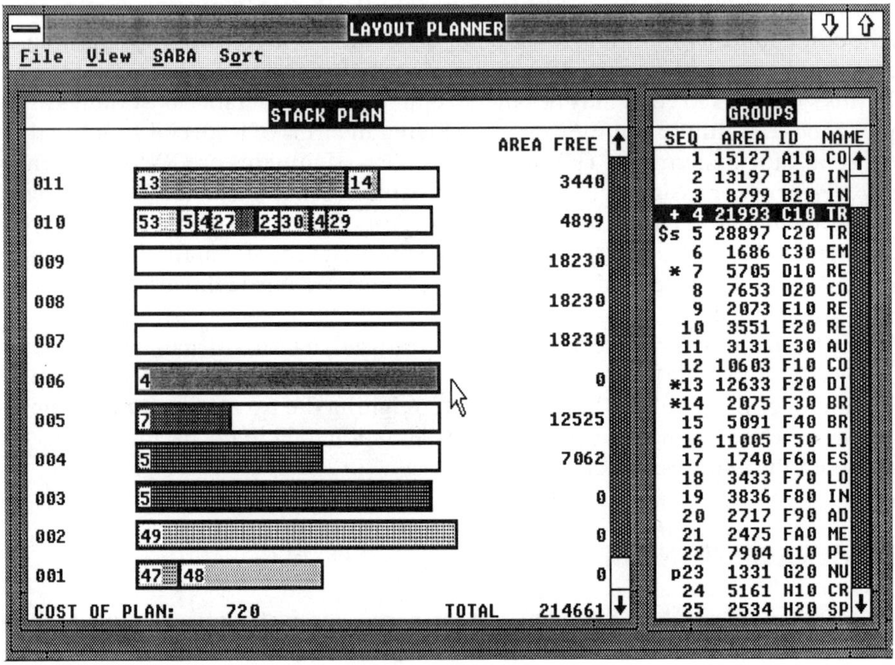

Figure 6.2. Screen windows.

Both the *plan* and *groups* windows can be scrolled and sized to show different portions of the plan or the group data. The stack shown in Figure 6.2 actually has 20 floors, so the user must scroll to see other floors of the plan. Figure 6.3 shows an alternative arrangement of windows. There are only three floors in this plan, so this arrangement allows a full-width *groups* window to be displayed. Two additional *pop-up* windows display more detailed information about a group or a zone. When a group is selected, either by selecting on the plan or selecting from the group list, a *group report* window is displayed. This report shows the status of the currently active group. If the group is already assigned to one or more zones, the area assigned to each zone is listed. The total assigned area, total required area, and total free area of the group are given. The report window for group 4 is displayed in Figure 6.3 and the group bar is highlighted on the plan.

There are also two action modes on the *group report* window: *move* and *free.* These become active if the group is already assigned, allowing the user to change the current assignment. If a group has not been previously placed, the program automatically places the group on one of the zones of the plan and activates the *move* mode. The actual interactive placement of groups on the plan differs slightly among plan types and will be discussed individually. A second *pop-up* window is the *zone report.* It displays the status of the selected zone with the groups currently assigned to the zone. The zone report window in Figure 6.3 shows the set of groups that are currently assigned to the second floor of the plan.

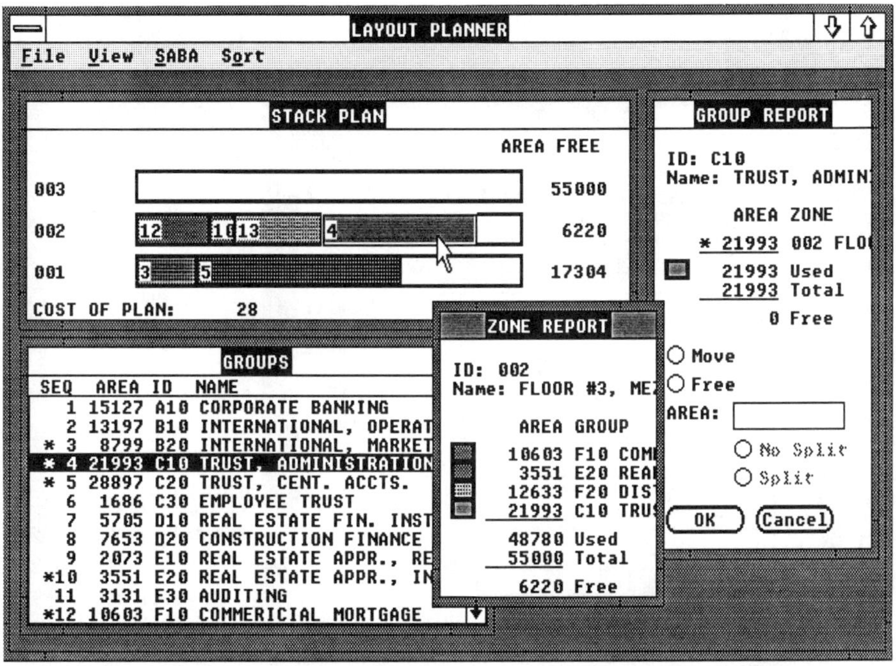

Figure 6.3. Alternative arrangement of windows with *pop-up* reports.

THE DESIGN PROCESS

The interactive design process is one of assigning the groups to locations on the plan. To make an assignment, a group is selected and tested on alternative locations until an assignment decision is made. This interactive process differs somewhat depending on the type of plan.

Stack or Zone Plan

A group is activated by selecting it from the group list or from the plan. By selecting a zone (i.e., rectangular bar representing a floor or location) on the stack window, the group will be placed on the zone. As a group is moved from zone to zone, the *free* area of the current zone is temporarily modified. The cost of the plan is also temporarily modified to reflect interaction costs associated with that group and others already on the plan. The group is placed permanently on a zone, when the *ok* button is selected on the *group report* window.

When a group is assigned to a zone, the entire area of the group is allocated to the zone. This area is displayed in the *area* edit box on the *group report* window. If the group area is greater than the free area on the zone, the group bar will extend over the end of the zone, and free area of the zone will be negative (see Figure 6.4). It is possible for the user

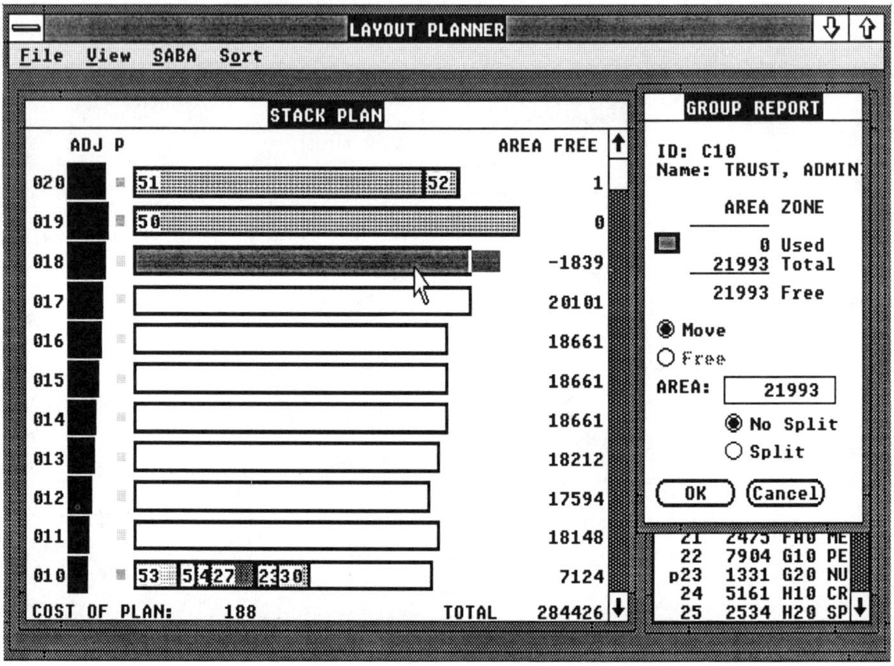

Figure 6.4. Placing a group on a stack plan.

to assign more area to the zone than may be available. The user can also split a group between zones. The exact area to be assigned to each zone can be determined by the system (i.e., whatever area is available on the zone) or can be entered by the user.

1to1 Plan

If a 1to1 problem is selected, outlines of the zones are required (this input can be provided by a CAD system). These outlines are shown in the *plan* window (see Figure 6.5). If a group has been assigned to a zone, the zone is filled with the color of the group (even if the group area requirement is actually smaller or larger than the zone area). The group ID is shown at the centroid of the zone.

When a group is activated for assignment, the group is displayed on the plan as a rectangle. The rectangle takes on the proportions of the zone to which the group is currently assigned, but with area equal to that required by the group. If the group area is less than the zone area, the group rectangle will be smaller than the zone rectangle, and vice versa. For example, the area of the group currently being placed in Figure 6.5 is larger than the zone area. Groups are moved from zone to zone by selecting on an empty zone. When the algorithm places groups in 1to1 mode, group area requirements are not considered. When operating in interactive mode, however, the graphic interface allows the user to consider these area requirements.

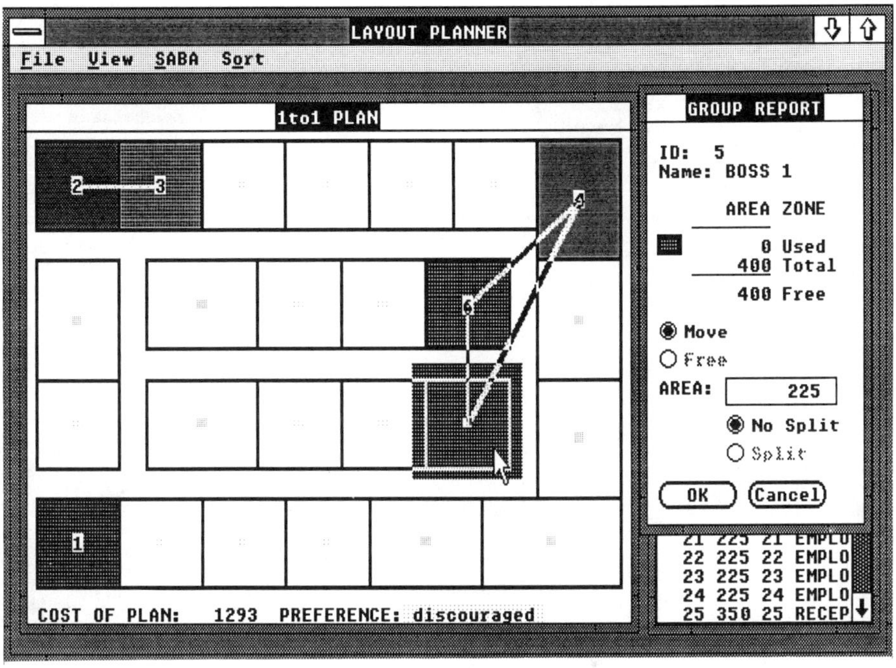

Figure 6.5. Placing a group on a 1to1 plan.

Block Plan

Block plan interaction is patterned after the approach taken by the automated algorithm. To solve large-scale problems (i.e., problems up to thousands of modules), SABA uses a two-tiered approach. The plan must be divided into zones by the user. These zones should generally be around 100 modules each. The algorithm first assigns groups to zones. Then on a zone-by-zone basis, the actual layout of each zone (generation of group perimeters) is determined by assigning groups to modules within zones. The process generates many small layout problems, where assignments to modules within zones are made considering the relative assignments of groups to other zones.

When creating a block plan, fixed costs are only relevant at the zone assignment level. For example, in the department store problem shown in Figure 6.6, the entrance zone at the lower left is preferred for high-impulse sale items, while the central and right zones are preferred for more major investment items. Thus zones can be used to partition the problem into architecturally meaningful space: for example, identifying entrance areas or separating exterior locations on the plan from interior areas. If there are no preference costs, zones just become artificial partitions that make the automated solution of a large problem feasible.

The *block plan* window shows the plan in terms of zone outlines and the underlying modular grid. As groups are placed, their perimeters are outlined and filled with system-

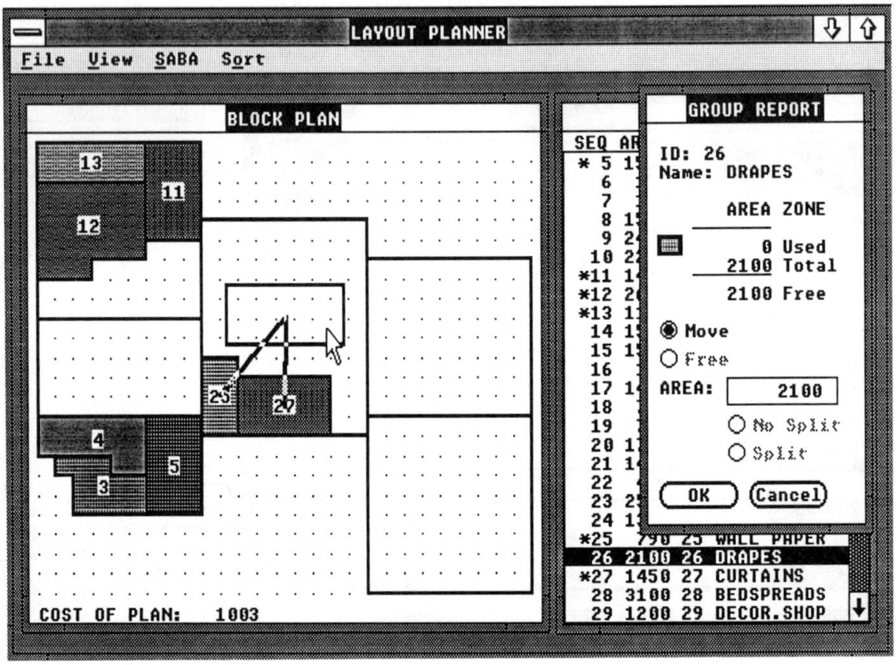

Figure 6.6. Placing a group on a block plan.

assigned colors. Group IDs are shown at the centroid of the group polygon (see Figure 6.6). When a group is activated, a rubber-band rectangle will appear at the centroid of a zone. The rectangle will have the area of the group and will be proportioned according to either the initial proportion specified in the input data or to the proportion of the bounding rectangle of the last shape the group took on the plan. Note the rubber-band rectangle shown for group 26 in Figure 6.6.

The rubber-band rectangle can be reproportioned or moved by grabbing the rectangle with the mouse. When the user selects the *ok* button on the *group report* window, all empty modules on the plan with centroids falling within the rubber-band rectangle are assigned to the group. A new polygon outline is calculated for these modules. If the area of the polygon is greater than the group area, it will be noted on the *group report* window. If the area of the polygon is less than the group area, a new rubber-band rectangle will appear at the centroid of the group on the plan. This rectangle will have an area equal to the free area of the group. The user can than move and stretch this rectangle to place it. Thus the actual group perimeter can take on any form as rectangles are added to rectangles. The group can also be split into multiple pieces.

DECISION AIDS

Selection of assignments can be made in a number of ways with varying amounts of input from the data and automated algorithm. These options are represented by the SABA *menu* available on the main menu you see in Figure 6.7.

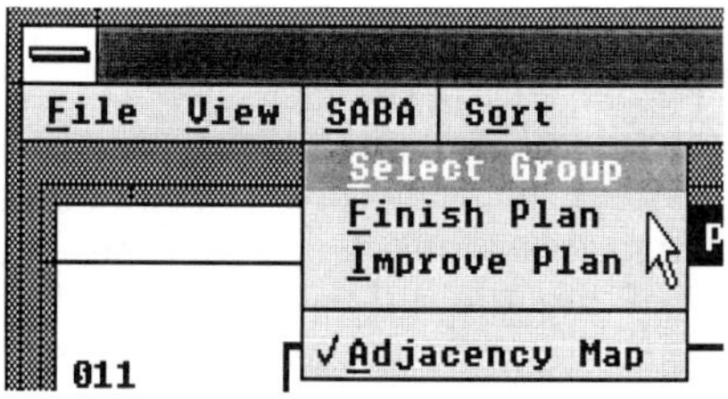

Figure 6.7. SABA menu.

Fully Automated Solution

At one extreme the user can request that SABA generate a complete solution and display the result. The only difference between this and the traditional automated layout approach is that the algorithm can be asked to generate the layout at any step in the process. The user can place as many groups as desired in the manual method and then ask SABA to "*Finish Plan.*" The groups already placed are considered preassigned by SABA and will not be moved. Thus it is an easy method for exploring alternative preassignment options in conjunction with the use of an automated algorithm.

After a plan has been generated by the algorithm, it is displayed for the user to evaluate and modify in interactive mode. At any point the user can ask the algorithm to attempt to improve the solution based solely on the value of the objective function ("*Improve Plan*" on the SABA menu). In this case a simple pairwise improvement procedure is employed. All pairwise group exchanges that meet area requirements and reduce the current value of the objective function are made.

Manual Mode With Aid from the Algorithm

At the other extreme, the user can use the system in a purely manual mode, placing each group by hand. The placement process can be aided by information provided by the algorithm. This information comes in two forms. First, when a group is selected to be placed or to be moved, an adjacency map showing levels of interaction with groups already on the plan is displayed. On the *stack* plan the adjacency map is shown as a bar chart to the left of the plan (see Figure 6.4). The wider the black bar the more desirable the floor (zone) is with respect to adjacency criteria (adjacency to groups already on the plan).

On the *1to1* and *block* plans, the adjacency map is shown as a network of interactions between groups (see Figures 6.5 and 6.6). The wider the link, the more important the adjacency requirement. The adjacency map shows only links between the group currently under consideration for placement and groups already on the plan. However, the entire adjacency network can be displayed as an option.

Fixed-cost information is displayed in a separate graphic format. Assuming that in most cases, the users will use the simple preference code for defining activity–space relationships, a colored graphic symbol is used to show location preference information. In the *stack* plan a small colored square is shown to the left of each zone: a *red* square means that the zone is prohibited for that group, a *yellow* square means that the zone is discouraged for the group, and a *green* square means a zone is preferred for the group (see Figure 6.4). Note that the user can override a prohibitive cost and place a group on a prohibited zone. The algorithm, however, will honor prohibited zones in a automated generated solution.

On the *1to1* plan, the preference key symbol is shown at the centroid of each zone (see Figure 6.5), and on the *block* plan, each module of a zone is colored with the preference key. While the automated algorithm considers both types of criteria simultaneously as fixed and interactive terms of the quadratic assignment objective function, displaying each type of information separately allows the user to make the appropriate trade-offs between the two.

At the bottom of the plan window a number representing the cost of the plan is displayed. This cost normally includes only interactive costs (i.e., the sum of the interactions between groups multiplied by distance or travel time between zones). If fixed-cost information has been specified as scaled data rather than with the preference codes, it will be included in the cost. This feature will allow a problem to be handled where fixed and interactive costs are on a comparable scale (i.e., dollars).

The single cost number generally must be viewed skeptically. Although it is another visual cue in the placement process, the user should recognize that it is not always a good measure for evaluating a plan. For example, in determining the value of the interactive costs for a block plan, distances are calculated on the basis of activity centroids. It is easy to see that the shape of the activity affects the value of the cost function. Long, thin, rectangular shapes may result in shorter distances between activities as compared to more compact square shapes, even though square shapes may be more desirable. It is much better to evaluate the plan by visually examining shapes and positions of groups. This visual inspection can be combined with the graphic representation of interactive costs and feedback on preference violations.

When a user selects a group for placement, additional information comes from the automated algorithm in the form of a suggested location for the group. The algorithm makes placement decisions on the basis of the value of the combined objective function (the fixed costs and interactive costs) considering both placements that have already been made and future requirements. When a group is selected, it is placed by the system at the algorithm's suggested location. The user can then use this as a starting point for manual interactive manipulation.

Another aid to the placement process is the "*Sort*" option. By default, groups are listed in the *groups* window by sequence number. They can be sorted, however, either by area or by the sum of their interaction scores to groups that are currently on the plan. This aids the user in group selection. In many cases it is desirable to place the largest groups first due to area constraints. Alternatively, the user may want to order the placement of groups by considering the level of interaction with groups currently on the plan. This allows the user to simulate the decision process used by a number of other automated algorithms. That is, select the next group to place as the one with the highest interaction to groups already on the plan. If no groups have been placed, the groups are sorted by total interactions, so the user could select the group with the most total interactions as a starting point.

Automated Mode with Aid from the User

An alternative approach lets the algorithm control the placement process with input from the user. If the user chooses "*Select Group*" from the SABA menu, the algorithm will determine the next group–zone assignment and display it on the plan. The user can accept this assignment and ask the algorithm to continue with the next selection, or he or she can modify the placement. The same graphic information available in the user-controlled selection process is displayed at each step. The only difference between this approach and the user-controlled method is the selection of the order of groups to place. Either the user or the algorithm may be used to control the order of group selection. Either way the suggested location from the algorithm will be given.

SOLUTION EVALUATION

While the three types of problems have different physical representations and slightly different evaluation criteria, the layout process is similar. For each of the three types of problems, four alternative solution modes were tested:

1. Solution generated by the automated algorithm
2. Solution generated by the automated algorithm and modified by the user
3. Solution generated interactively with the algorithm controlling the order of group placement, but with the user modifying activity location and shape
4. Solution generated interactively by the user with graphic aids from the algorithm

Although not anticipated, it is not surprising in retrospect that a different solution mode was best suited to each of the three types of problems.

The *stacking problem* tested was quite complex. Fifty-four groups were assigned to 20 floors of a high-rise bank building. Some of the groups were preassigned. The executives were preassigned to the top two floors, a number of service groups such as reproduction and security were preassigned to the transfer floor (floor 10), and the public banking operations were preassigned to the lower floors. Preference costs were used to discourage or prohibit a number of assignments. For example, most groups were discouraged from the two floors closest to the executives and prohibited from the transfer floor. Group 5 (Central Accounting) was discouraged from all except the third floor. However, it was too large to fit on a single floor, so some fixed cost had to be incurred.

The best results for this stacking problem were generated interactively. Surprisingly similar results were obtained when the algorithm controlled the order of group placement as when the user was in control. In fact, the same value for the cost of the plan (3308) occurred for both methods, although the actual arrangement of groups on the zones differed slightly. One arrangement is shown in Figure 6.8. In these solutions only those groups that were too large to fit on a single zone were split between floors, and the only preference criterion that was violated was for group 5—a cost that could not be avoided.

Both of the solutions generated interactively seemed significantly better than the SABA-generated solution, which is shown in Figure 6.1. The cost was higher for the SABA plan (4500) and there were more split groups. While SABA generated the plan very quickly (a few seconds on an IBM PS/2 model 80), it was actually more time consuming to improve this plan manually than to create the plan interactively using the graphic aids.

For the *one-to-one problem* the algorithm generated a good solution very quickly with a cost of plan of 5080. This plan was improved slightly by hand (a lateral shift of three

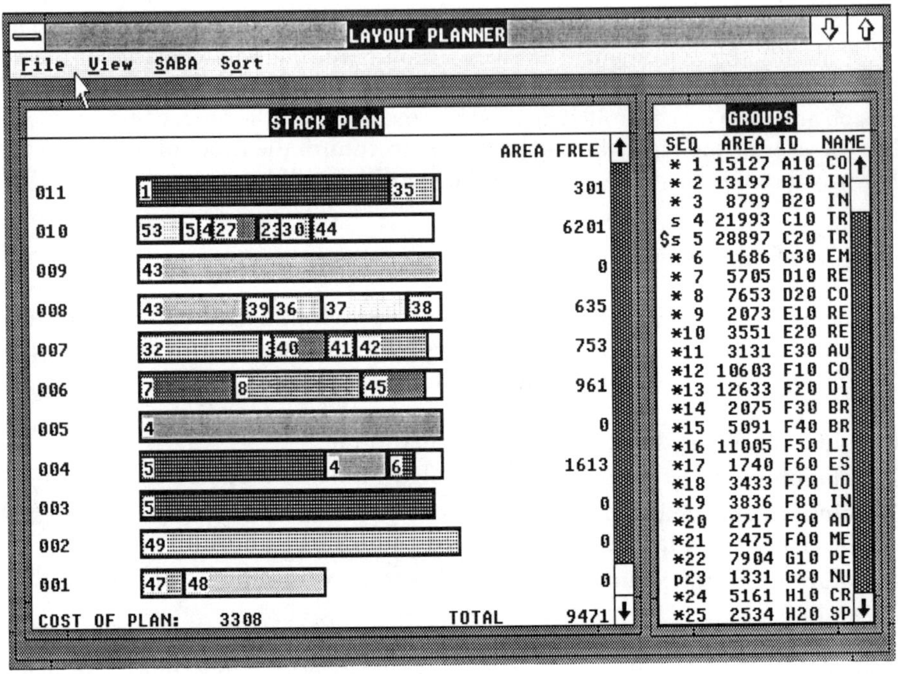

Figure 6.8. The *best* stack plan.

groups), resulting in a reduced cost of 5031 (Figure 6.9). Solutions generated manually were not as good either in terms of the cost of the plan (5587) or by visual examination of the interaction network. This problem, which involved the assignment of employees to individual offices, is the ideal application for the automated method. There were no preference costs and no problems of maintaining group contiguity. Group area requirements were ignored. Only interactive costs were considered.

In the actual problem, however, group and zone areas are not equal. For a realistic solution, groups with larger areas should be assigned to larger offices. Area considerations can be incorporated in the 1to1 model using preference costs. The assignment of large groups to small offices may be prohibited or discouraged. In manual placement mode, the size of the group rectangle provides a visual que for the user to consider area requirements. Working with the area constraints, the pure algorithmic solution is not as good as the user-generated solution. The algorithm is not as capable as the human being in making the trade-offs between the interactive costs and the fixed costs now included due to area constraints. Once these additional constraints are considered, the manual methods with the graphic aids are the preferred solution modes as in the stacking problem.

The *block plan* is the most difficult type of problem to solve both for the automated algorithm and for the manual solution mode. For the department store block plan, a better solution resulted when the output from the automated algorithm was used as a starting point. A significant amount of user manipulation was still required to improve the plan, but it took less time than building the solution from scratch manually. Figure 6.1 shows the original plan generated by SABA, while Figure 6.10 shows the manually

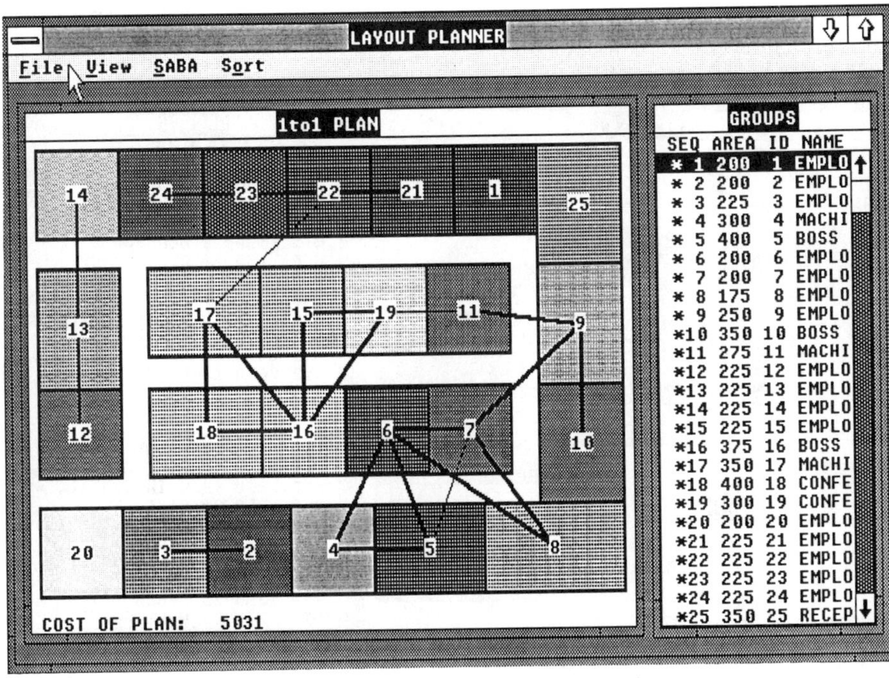

Figure 6.9. The *best* 1to1 plan.

improved plan. The starting solution generated by the algorithm provided a valuable visual representation of the problem criteria. It would be ideal to use such a starting solution as a template on a separate layer (as in a CAD system) on top of which the manual solution could be constructed.

In all cases, except for the pure one-to-one problem where area constraints were ignored, some method of combining the algorithm and information about the criteria with user interaction resulted in the best solutions. *Best* is defined in terms of the numeric cost associated with the plan; a visual examination of group location, contiguity, and shape; and the time required to generate the solution.

CONCLUSION

Algorithmic solution techniques have in the past proved to be useful tools for generating solutions to design problems such as facility layout, where some of the criteria can be specified in quantitative form. Experience has shown these tools to be most effective when employed in an iterative fashion. For example, in a typical layout problem, the design process might start by generating a plan where only information on group interactions is considered. Generally, the result produced will be unsatisfactory, prompting the designer and/or client to make their design requirements more explicit. This can be accomplished by adding information on group–location preferences and group preassignments. The consequences of these additional criteria can be identified by studying the resulting plan. Appropriate trade-offs can then be made as the problem is gradually

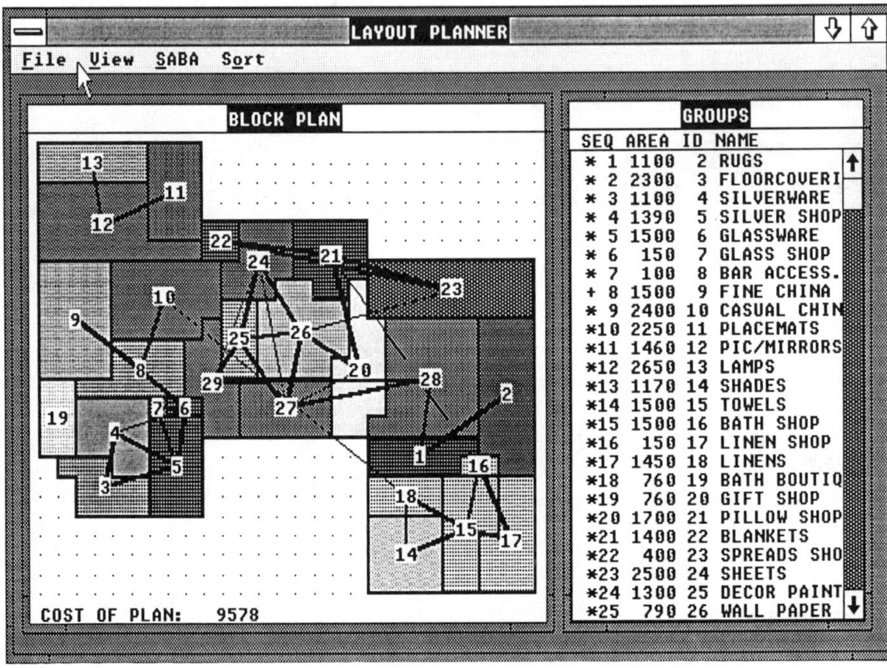

Figure 6.10. The *best* block plan.

transformed from one in which interactive costs clearly dominate the assignment process to one in which preassignments and preference costs are dominant.

Such an iterative approach to the design process was possible in an ad hoc way with older *batch* versions of automated programs. The addition of an interactive graphic interface, however, not only facilitates this process but can change the way in which algorithmic solution techniques are used by allowing a more direct designer–algorithmic partnership during the design process.

Information generated by the algorithm can be displayed graphically to aid the user in making trade-offs between competing criteria. Rather than generating one least costly plan, an understanding of the cost consequences of location decisions can be provided, such that the search for a solution responds to a broad spectrum of complex and often ill-defined criteria. It seems clear that real-time, graphically compelling feedback combined with human intuitive reasoning and visualization skills helps overcome the shortcomings of the computerized representation of a design problem.

REFERENCES

Gaskell, J. M., and L. G. Martin, (1989). "Study Examines AGV Movements," *IFMA Journal*, June, pp. 30–35.

Graves, G. W., and A. Whinston, (1970). "An Algorithm for the Quadratic Assignment Problem," *Management Science*, **17**(3): 453–471.

Hamer, J. M. (1988). *Facility Management Systems*, Van Nostrand Reinhold, New York.

Koopmans, J. C., and M. J. Beckmann (1957). "Assignment Problems and the Location of Economic Activities," *Econometrica*, **25:** 53–76.

Liggett, R. S. (1980). "The Quadratic Assignment Problem: An Analysis of Applications and Solution Strategies," *Environment and Planning B*, **7:** 141–162.

Liggett, R. S. (1985). "Optimal Spatial Arrangement as a Quadratic Assignment Problem," in *Design Optimization*, J. S. Gero (ed.), Academic Press, New York, pp. 1–40.

Liggett, R. S., and W. J. Mitchell (1981). "Optimal Space Planning in Practice," *Computer Aided Design*, **13**(5): 277–288.

7

MULTIPLE OPERATIVE AND INTERACTIVE MODES IN KNOWLEDGE-BASED DESIGN SYSTEMS

Rivka E. Oxman

Faculty of Architecture and Town Planning
Technion–Israel Institute of Technology
Haifa, Israel

A conceptual basis for the development of an expert system that is capable of integrating various modes of generation and evaluation in design is presented. This approach is based on two sets of reasoning processes in the design system. The first enables a mapping between design requirements and solution descriptions in a generation mode of design; and the second enables a mapping between solution descriptions and performance evaluation in an evaluation and prediction mode. This concept supports a formal framework necessary for a knowledge-based design system to operate in a design partnership relation with the designer. Another fundamental concept in expert systems for design, dual-direction interpretation between graphic and textual modes, is presented and elaborated. This encoding of knowledge behind the geometrical representation can be achieved in knowledge-based design systems by the development of a "semantic interpreter" which supports a dual direction mapping process employing geometrical knowledge, typological knowledge, and evaluative knowledge. An implemented expert system for design, PREDIKT, demonstrates these concepts in the domain of kitchen design. It provides the user with a choice of alternative modes of interaction: as a "design critic" for the evaluation of a design, as a "design generator" for the generation of a design, or as a "design critic-generator" for the completion of partial solutions.

INTRODUCTION

During the past decade there has been general enthusiasm regarding a potential breakthrough in computer-aided design due to the introduction of artificial intelligence technologies. The successful development of knowledge-based systems has been largely responsible for this enthusiasm, given its applicability in the domain of design. The field of knowledge-based design (KBD) has emerged in recent years from the application of artificial intelligence (AI) concepts and methods to design. As is by now well known, such systems are inherently different in computational approach from conventional systems. The inferencing of knowledge in knowledge-based systems relies on reasoning rather than on traditional algorithmic calculation. In addition to reasoning techniques, the approach provides new methods and tools of computational representation which may be employed in the computation of design processes. These new technologies have also enhanced the formalization of the various kinds of knowledge employed in design in order that such knowledge may be represented as part of a system's "knowledge base."

The concept of the knowledge base is among the unique characteristics of these systems. Knowledge bases employ facts as constituent elements analogous to data in a database. The knowledge in a knowledge base is both the factual content and the means whereby it may be interpreted. It may be defined as a collection of facts (data) and knowledge (rules) about the facts. As opposed to data in conventional systems, this encoded knowledge captures in some way the symbolic content of a decision-making process. In knowledge-based systems one of the key factors is the separation of the knowledge base from its reasoning mechanism. An *inference mechanism* is a tool that builds derivations from both facts and knowledge. It is also possible in such systems to deduce and induce new facts from a facts base. These and other characteristics offer promise for encoding knowledge in the computerization of design, for providing reasoning capability in design systems, and for integrating design knowledge and reasoning with existing CAD capabilities such as visualization and evaluation.

A first generation of KBD has been associated with applications of the best known form of knowledge-based systems, the expert system. Among examples of systems that have successfully applied knowledge-based system technology to architectural design are a system that utilizes the requirements of a building code to provide information for design [Rosenman and Gero 1985]; a system that interprets the description of window types [Thomson 1987]; and ALEX, an expert system that behaves like a design apprentice [Kalay et al. 1985]. This technology has also been utilized in architectural spatial synthesis in systems such as PREDIKT [Oxman and Gero 1987], JANUS [McCall et al. 1990], and a building layout system [Flemming et al. 1988].

In this paper an approach to the application of KBD technology in architectural design is developed through the explication of the implemented system PREDIKT. We elaborate on the conceptual framework for the development of the system and describe the tools and techniques by which it was implemented. A potential integration of multiple, alternative modes of design in the performance of evaluation and generation tasks is proposed as one of the significant features of PREDIKT's utilization of knowledge-based technology. The system provides a range of operative and interactive modes in a design partnership relationship with the human designer. In conclusion we consider the future potential of such *design partnership knowledge environments for design* in which both CAD and KBD capabilities are integrated in design support systems.

THE CONCEPT OF INTERPRETATION IN KNOWLEDGE-BASED DESIGN SYSTEMS (KBDS): TOWARD AN INTEGRATION OF PERFORMANCE EVALUATION AND FORMAL GENERATION

The general characteristics of KBS technology have been briefly reviewed with respect to their relevance to the computation of design. In the following sections a specific approach to KBS in which design is considered *an interpretative process* of mapping between formal representations and meanings [Mitchell 1989, Coyne 1988] is presented and its application in architectural design demonstrated. In such a conception, architectural design is characterized by successively and reiteratively cycling through modes of generation and evaluation. Generation is here regarded as a process whereby the *interpretation* of a description of design goals leads to the production of a a geometrical description. Evaluation may be regarded as the process of interpreting the attributes of that description as to its performance and, possibly also, testing this performance against design norms. In this view of knowledge-based design systems, interpretation plays a role as a reasoning process that connects between different descriptions.

Current CAD systems provide modalities such as *generate*, a change in the graphic representation, and *test*, through the analytical procedure for performance evaluation. These processes are highly developed in CAD systems in certain domains, such as the evaluation of structural and energy performance, or in construction and life-cost estimation. Despite this high level of technical development, these analytical processes are not embedded in an automated generation–evaluation cycle as part of a continuous and reiterative process. The procedural inference in current CAD systems appears to be limited in the ability to record, maintain, and infer new knowledge generated through interpretation and evaluation *during a design process*, as compared to *subsequent to* a design generation process. KBD systems, on the other hand, which are essentially characterized by their reasoning capability, can potentially support the inference of such new knowledge at any stage of the design process in a continuous rather than sequential process.

In this paper we present an approach based on KBD technology for applying knowledge by augmenting both generative and evaluative tasks in the architectural design process. In this approach interpretation provides a mapping between design descriptions. This mapping relates goal descriptions to geometrical descriptions and geometrical descriptions to performance evaluations. In the following sections, the conceptual basis for the development of a system that realizes these concepts in KBD is described, and a method for the representation and control of interpretive knowledge in design evaluation and design generation is presented. The implementation of the concepts is presented through the PREDIKT system. This is a KBD system that provides the designer with a *design knowledge support environment* in a well-defined design domain. This environment includes a variety of operative modes (*modes of operation*) up to the final generation of a design. These modes include full manual generation with automated checking, partial manual generation with the system completing the design, and automated generation in which the system carries out the entire design. This system's operation also reflects the integral importance of both graphic and textual modes (*modes of interaction*) in design. It provides dual-mode interaction, including interpretation of both graphical and textual descriptions. This is achieved by a simultaneous mapping between symbolic text interpretation and graphic interpretation.

Knowledge-Based Design: The Role of Interpretation

Design may be considered a process of transformation of representations from one form of abstraction to another. At different stages, different representations provide descriptions of the design. At an initial stage, design is usually described and represented as a set of requirements to be satisfied and goals to be achieved. At a final stage it is usually represented as the graphic configuration of a specific design. Each of these different forms of representation refers to the same design. Designers interpret requirements and goals in order to produce a specific design artifact. The process of mapping from one form of abstraction to another is achieved by what is referred to here as the *interpretation of the design description*. One of the key factors in a KBDS is the manner in which interpretation operates to support the mapping process between one form of description and another. Interpretive knowledge is not always explicit in the process of design. A knowledge-based view of design renders the knowledge through which design interpretation is made explicit and amenable to the process of computation.

In the processes of design generation and design evaluation, design interpretation is supported by reasoning. Design interpretation can be modeled in knowledge-based systems as deductive or inductive inference, and the *derivation of an implicit design description* through reasoning may be regarded as interpretation. In order to interpret and produce *new knowledge statements* from design descriptions, we require interpretive knowledge. The explicit description of a design may be encoded computationally in a KBDS as "facts," and the knowledge that supports the interpretation of these facts may be represented as "rules."

To explain this idea, let us consider a simplified rule representation in the form "if A, then B." The rule itself may be considered as the interpretive knowledge. The rule consists of two facts, A and B; it means that if A is an explicit attribute of the design description, then attribute B of that design may be inferred.

Interpretation in a Knowledge-Based Design System: Mapping Between Design Descriptions

Interpretation is defined as a process that enables the mapping of the representation of a design description into another form of description of that design. Interpretation maps among formal, typological, and performance descriptions:

1. A formal description including:
   a. Graphical description (coordinates)
   b. Geometrical description (geometrical elements and their attributes)
   c. Topological description (relations between elements)
2. A typological description (attributes of the type)
3. A performance description (performance attributes)

The role of interpretation in design is to derive properties of each of these forms of the design that may not be explicit in the design description. This is enhanced when descriptions can be represented as a hierarchical schema in which inherent knowledge about the way these facts are related is embedded in the representation. Such *structured representations* are significant in providing the medium to encode the different kinds of knowledge that may be employed during evaluation.

Interpretive knowledge may be represented as rules that map design descriptions of one form to another until it reaches performance description. It is important to define implicit attributes that will support the interpretation from a graphical description to a performance description. The process of evaluation occurs at various stages of design. Design descriptions are evaluated after being interpreted in sequential order, from graphical description to performance description. Each description, after being interpreted, must be checked against attributes of the next form of knowledge.

A knowledge-based system may be seen to consist of interpretation–evaluation tasks in stages of design development. Let us consider *geometry–typology*. The *geometrical attributes* of a design are interpreted into the next form of *typological descriptions* as the *topological attributes of that design*. At this stage an evaluation process checks the validity of geometrical and topological attributes against the attributes of the type. For example, Figure 7.1 illustrates a simple typological evaluation of a geometrical description of certain architectural elements, such as sink, counter, window, wall, in various relationships in which configuration *a* is acceptable because of appropriate relationships between elements, and configuration *b* is nonacceptable.

Computationally, evaluation may be represented by rules which state that the attributes of the design are satisfactory or unsatisfactory by comparing existing attributes against attributes of the type. At the level of *typology–performance*, the interpretation process can derive the performance attributes of a description. Computationally, evaluation may be represented in this case by rules that compare desired performance attributes with the performance of derived attributes.

In *design generation*, the interpretation process is reversed. Design starts from goal descriptions and maps to object descriptions. Design descriptions are derived from goal statements through the processes of knowledge-based interpretation. This is possible only when the design domain is based on certain "closed-world" assumptions. In such domains exact mapping between design requirements and the design description is possible because of the existence of a well-accepted tradition of normative good practice. It is typical of design domains in which there is well-accepted and well-formulated knowledge. For example, bathroom design, kitchen design, or detailing may be considered examples of such domains. Under such closed-world assumptions, deductive knowledge can be manipulated reversely as abductive rules [Coyne 1988].

Interpretation Through Backward and Forward Reasoning Mechanisms

Interpretation as a reasoning process in a knowledge-based system may be represented by a set of rules which map from the formal representation to the performance level in the following form:

If	<existing geometrical design attributes>
then	<derived typological design attributes>
If	<existing typological design attributes>
then	<derived performance design attributes>
If	<derived performance attributes meet desired performance attributes>
then	<design is satisfied>

The reasoning mechanisms of *backward chaining* and *forward chaining* are the two methods of deriving new facts from known facts in automated reasoning processes. In a

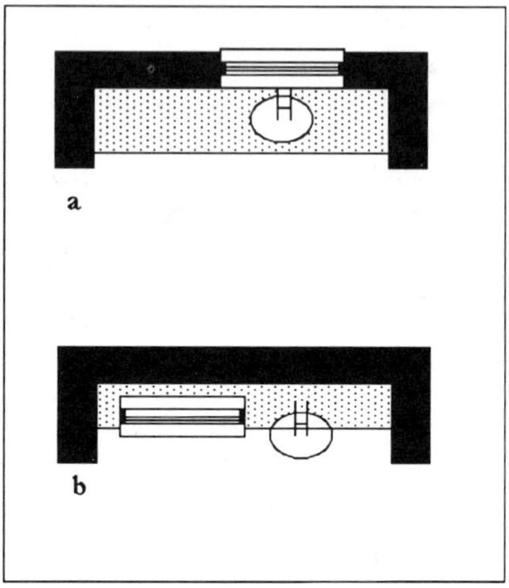

Figure 7.1. Typological evaluation in PREDIKT.

design context both of these mechanisms make it possible to derive implicit design attributes from explicit ones. Reasoning in design works either forward or backward from form to performance, or vice versa.

The *backward-chaining mechanism* tests the derivation of design attributes against existing formal descriptions. In the rule illustration above you work backward, and goals are tested against descriptions. For example, derived performance attributes may be tested for their compliance with typological characteristics; or, in effect, testing whether a certain performance is acceptable in a certain class of design. A set of attributes to be satisfied constitutes a goal. The truth of a goal statement is proven either by matching it against one of the facts in the design description base or by matching it against the goal of a prior rule in the knowledge base. This is backward chaining. The process continues recursively until a subgoal is matched to the attributes of a suitable design description.

Forward chaining starts from existing known facts about the design and the interpretations are inferred from these facts. Descriptions lead to interpretations and in the rule example above you move in the forward direction. In such reasoning in design, interpretations are direct derivations.

Semantic Interpretation of Syntactical Descriptions

In regarding design as an interpretive process that maps between descriptions, generative and evaluative processes must also be considered as the *mapping between syntactical and semantic descriptions through interpretation.* To fully understand the significance of this seminal concept in knowledge-based design systems, we must address the following two meanings of the terms *syntax* and *semantics* in KBDSs:

1. The semantical interpretation of syntactical descriptions in design generation and design evaluation
2. The textual and graphical interpretation.

Semantical Interpretation of Syntactical Descriptions in Design Generation and Design Evaluation. In configurative spatial design, elements such as points, lines, labels, and their properties provide syntactical descriptions. Interpretive knowledge is the knowledge that derives the geometrical, typological, and performance attributes of these syntactical elements, or which may be said to *map a syntactical description into its semantics.* The possibility of mapping requires the existence of a formulation of hierarchical schema of descriptions in the order such as the one suggested above, as well as well-formulated knowledge. In such an approach, the semantic description at one level provides the syntax for the next level.

In design generation, goals and requirements constitute a syntactical description for the derivation of their semantics as geometrical attributes, which are then converted to a graphical representation. Conversely, in design evaluation, a graphical design description constitutes a syntactical description for the derivation of performance attributes. Semantics in evaluation is concerned with the performance derivation of that description.

Textual and Graphical Interpretation. Among the research subjects of importance in the application of this technology to design is the integration of graphic capability in the systems. Designers express their thoughts, concepts, and designs graphically as well as verbally. Both methods of communication are characteristic of, and integral to, the process of design. Sketching in design, for example, offers an exploratory medium through which unanticipated attributes of designs emerge. Traditionally, a geometrical description of a design is represented in a graphical form while the performance attributes of that design are represented symbolically. In design, syntax and semantics are also concerned with the mappings between textual and graphical representations. A graphical representation may provide a syntactical description to be interpreted into a textual representation that describes its semantics; and vice versa, a syntactical textual description may be interpreted into a graphical representation that describes its semantics. A KBDS must provide for this interdependency of media. The knowledge base must operate to support both of these media, which are referred to here as *modes of interaction.* This duality of interaction mode is an essential characteristic of a KBDS for architecture.

Various research works have explored diverse aspects of integrating computer graphics techniques with knowledge-based methodologies. Among these are *geometric reasoning* as the systems capability of understanding shapes and their attributes [Woodbury and Oppenheim 1988], *object modeling* with logic expressions [Szalapaj and Bijl 1985], and *semantic interpretation as a means for interfacing and integrating* with CAD [Manago and Gero 1986; Gero et al. 1986]. These issues contribute to a theoretical foundation for the development of an approach to the implementation of graphical interpretation in KBDSs. The implementation of these concepts is presented below. The PREDIKT system provides for both the *interpretive mapping between syntactical and semantical descriptions* and *dual-mode interaction* from a common knowledge base.

THREE OPERATIVE MODES FOR A KBDS: DESIGN GENERATOR, DESIGN CRITIC, DESIGN GENERATOR AND CRITIC

With respect to the *operative* as well as *interactive characteristics* of KBDSs, there exist various theoretical possibilities. In the first of these the system acts as a "design generator." In the second the system acts as a "design critic." The third possibility combines the characteristics of the first two and acts as both "design critic and design generator." Each of these different modes of operation of the system also provides the different modes of interaction.

Design Generator

One possibility of applying expert systems technology in the processes of computerized design generation is that in which the expert system is capable of producing, or generating, the formal description of a design. Design generation is elaborated here as a goal-driven process of formal generation, as opposed to a purely syntactical approach. The best known example of the formal approach is the shape grammar formalism [Stiny 1980], in which design knowledge is represented by transformational rules for the manipulation of formal patterns. In the proposed system, design is considered as the goal-directed search successively from one state to another. It starts with an interpretation of the design goals description as operative subgoals, which leads in the end to the production of a formal design description. The elements of this formal description may be derived from a formalism that describes *typological knowledge of architectural typological classes* [Oxman 1990]. The formal representation consists of the syntactical descriptions of the types and subtypes. In addition to this, the representational formalism may also include a *formal refinement schema*, the operation of which can be goal directed. We refer to this type of formalism as the GPRS (Generative Prototype Refinement Scheme).

Design Critic

Another possibility for the application of expert system technology in design exists in processes of evaluation. In this case an expert system can be made to function as a design critic. This utilization of the technology in design is, in fact, analogous to that of the common diagnostic function of expert systems. In this application, the design critic "diagnoses" the faults of a design given a description of design goals. The expert system can evaluate, provide criticism, and recommend corrections in design. During this procedure, the logical basis of the system's reasoning is made available to the user, usually as a screen display. This provides for interaction with the system, through such mechanisms as override. The design critic mode may support evaluative analyses, provide an explanation, and even provide suggestions for the improvement of the design.

Design Critic and Design Generator: The Significance of Integrating Alternative Modes of Design

The ideal situation would be one in which the system operates in both modes interchangeably. The system would integrate the approaches and would operate equally well as an expert system in either mode. Such a system may support an important function in a design partnership conception, that of the capability of *completing partial design solutions*. Partial solutions once represented manually can be evaluated against criteria and desired attributes. The knowledge of the partial solution can provide a context to guide the completion of a prototypical solution.

Until now CAD programs have generally supported either mode of operation separately. Usually, the sequence of operation is generation followed by evaluation. Evaluation is performed upon completed designs and the results of the evaluation process are difficult to incorporate back into design generation. It is also generally difficult to embed this process in partial solutions. Yet the ability to evaluate and modify partial solutions is one of the characteristic processes of architectural design. KBDSs provide the "explanation facility" of expert systems, which can be associated with the process of evaluation. But in a more significant sense, KBDSs provide the possibility to develop integrated systems in which evaluation can be tied back directly into the generation process, thus modifying the normal sequence of operations characteristic of current CAD systems.

PRIMARY METHODOLOGICAL ISSUES IN THE IMPLEMENTATION OF KNOWLEDGE-BASE DESIGN

We can summarize the methodological issues that have been raised in the previous sections. We then describe the operation of a system that attempts to address these issues in architectural design.

1. Provide the three modes of operation in an architectural KBD system. Design evaluation and design generation should be integrated, should be applicable to partial solutions, and should interact in the process of design refinement until a completed design state is reached. This requires a systems operation in which a knowledge base supports both design evaluation and design generation from the same systems architecture; and the possibility of inference from the knowledge base by both forward chaining and backward chaining.

2. Provide dual-mode interaction for both graphical and textual input. Both modes must initiate responses with the traditionally interactive nature of the expert system dialogue. This is achieved by the construction of a semantic interpreter that converts the verbal representation to a graphical representation and vice versa. A suitable representational media for architectural design must be developed to support the complex representational aspects and reasoning of the design process as well as to support the integration of evaluation and generation.

PREDIKT: AN EXPERT SYSTEM AS DESIGN GENERATOR, A DESIGN CRITIC, AND A DESIGN CRITIC AND GENERATOR

PREDIKT (PREliminary Design of KITchens) is an icon-oriented graphics system that captures knowledge in the domain of kitchen design. As a knowledge-based system it elaborates the theoretical foundations that have been presented and addresses certain of the key methodological issues relevant to the implementation of KB technologies to design in general and to architectural design in particular.

Modes of Performance

Two basic requirements in the development of a partnership approach in knowlege-based design are demonstrated by PREDIKT. The first of these is the integration of design generation and design evaluation. Due to this integration, the system is capable of providing the user with a choice between alternative modes of operation described above. This is one of the essential differences between the operational characteristics of PREDIKT and other expert systems in the domain of floor plan layout. JANUS [McCall et al. 1990] operates as a design critic. Other expert systems, such as that of Flemming et al. [1988], which is based on a shape grammar approach, are generative. PREDIKT provides these and a third mode that integrates both design evaluation and design generation to develop and complete partial design solutions. PREDIKT also provides this choice of mode to the user.

The second is the provision of a rich interactive environment. PREDIKT offers the user a choice between modes of graphic and textual interaction. By comparison, JANUS provides connection in one direction between the CAD representation and Hypertext technology. PREDIKT provides both textual mode and graphical mode during a working session and allows dual-direction interaction, graphical–textual or textual–graphical, both of which operate through a semantic interpreter. It maps simultaneously between these two sets of communications during both generation and evaluation. These characteristics of the system are discussed below.

PREDIKT as Design Critic in the Mode of Design Evaluation. Using PREDIKT as a design critic, the designer constructs a design description and the system, containing evaluative and interpretive rules, checks the design as it proceeds, makes recommendations for improvements through its explanation facility, and carries on a dialogue with the user regarding design performance. In this mode the designer interacts with the system graphically using the graphic window.

The user asks the system to check the design display. The system interprets its attributes and evaluates the design. The semantics implicit in the graphics are interpreted by activating an interpretive knowledge base, which derives attributes from the graphical description. These are then represented as new deduced facts about the current state of the design description and placed in a facts base. PREDIKT evaluates the attributes by activating its rule-based evaluative knowledge and subsequently textually criticizes the design created by the user in a textual window. This is in the form of evaluating the particular aspect of the design as satisfactory or unsatisfactory. Figure 7.2 illustrates an example of this mode in a screen display of a design critic session in which a graphical display of an area of a kitchen has been checked and evaluated relative to norms.

PREDIKT as a Design Generator. As design generator the system automatically generates a design solution from an initial state, given the information on hand. The system interacts with the user textually and requests certain required information. For example, to interpret a given design description of a floor area of a specific kitchen, the system poses questions about required activities and types of movement. As a result, the system can derive an alternative of the layout of counter shapes for the kitchen. Figure 7.3 presents a screen display during a generator session in which the user employs the system to satisfy design goals and generate a layout of the kitchen counter. The system reasons from the given facts about requirements and generates a graphic design description that satisfies the textual goal.

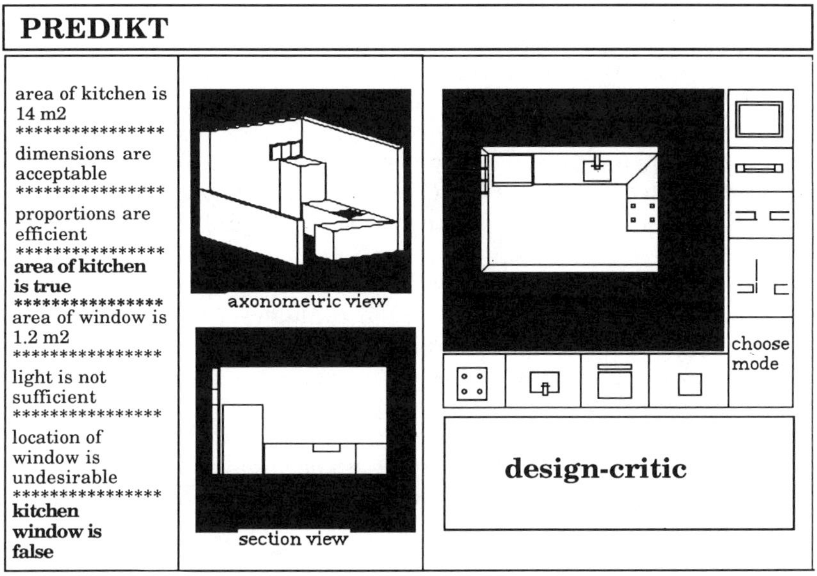

Figure 7.2. Design critic session in PREDIKT.

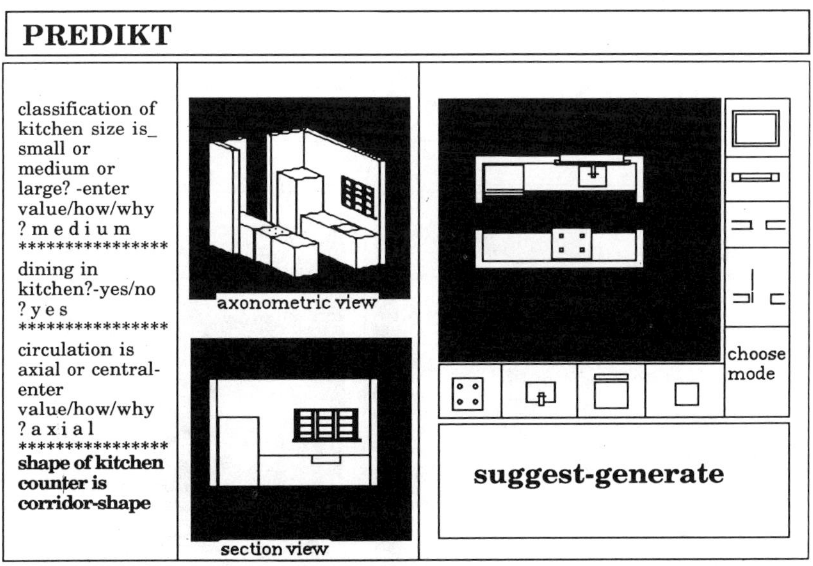

Figure 7.3. Design generator session in PREDIKT.

PREDIKT as Design Developer. During the development of the design, the designer may generate a partial design. For example, the designer may provide the counter shape, and the system will evaluate it and complete the design by locating other elements, such as the sink, stove, and refrigerator. The system first checks the validity of the partial design in the same manner as in the design evaluation mode. It then changes the mode of operation and switches to the generation mode. Thus a partial design solution provides facts that the system can derive and it acts upon them as an initial design description in the mode of design generation in completing the design. Figure 7.4 and 7.5 illustrate the process of such a completion of design. The user has asked the system to complete and refine a design. The system interprets the given partial design state in an evaluation mode (illustrated in a screen display in Figure 7.4), evaluates its correctness, and applies its knowledge of kitchen design in a generation mode to complete that design (illustrated in a screen display in Figure 7.5).

In all three modes, the system infers from the given facts of a design description. When it requires additional information to complete a derivation process, it requests this information from the user. The system can be asked at any stage to justify its conclusions or to explain through its explanation facility why certain facts have been proven true or untrue.

Methods and Techniques

Due to its utilization of the same knowledge base for both design evaluation and design generation, PREDIKT is capable of supporting a multiple interactive mode and integrated operational approach. This is achieved by using a knowledge-based systems

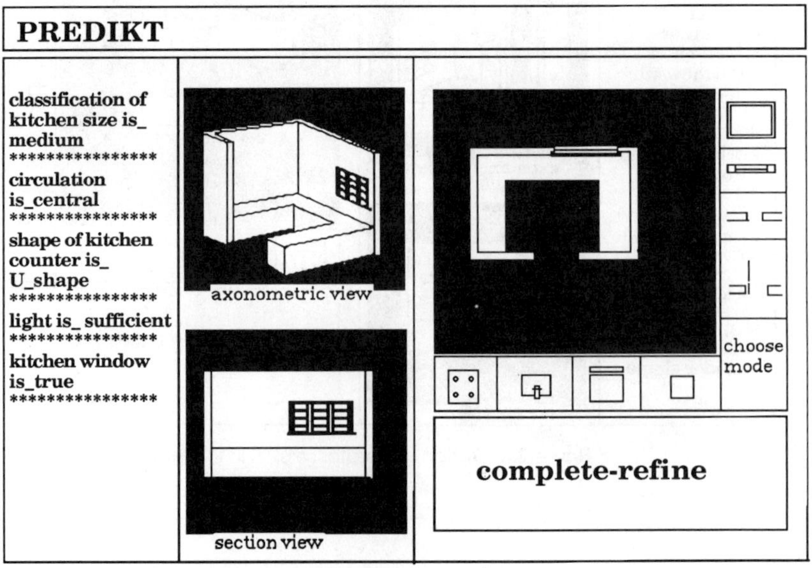

Figure 7.4. Critic-generator session in PREDIKT: a critic mode.

architecture that provides the capability of *a semantic interpreter and both forward* and *backward chaining*. The system is written in Quintus Prolog running on a SUN Microsystems workstation. It consists of three modules:

1. An expert system shell with both forward chaining and backward chaining
2. A knowledge base in the form of design rules
3. A semantic interpreter that converts graphical representations to textual representations, and vice versa

Figure 7.6 illustrates the components of the PREDIKT system. In the following section, each of these modules is described briefly.

The Expert System Shell. The shell BUILD [Rosenman 1986] contains the basic components of an expert system: an interface facility, an inference mechanism, an explanation facility, and a state description. In the PREDIKT application, backward chaining and forward chaining accommodate design evaluation and design generation.

A Knowledge-Base in the Form of Design Rules. A knowledge base of design rules provides the body of interpretive and evaluative knowledge of the system's schema of hierarchical levels of design descriptions which have been illustrated previously. An integrated approach requires the implementation of both generative and evaluative rules within the knowledge base. An example of evaluative and interpretive knowledge is given by rules listed below. The first rule interprets; the second and third evaluate.

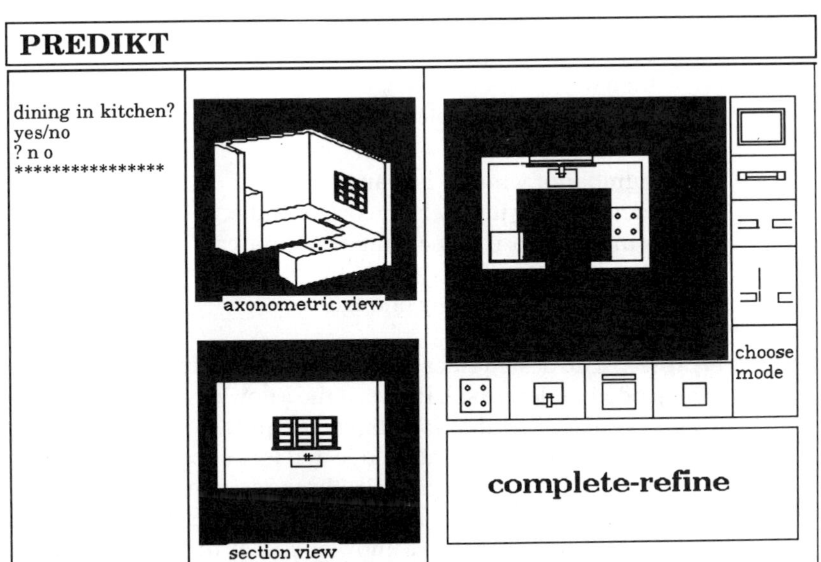

Figure 7.5. Critic-generator session in PREDIKT: a generator mode.

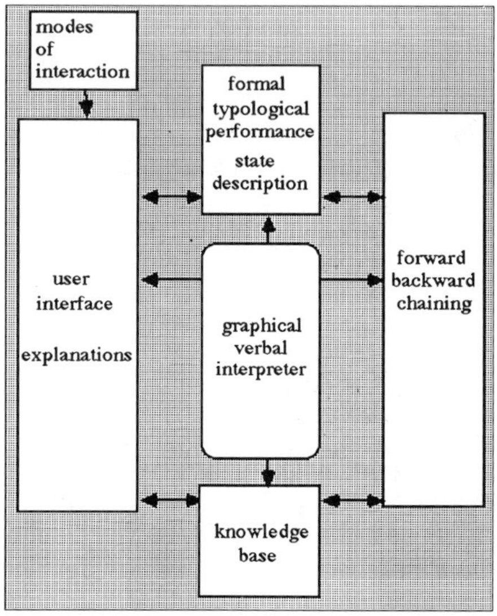

Figure 7.6. System components in PREDIKT.

If
area of kitchen in sqm is greater than 30.0 and
width of kitchen in cm is greater than 210
then
classification of kitchen size is large.
If
classification of kitchen size is large and
eating is to be provided in kitchen and
counter is L__shape
then
counter is acceptable.
If
number of window is 1 and
location of window is in corner and
window area is 40 percent of floor area
then
size of window is big enough.

A Semantic Interpreter. The design rules that are related to the domain of residential kitchen design are encoded as described. Between the knowledge base and the graphic display there is a *semantic interpreter* that can convert graphics to facts, and vice versa. It employs the evaluative and interpretive knowledge that is needed to carry out these conversions. When the user displays a design graphically, the system interprets the syntactic description and converts it to facts that are checked against rules in the knowledge base. The utilization of evaluative and interpretive knowledge may be demonstrated by a simple predicate written in Prolog:

```
check_execute:
     locate_object
     generate_facts
     interpret - evaluate (find 'the approval of the object')

locate_object:
     execute and display the syntactical relations between
     entities of an object

generate_facts:
     derive topological and typological attributes of the
     syntactical display

interpret_evaluate
     evaluate the validity of derived attributes against
     desired attributes in the system knowledge base
```

The following example illustrates the function of the semantic interpreter within the systems operation. The designer wishes to input a kitchen plan graphically. This is performed syntactically by specifying the x- and y-coordinates of two opposing corners of the plan (assuming that it is rectangular), by pointing with a mouse. These two corners have coordinates (X1,Y1) and (X2,Y2); then the transformational knowledge can be written in Prolog as

```
locate_area:
     Dx is X2 - X1,
     Dy is Y2 - Y1,
     draw_area (Dx,Dy),
     assert (length of kitchen is_Dx),
     assert (width of kitchen is_Dy),
     assert (area of kitchen is Dx*Dy).
```

As illustrated, this graphical–textual transformation results in semantic facts being asserted and encoded in the facts base. In fact, syntax–semantics conversions require much more complex transformations than those of this simple example.

When the user describes goals textually, the interpreter derives their semantics and displays the resultant syntactic relations of elements graphically. These two modes are employed in the process of design evaluation and design generation. Figure 7.7 illustrates a typical screen display of the interpreted facts, which have been derived from a graphical display.

TOWARD THE REPRESENTATION OF GENERIC KNOWLEDGE IN KNOWLEDGE-BASED DESIGN

Through its organizational structure, the schema of a database implicitly describes knowledge about the meaning of the data in the database. In a knowledge-based system

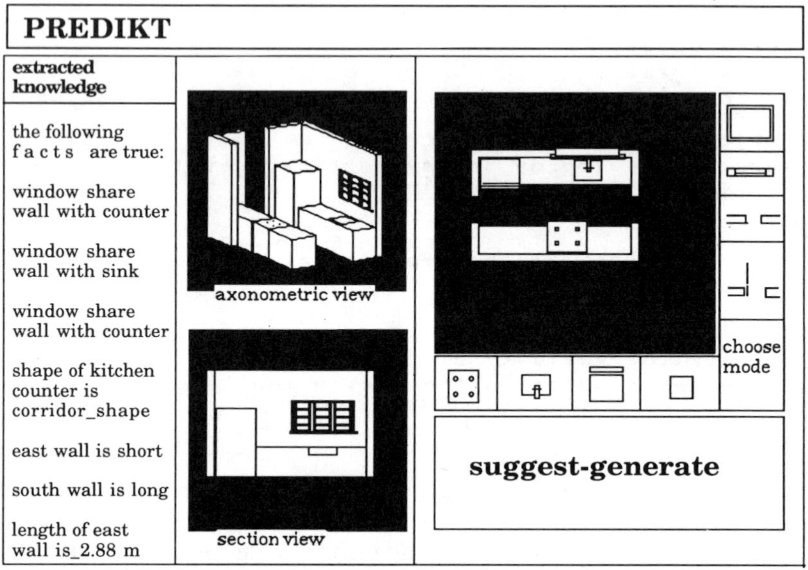

Figure 7.7. Interpreted facts after a graphical display.

the organization of the facts in the knowledge base may inherently represent a certain schema of knowledge. In a KBDS such *structured representational schema* may serve as a model that provides implicit knowledge about the thing being designed. The way such schema function as a *central representation* in KBDS is of major significance to the augmentation of evaluation and generation in KBDS.

Certain of these ideas have been demonstrated in the PREDIKT system. In the current version of PREDIKT the facts base is employed as a central representation of all of the derived facts of the various design descriptions that support design evaluation and design generation. A schema that contains a formal description, a typological description, and a performance description may, in fact, provide a model of the artifact being designed. The current facts base lacks such an organizational schema, which can effectively represent inherent knowledge of the design world. Figure 7.8 illustrates diagramatically a facts base without an underlying organizational schema. Such a facts base provides a low level of inherent knowledge about the design objects.

In an enhanced representation, the system contains the knowledge of design prototypes from which instances can be derived. It possesses knowledge in a generic sense by being capable of providing properties, variables, and methods for interpretation and evaluation. The instance should be derived by inheriting these from a generic prototype implicit in the kitchen description. Figure 7.9 illustrates diagrammatically the concept of a generic description based on a schema. In contrast to the preceding diagram, here inherent knowledge exists in the representational scheme itself. This is represented in the diagram by the linkage of facts on the left side, which is the scheme of knowledge of prototypes. In this case, any facts base that is created during a session is an instance of that schema.

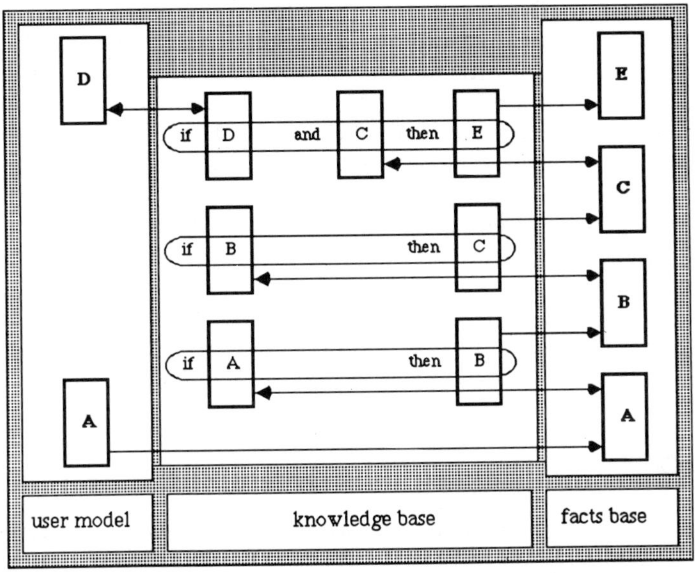

Figure 7.8. Fact base lacking schema.

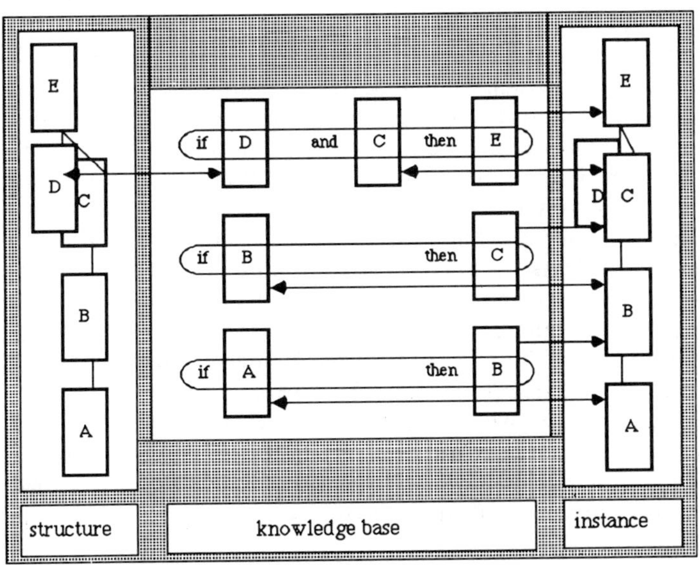

Figure 7.9. Generic schema behind a fact base.

Systems having underlying schema in the facts base have important operative characteristics. During design, a prototype is refined and instantiated according to various goals and constraints. In such systems certain information that had to be requested formally from the user may now be derived directly from the schema.

The development of representational schema capable of encoding these representations is a significant theme in some research work in knowledge-based design systems. Currently, work in constructing such a model of a prototype schema [Gero 1987, Oxman and Gero 1988] and applying it to architectural design [Oxman 1990] is being undertaken. In such an approach, a rich *hybrid knowledge representation environment* is required to encode, access, and manipulate various kinds of knowledge in the design process. From the computational point of view, the role of knowledge-based interpretation in such an approach is to support communication of the different representational methods with one another through the central representational scheme.

CONCLUSIONS

In this paper we have demonstrated the potential of knowledge representation and automated reasoning techniques for an approach to KBDS that supports the integration of alternative modes of design evaluation and design generation in knowledge-based design. A partnership approach with an *integrated operational mode* and *a dual interaction mode* in integrated evaluation–generation has been presented. Certain aspects of the implementation of such an approach were demonstrated through the PREDIKT system. Its integrated performance is based on an interpreter that provides communication with the graphic system and enables the knowledge bases to communicate with the database of the graphic system in both design evaluation and design generation.

The development of representational schema that are capable of encoding geometrical representations in addition to other types of knowledge is an important area of research work on which the success of the next generation of implementations of knowledge-based design systems will depend. It appears that an integrated combination of current geometry-based and knowledge-based technologies, providing communication between CAD and KBDS capabilities, can significantly enhance the performance of CAD systems as well as the role of design partnership in CAD. These concepts will have an impact on design supportive computer systems by enhancing our current capability to integrate various types of knowledge, such as formal, typological, and performance knowledge. In effect, this will combine geometric-based CAD and knowledge-based design concepts and will move us a step closer to intelligent, integrated CAD systems as a design support environment. A significant contribution to this advancement will be the enhancement of the function of integrated design generation and evaluation in interactive design support systems.

REFERENCES

Coyne, R. D. (1988). *Logic Models of Design*, Pitman, London.

Flemming, U., R. Coyne, T. Glavin, and M. Rychner (1988). "A Generative Expert System for the Design of Building Layouts, Version 2," in *Artificial Intelligence in Engineering: Design*, J. S. Gero (ed.), Elsevier, New York.

Gero, J. S. (1987). *Prototypes: A New Schema for Knowledge-Based Design*, Working Paper, Architectural Computing Unit, Department of Architectural Science, University of Sydney, Sydney, Australia.

Gero, J. S., R. E. Oxman, and C. Manago (1986). "Graphics and Expert Systems," in *AUSGRAPH '86: 4th Australian Conference on Computer Graphics.*

Kalay, Y. E., A. Harfmann, and L. Swerdloff (1985). "ALEX: A Knowledge-Based Architectural Design System" in *ACADIA '85 Workshop Proceedings*, P. G. McIntosh (ed.), Arizona State University, Tempe, Ariz.

Manago, C. and J. S. Gero (1986). *Some Aspects of Communication Between Expert Systems and a CAD Model,* Working Paper, Architectural Computing Unit, Department of Architectural Science, University of Sydney, Sydney, Australia.

McCall, R., G. Fischer and A. Morch (1990). "Supporting Reflection-in-Action in the JANUS Design Environment," in *The Electronic Design Studio*, M. McCullough, W. J. Mitchell, and P. Purcell (eds.), MIT Press, Cambridge Mass, 1990.

Mitchell, W. J. (1989). *Logic of Architecture*, MIT Press, Cambridge, Mass.

Oxman, R. E. (1990). "Architectural Knowledge Structures as 'Design Shells,' " in *The Electronic Design Studio*, M. McCullough, W. J. Mitchell, and P. Purcell (eds.), MIT Press, Cambridge Mass.

Oxman, R. and J. S. Gero (1987). "Using an Expert System for Design Diagnosis and Design Synthesis," *Expert Systems,* **4:** 4–15.

Oxman, R. and J. S. Gero (1988). "Designing by Prototype Refinement in Architecture," in *Artificial Intelligence in Engineering: Design*, J. S. Gero (ed.), Elsevier, New York.

Rosenman, M. A. (1986). *The BUILD Manual*, Architectural Computing Unit, Department of Architectural Science, University of Sydney, Sydney, Australia.

Rosenman, M. A., and J. S. Gero (1985). "Design Codes as Expert Systems," *Computer-Aided Design*.

Stiny, G. (1980). "Introduction to Shape and Shape Grammar," *Environment and Planning B*, **7:** 343–351.

Szalapaj, P., and A. Bijl (1985). "Knowing Where to Draw the Line," in *Knowledge Engineering in Computer Aided Design*, J. S. Gero (ed.), North-Holland, Amsterdam.

Thomson, J. V. (1987). "A Water Penetration Expert System Using PROLOG with Graphics," in *Applications of Expert Systems*, J. R. Quinlan, (ed.), Addison-Wesley, Reading, Mass.

Woodbury, R., and I. Oppenheim (1988). "Geometric Reasoning: Concepts and Demonstration," in *Artificial Intelligence in Engineering: Robotics and Processes*, J. S. Gero (ed.), Elsevier, New York.

8

CONSTRAINTS FOR GENERATING BUILDING ENVELOPE DESIGN ALTERNATIVES

Paul Fazio
Claude Bédard

Centre for Building Studies
Concordia University
Montreal, Canada

Krishnan Gowri

SIRICON Inc.
Montreal, Canada

The building envelope design process involves selecting materials and constructional types for the envelope components. Many different materials need to be combined for wall and roof assemblies to meet various performance requirements, such as thermal efficiency, cost, acoustic, and fire resistances. The number of performance attributes to be considered in the design process is large. Lack of information, time limitations, and the large number of feasible design alternatives generally force the designer to rely on past experience and practical judgment to make rapid design decisions.

Current work at the Centre for Building Studies focuses on the development of knowledge-based synthesis and evaluation techniques for reducing the problems of information handling and decision making in building envelope design. The generation of design alternatives is viewed as a search process that identifies feasible combinations of building envelope components satisfying a set of performance requirements. These are constraints established from user needs, design context, building code requirements, material compatibility, practicality of design, and so on. In this paper we discuss knowledge acquisition and representation issues involved in the definition of constraints to guide the generation of feasible combinations of envelope components.

INTRODUCTION

During the preliminary stages of a building project, the architect–designer prepares schematic drawings consisting of floor plans and elevations. Inherent in this process is the selection of materials and systems for the structure and envelope of the building. Decisions made at this stage greatly influence construction time, quality, operating costs, and so on. But these decisions are based on minimum information regarding building location, type, occupancy and budget specified by owner and developer. Often time pressures and lack of adequate information force the designer to rely on experiential knowledge, heuristics, and judgment for making rapid design decisions. Hence a systematic methodology needs to be developed for reducing the problems of information handling and to provide a rational approach to design decision making.

Recent developments in design research and knowledge-based system techniques show that it is possible to develop tools for the preliminary design stage [Gero and Coyne 1984, Sriram et al. 1985, Jayachandran and Tsapatsaris 1988]. Design in general is a goal-directed activity for developing detailed descriptions of artifacts satisfying the functional specifications and life-cycle requirements of a product. Architectural and building design problems are radically different from other domains because of the subjective nature of space planning, and selection of materials and systems that are knowledge intensive, project specific, and involve little computation [Willey 1976]. Building envelope components have a large number of attributes that are to be considered in sufficient detail for successful performance.

The primary objective of the present study is to formalize the building envelope design process, enabling the integration of the many different types of information, including performance standards, material properties data, constructional types, design heuristics, and evaluation procedures. Problem decomposition and constraint-based search techniques in a knowledge-based environment are utilized in this study to develop a building envelope analysis and design tool known as BEADS. We describe the design constraints, knowledge sources, and representation methodology for generating building envelope design alternatives.

BUILDING ENVELOPE DESIGN PROCESS

External walls, roof, and fenestration are the major components of the building envelope that protect the indoor environment from rapidly fluctuating outdoor climatic conditions. Successful performance of an envelope assembly depends on the judicial selection of materials and constructional systems to meet the various functional and user requirements of the building. Thermal resistance, moisture protection, structural strength, fire and acoustic resistance, cost, buildability, and maintainability are some of the important performance requirements that need to be considered in design. A more complete list of requirements and a detailed discussion of each can be found elsewhere [Corkan 1985, Hutcheon 1963]. The nature of building envelope design is further complicated, as these performance attributes often conflict with each other. Research in building envelope design has focused on issues such as energy efficiency or cost which can be modeled into mathematical problems. Most other performance requirements have either been neglected or considered on the basis of previous experience.

The AJ handbook of building enclosure [Elder and Vandenberg 1974] presents the sequence of design decisions as establishing the user needs, performance requirements, design constraints, and selecting materials and components for the assembly. The identification of user needs and establishment of design constraints are complex processes requiring expertise and knowledge to determine the expected level of performance for all the various functions of the envelope assembly. Decisions on the selection of materials and components require information on material properties, cost, and compatibility between materials. It is difficult for the designer to consider more than a few alternative combinations of materials and components because of the large number of variables in the process. Although the large number of available materials and constructional types guarantee a designer's freedom and creativity, they also complicate the search for an optimal solution with an infinite number of possible design alternatives.

A systematic methodology for design decision making needs to be established to assist the designer in considering more performance attributes and in exploring more design alternatives. This is a problem of information processing and decision making that may be solved using computer-aided techniques. In an effort to automate the building envelope design process, the following three distinct stages have been identified:

1. Establish the design context.
2. Generate design alternatives.
3. Rank and select alternatives.

Earlier research attempts have adopted mathematical optimization and decision-making techniques [Mattar et al. 1978, Murthy et al. 1986] for evaluating building envelope design alternatives. These techniques can help at the final stage but lack the ability to assist in establishing the quantitative requirements for the design context and in the synthesis of design alternatives. These two issues can now be addressed adequately with reference to recent developments in artificial intelligence and knowledge representation methodologies.

CONSTRAINT-BASED SEARCH: A REVIEW

Design in general can be viewed as a search process to find a solution that satisfies a set of constraints defining the problem context. Finger and Dixon [1989] have reviewed the various models of computer-aided design. The basic problem types in design are classified into the following models: (1) parametric, (2) configuration, and (3) conceptual (or preliminary). All these models use constraints either explicitly or implicitly for evaluating design parameters, and for generation or formulation of design alternatives. A constraint is defined as the description of a desired relationship among one or several variables or objects [Morganstern et al. 1988]. Explicit representation and use of constraints enable the computer modeling of the designer's search process in identifying design descriptions. A comparative evaluation [Kalay et al. 1990] of the various search techniques for architectural design shows that constraint-based search is suitable for synthesis–evaluate problem solving and in providing a partnership approach to design. Gross et al. [1987] have discussed the use of a "constraint model" for demonstrating and exploring the computability of design. Chan and Paulson [1987] have shown that the explicit representation of constraints can be used in exploratory design to devise design descriptions in a way that guarantees the result to be consistent with constraints instead of using constraints merely to check design descriptions.

Constraints in design can be classified in many different ways. Sriram and Maher [1986] present a classification of five constraint types: (1) synthesis constraints, (2) interaction constraints, (3) causal constraints, (4) parametric constraints and, (5) evaluation constraints. These essentially represent the constraints used in the various stages of design. Sapossnek [1989] classifies constraints as (1) implicit constraints, (2) place constraints (3) inherited constraints, and (4) accumulation constraints, based on the implementation of generic constraint-based systems. Another classification [Gross et al. 1987] is concerned with the design objects as definitional, functional, and preferential constraints. All the classifications above are somewhat artificial but certainly useful in understanding the role of constraints and representation issues.

In building envelope design, constraints can be used for expressing the functional relationship between envelope components, compatibility between materials and systems, interaction between performance requirements, designer preferences, and so on. One of the most important tasks in using these constraints is to assist the designer with the identification of feasible alternatives that can be considered in the early stages of design. Explicit use of the above-mentioned building envelope constraints along with design heuristics will enable the efficient use of constraint-based search techniques in the alternative generation process. In the context of the present study, constraint-based search may be seen as a combination of problem decomposition and multiple generate–test paradigms. Building envelope components such as walls, roof, and windows can be designed separately and combined to satisfy the constraints of the problem context. The design of envelope components may be logically sequenced so that the compatibility between components are ensured in the generation process. The remainder of this paper deals with how building envelope design constraints are identified and used for the generation of design alternatives. The knowledge sources, knowledge acquisition process, and implementation details are described briefly.

BUILDING ENVELOPE DESIGN CONSTRAINTS

Constraints can represent performance requirements, functional relationships, and design preferences in identifying materials and systems for building envelope components. Quantitative requirements can be defined for thermal, structural, and other performance based on standards and codes that stipulate minimum acceptable levels of performance. For example, ASHRAE Standard 90A-80 [ASHRAE 1980] provides tables and graphs to determine the recommended thermal resistance values for external walls, roof, and fenestration. Constraints such as these can be used in defining the boundaries of the design search space. The problem here is to identify all the necessary constraints, the source of information, and the user input required to establish the design context. In the case of thermal resistance, ASHRAE Standard 90A-80 requires the degree days of the building location and envelope areas to determine the recommended energy budget. The degree days of various locations can be obtained from weather data tables as available in the *Supplement to the National Building Code of Canada* [NRCC 1985]. Hence it is possible to extract all the necessary information simply by knowing the building location and the envelope areas from the user. Figure 8.1 shows an example definition of the design context schema in the BEADS prototype implementation. Also identified in this figure are the sources of information and the relations between performance attributes in establishing the design context. The weather data table and ASHRAE Standard 90A-80 requirements are incorporated in the knowledge base as schemas and procedures (demons), respectively. The relationship between performance attributes are dynamic in nature for maintaining the consistency of the design context and the logical dependency of performance requirements. For example, the energy budget value is calculated by accessing the values corresponding to thermal transmittance (U_0) for wall and roof, and areas of envelope components. The U_0 values depend on the degree days of the location. Thus changing a location or any other user input parameter will automatically reassert the value corresponding to the attributes affected. This representation is efficient for quick redesign and examining the effect of various design parameters. Most of the performance-related constraints established by the design context are used in the alternative generation process to check the suitability of materials, systems, and their combinations. These checks may be simple predicate relationships or more complicated analytical computations and verifications.

The functional relationships between building envelope components are represented in a hierarchy shown in Figure 8.2. A feasible design alternative will consist of a

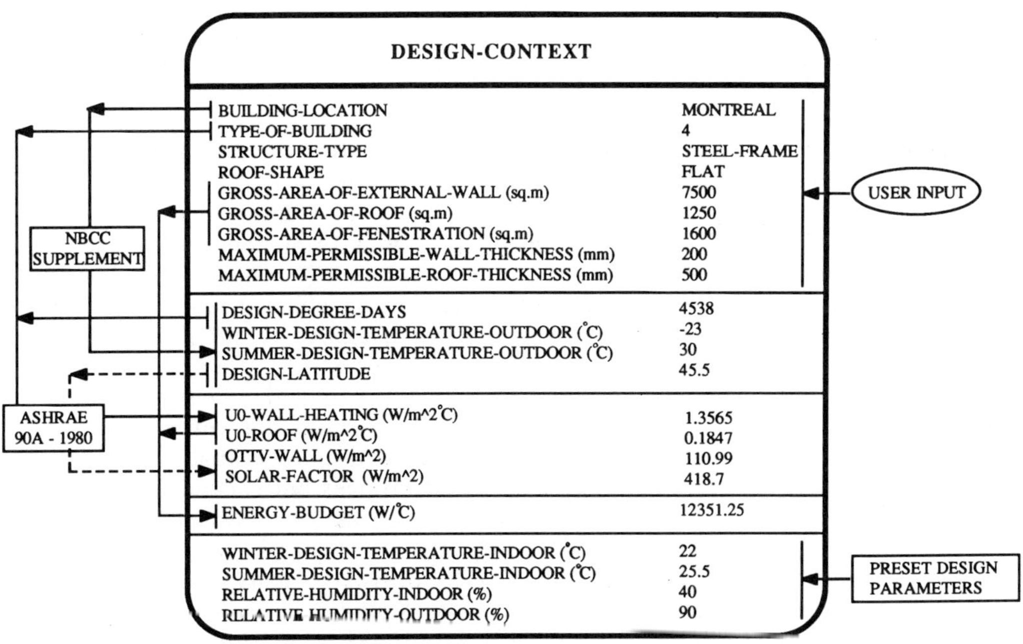

Figure 8.1. Example of a design context and information sources.

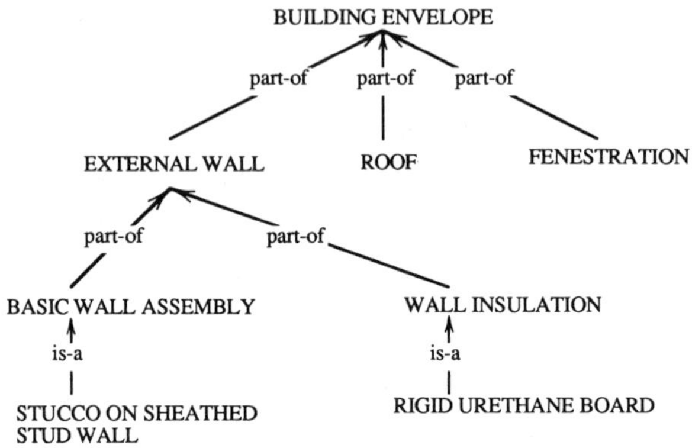

Figure 8.2. Hierarchy of building envelope components.

wall, roof, and glaring type, each satisfying some local constraints, such as thickness and R-values, and together meeting the energy-efficiency requirements. A typical wall used in the North American climate consists of many layers of different materials, such as exterior cladding, insulation, vapor barrier, air barrier, and interior finishes. Building materials for each one of these functions are available in large number with competitive performance and cost. Although materials satisfactory for a particular function may be easy to identify, the problem of combining them together as a system or assembly is more complicated.

Functional constraints can be used in generating design alternatives but are found to be difficult to implement at the system level. It is possible at the component level and has been reported elsewhere [Fazio and Gowri 1989] for the design of roof systems only. Instead of relating all the materials by functional constraints, the present study uses basic constructional types and insulation materials separately. These basic types for walls and roofs are widely used in practice and can be found in design literature [MCC 1982, AIA 1982, AMA 1986]. In terms of energy efficiency, the prime concern is to find the appropriate insulation types and thicknesses for walls and roofs. Other constraints on suitable structure for each constructional type, such as the effect of roof slope in the choice of roof finish materials, also need to be considered.

Figure 8.3 shows an example of basic wall type details which is an instance of the BASIC-WALL-TYPE schema. The suitable building type and suitable structure type information for each basic wall type are obtained by consultation with a practicing designer. Currently, the building type classifications correspond to that of ASHRAE Standard 90A-80. The only structure type considered is frames, but a distinction is made among wood, concrete, steel, and composite constructions. These two slots were introduced after a preliminary version of the prototype was examined by a few practitioners. The slot NAME-OF-COMPONENTS consists of schema names corresponding to the materials used for various layers of the wall, except insulation. The order of values in this slot corresponds to that of the layers from exterior to interior of the wall. Each material name is an instance of a schema representing building material properties. Thus the R values and thickness slots of walls are automatically asserted by retrieving the corresponding values from the constituent materials. The value INSULATION in the NAME-OF-COMPONENTS slot refers to an insulation material that is permitted for use in wall assemblies.

The representation of building materials information is different from that of insulation materials. This is necessary because the thickness of insulation is determined during the design process, whereas this value is constant for all other materials present in the basic construction type. An example of roof insulation material and its representation is shown in Figure 8.4. All insulation materials are instances of the INSULATION-TYPE schema, but only those suitable for walls and roof are related to the WALL-INSULATION and ROOF-INSULATION schemas, respectively. This semantic relationship is used during the alternative generation process to identify selectively the appropriate insulation materials for the envelope component under consideration. *R* values, vapor resistance, available thickness, and cost are obtained from product literatures. The slot IN-COMPATIBLE consists of building material names, which are not compatible with the insulation material. This information was obtained from design handbook and technical notes [Hardy 1985].

There are 17 basic wall types, 10 basic roof types, four glazing types, six insulation types, and about 30 building materials available in the BEADS prototype. The basic roof types and glazing types have a representation similar to that of the wall types discussed above but with different attributes and functions.

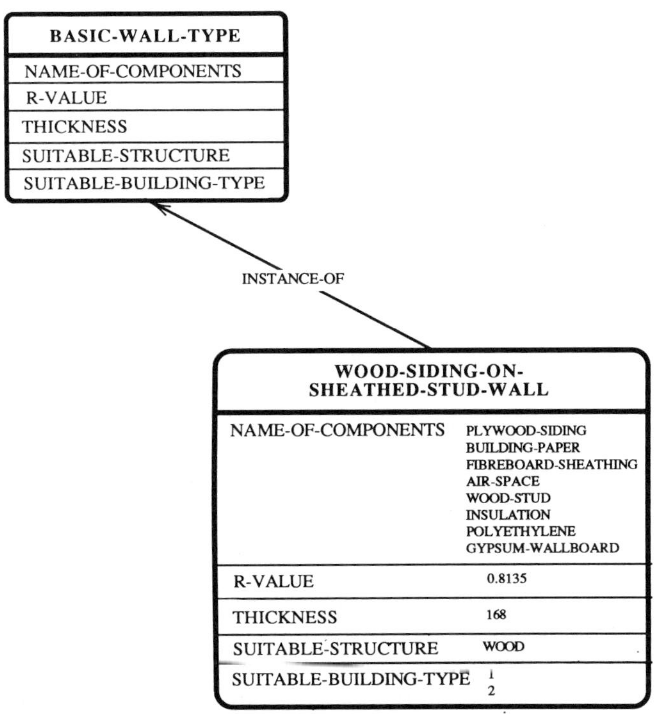

Figure 8.3. Basic wall type description.

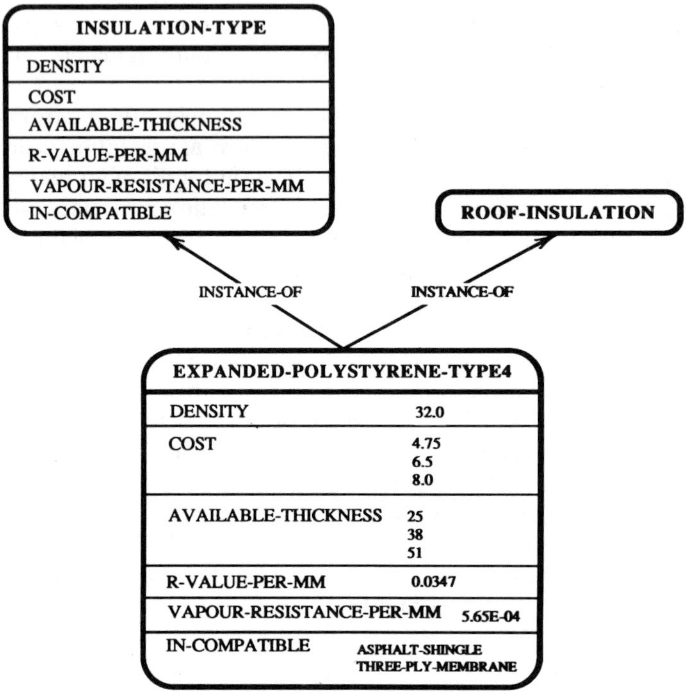

Figure 8.4. Insulation material description.

DESIGN ALTERNATIVE GENERATION PROCESS

The generation of building envelope design alternatives is a process of identifying basic construction types and insulation materials for walls, roof, and glazings. The semantic relationships and constraints on compatibility between materials and construction types are resolved during the generation process. Figure 8.5 shows an example of how a basic wall type schema description is related to building and insulation material schemas. Once a basic wall type such as CONCRETE-BLOCK-WALL is selected upon satisfying the constraints on structure, building type, and basic thickness requirement, the design alternative generator tries to identify an insulation material such as EXPANDED-POLYSTYRENE-TYPE2 that is suitable for use in wall constructions. In selecting the insulation material, constraints on material compatibility, total wall thickness, and basic R-value requirement are verified. Three lists of thicknesses, R values, and vapor resistance values, corresponding to the various layers of the wall assembly, are created by retrieving them from the corresponding building and insulation material schemas. These lists are used later for checking the possibility of interstitial condensation in the wall assembly. After identifying a suitable wall assembly, the alternative generator proceeds to find a roof type, roof insulation, and glazings for the envelope assembly. A sequential scheme is implemented for systematically searching the design space for identifying all the feasible combinations of envelope components. The following is a list of operations and constraints checks performed in the generation process:

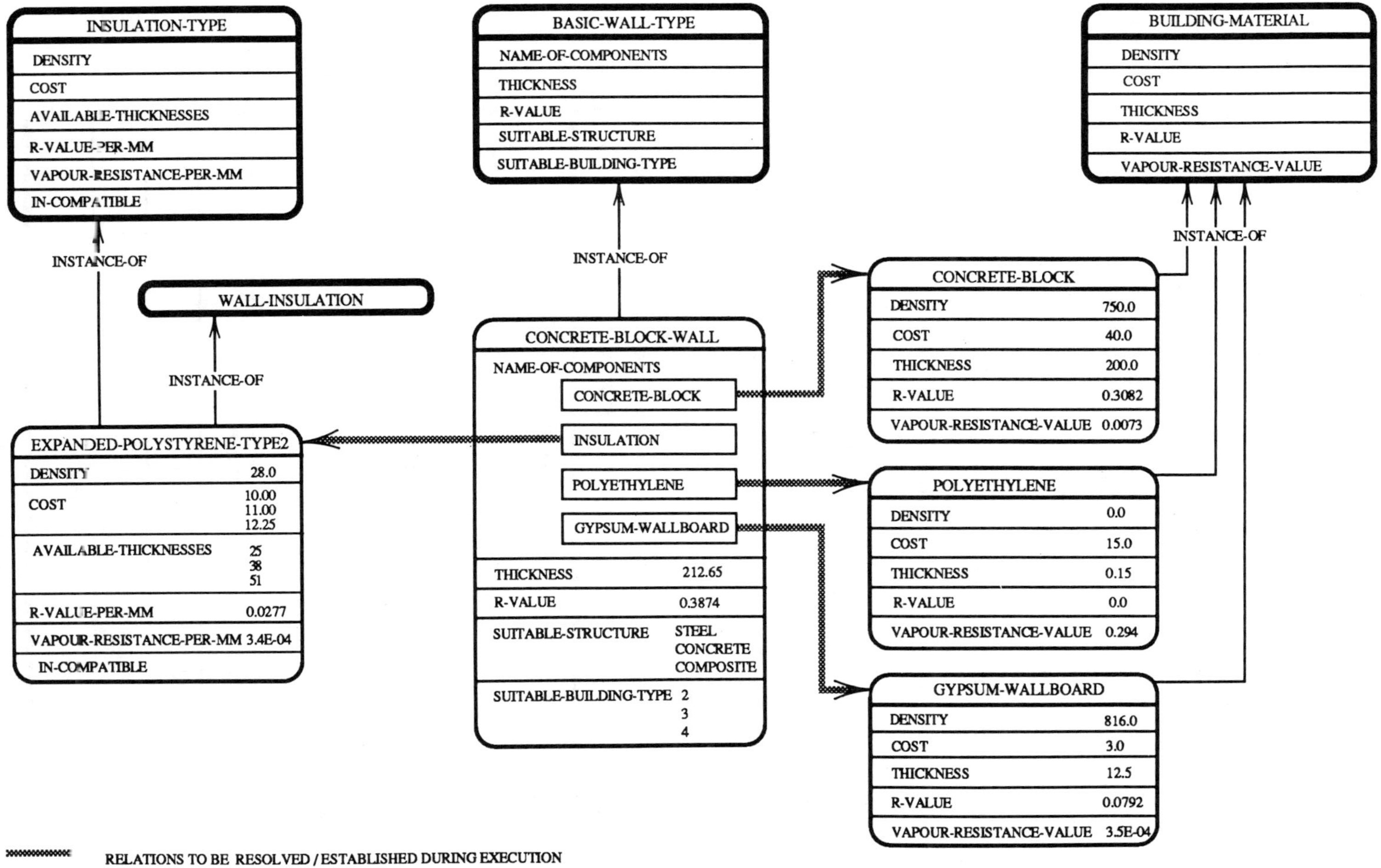

Figure 8.5. Example of relationship between various schemas.

Stage 1: A basic wall type suitable for the structure type and building type and satisfying the thickness requirement is chosen.

Stage 2: An insulation type suitable for the wall is chosen satisfying the thickness requirement and R-value requirement.

Stage 3: A static condensation analysis is performed to determine the suitability of the wall assembly.

Stage 4: A basic roof type suitable for the structure type and building type, satisfying the roof slope requirement and permissible roof thickness, is chosen.

Stage 5: A roof insulation is selected satisfying the thickness requirement.

Stage 6: A glazing type suitable for the building type is chosen.

Stage 7: The combination of wall, roof, and glazing is checked for overall energy consumption for heating, and if required, a cooling check is also made. The details of a successful combination are asserted in an instance of the FEASIBLE-ALTERNATIVE schema, as shown in Figure 8.6.

The process above is repeated until all feasible design alternatives are identified. Whenever a constraint is violated, the alternative generator backtracks to the preceding level. For example, if the condensation check for a wall assembly fails, other available thicknesses of the insulation are tried until a successful one is found. If there are no suitable thicknesses in this insulation material, another wall insulation material is chosen

FEASIBLE-ALTERNATIVE-54		
WALL-TYPE	STUCCO-ON-CONCRETE-WALL	
ROOF-TYPE	BUILT-UP-OR-SINGLE-PLY-ON-METAL-DECK	
INSULATION-TYPE-FOR-WALL	GLASS-FIBRE-BOARD	
INSULATION-TYPE-FOR-ROOF	RIGID-GLASS-FIBRE	
GLAZING-TYPE	DOUBLE-GLAZING	
ENERGY-CONSUMPTION	2701.26	(W/°C)
MATERIAL-COST	594982.0	($)
TOTAL-THICKNESS-OF-WALL	281.5	(mm)
TOTAL-THICKNESS-OF-ROOF	140.0	(mm)
UTILITY-VALUE	32.3 9.8 1.4 10.6 10.5	

Figure 8.6. Instance of a FEASIBLE-ALTERNATIVE schema.

and the process continues. Finally, if there are no other wall insulations, a different basic wall type is used and the process above repeated until all the materials, systems, and combinations are explored.

The foregoing process follows a top-down refinement of performance requirements and combinations of envelope components. Once generated, design alternatives are evaluated and ranked according to preferences on construction types and priorities of performance attributes specified by the designer. Hence the designer has the freedom to express preferences at the evaluation stage and to examine the performance of specific design solutions. This can be viewed as a form of bottom-up selection process that is available to the designer once all the feasible design alternatives have been generated. The designer may also define the details of new construction types for envelope components in the knowledge base and explore their feasibility.

Since the alternative generation process is presently implemented using object-oriented programming techniques, this system can readily be extended for considering other constraints and performance attributes. A more detailed description of the BEADS prototype system, its architecture, and an example consultation session can be found in Fazio et al. [1989].

CONCLUSIONS AND FURTHER WORK

The present study demonstrates the use of constraints derived from performance requirements and functional relationships for the generation of building envelope design alternatives. Schema representation of building objects enable heuristic knowledge to be coded in a structured form for efficient use in the generation process. Control techniques for constraint checking can be implemented easily because of the versatile knowledge representation for constraint formulation, representation, and retrieval. The BEADS system is a useful tool for considering a large number of alternatives during the preliminary design stage when the available information is limited to geometric and user requirements. Further work is under way to extend the number of performance attributes to include fire resistance, acoustic resistance, air-barrier requirements, and so on, and to develop a graphic interface for examining the cross-section details of building envelope components. Also, field testing of the prototype is being carried out to identify any revisions and improvements required by designers.

REFERENCES

AIA (1982). *Analysis: Thermal Transfer Through the Envelope*, Architects Handbook of Energy Practice, American Institute of Architects, Washington, D.C.

AMA (1986). *Manual for Masonry Walls and Marbles, Tiles and Terrazo Installations in Atlantic Canada*, Atlantic Masonry Association, Fredericton, New-Brunswick, Canada.

ASHRAE (1980). *ASHRAE Standard 90A-1980: Energy Conservation in New Buildings*, American Society of Heating, Refrigerating, and Air-Conditioning Engineers, Inc., Atlanta, Ga.

Chan, W.-T., and B. C. Paulson, Jr. (1987). "Exploratory Design Using Constraints," *Artificial Intelligence in Engineering Design, Analysis and Manufacturing*, **1**(1).

Corkan, J. (1985). "Check-list for Design to Overcome Building Envelope Failures," *Building*

Envelope Failures: Causes and Remedial Measures, Seminar 1: Walls, Alberta Building Envelope Council, Calgary, Alberta, Canada, March.

Elder, A. J., and M. Vandenberg (1974). *AJ Handbook of Building Enclosure*, Architectural Press, London.

Fazio, P., and K. Gowri (1989). "A Knowledge-Based System for the Selection and Design of Roof Systems," *Proceedings of the 11th International Congress CIB '89,* Paris.

Fazio, P., C. Bédard, and K. Gowri (1989). "Knowledge-Based System Approach to Building Envelope Design," *Computer-Aided Design*, **21**(8).

Finger, S., and J. R. Dixon (1989). "A Review of Research in Mechanical Engineering Design, Part I: Descriptive, Prescriptive and Computer-Based Models of Design Processes," *Research in Engineering Design*, **1**(1).

Gero, J. S., and R. Coyne (1984). "The Place of Expert Systems in Architecture," *Proceedings of Computer Aided Design '84*, Butterworth, Sevenoaks, Kent, England.

Gross, M., S. Ervin, I. Anderson, and A. Fleisher (1987). "Designing with Constraints," in *Computability of Design*, Y. E. Kalay (ed.), Wiley, New York.

Hardy, S. (1985). "Evaluating Roof Insulation Systems", Architecture, August.

Hutcheon, N. B. (1963). "Requirements for Exterior Walls," *Canadian Building Digest 48*, Division of Building Research, National Research Council of Canada, Ottawa, December.

Jayachandran, P., and N. Tsapatsaris (1988). "A Knowledge-Based Expert System, for the Selection of Structural Systems for Tall Buildings," *Proceedings of Microcomputer Knowledge-Based Expert Systems in Civil Engineering*, H. Adeli (ed.), American Society of Civil Engineers, New York.

Kalay, Y., L. Swerdloff, and B. Majkowski (1990). "Process and Knowledge in Design Computation," *Journal of Architectural Education,* **43**(2): 1–7.

MCC (1982). *Guide to Energy Efficiency Masonry and Concrete Buildings*, Masonry Council of Canada, Downsview, Ontario.

Mattar, S., W. Bitterlich, P. Manning, and P. Fazio (1978). "A Decision Model for the Design of Building Enclosures," *Building and Environment*, **13:** 201–216.

Morgenstern, M., A. Borgida, C. Lassez, D. Maier, and G. Wiederhold (1988). "Constraint-Based Systems: Knowledge About Data," *Proceedings of the 2nd International Conference on Expert Database Systems*, L. Kerschberg (ed.), Benjamin-Cummings, Menlo Park, Calif.

Murthy, N. S. H., J. S. Gero, and A. D. Radford (1986). "Multifunctional Material System Design," *Journal of Structural Engineering* **112**(11).

NRCC (1985). *Supplement to the National Building Code of Canada*, Associate Committee on the National Building Code of Canada, National Research Council of Canada, Ottawa.

Sapossnek, M. (1989). "Research in Constraint-Based Design Systems," in *Artificial Intelligence in Design*, J. S. Gero (ed.), Computational Mechanics, Southampton, England.

Sriram, D., and M. Maher (1986). "The Representation and Use of Constraints in Structural Design," *Proceedings of AIENG'86*, Computational Mechanics, Southampton, England.

Sriram, D., M. L. Maher, and S. Fenves (1985). "Knowledge-Based Expert Systems in Structural Design," *Computers and Structures*, **20**(1–3).

Willey, D. S. (1976). "Approaches to Computer-Aided Architectural Sketch Design," *Computer-Aided Design*, **8**(3).

PART THREE

MULTICRITERIA EVALUATION

9

PERFORMANCE EVALUATION IN AN INTEGRATED SOFTWARE ENVIRONMENT FOR BUILDING DESIGN AND CONSTRUCTION PLANNING

S. J. Fenves
U. Flemming
C. Hendrickson
M. L. Maher

Engineering Design Research Center
Carnegie Mellon University
Pittsburgh, Pennsylvania

In this paper we describe the role of performance evaluation in the Integrated Software Environment for Building Design and Construction Planning (IBDE), which is a testbed for examining integration issues in the same domain. Various processes in IBDE deal with the spatial configuration, structural design, and construction planning of high-rise office buildings. Performance evaluations occur within these processes based on different representation schemes and control mechanisms for the handling of performance knowledge. Within this multiprocess environment, opportunities also exist for performance evaluation across disciplines through design critics.

INTRODUCTION

The Integrated Software Environment for Building Design and Construction Planning (IBDE), under development in the Engineering Design Research Center (EDRC) at Carnegie Mellon University, is a testbed for examining integration issues in the domain identified in its title. The major issues examined are (1) the representation of project information as a project progresses; (2) communication of that information among the participating disciplines, which are represented through computer-based processes; and (3) control of the overall process. The test and demonstration domain is that of the spatial configuration, structural design, and construction planning of high-rise office buildings.

The architecture of the environment is based on the blackboard model [Nii 1986], in which various knowledge-based processes communicate through a shared memory referred to as the blackboard. IBDE consists of the following components: knowledge-based processes, controller, datastore and datastore manager, message and constraint blackboards, and user interface. The architecture is illustrated in Figure 9.1. The current version of the system integrates the following knowledge-based processes:

- ARCHPLAN: an interactive architectural planner to determine basic design characteristics
- CORE: a space planner for the service core of the building
- STRYPES: a structural system configurer
- STANLAY: a structural layout and approximate analysis system
- SPEX: a structural component designer
- FOOTER: a foundation designer
- CONSTRUCTION PLANEX: a construction planner, estimator, and scheduler.

ARCHPLAN produces a starting point for the other processes by determining the overall configuration of the building under consideration. CORE generates the spatial arrangement of the service core. STRYPES, STANLAY, and FOOTER provide a preliminary structural system and foundation design using the design synthesis shell EDESYN; therefore, these three processes share a common representation scheme for performance evaluation. SPEX was developed independently of IBDE and was integrated to provide structural component design. CONSTRUCTION PLANEX was also developed independently of IBDE and provides the construction planning knowledge for IBDE. Thus there are five different mechanisms for encoding performance knowledge, one each for ARCHPLAN, CORE, EDESYN, SPEX, and PLANEX.

Each process is a knowledge-based system in its own right and has a strong generative flavor; that is, it advances the state of the design. Performance evaluations are an integral part of both the individual processes and the overall process and play multiple roles in the development of a design:

1. Those processes that examine alternative solutions make their choice based on performance criteria or constraints which reflect the particular domain knowledge relevant for the type of problem handled by the process.
2. In advancing the state of the design, each process generates constraints for subsequent (or even parallel) processes. When encountering an unacceptable decision made by another process, a process turns into a *critic* of that process and posts its criticism on the blackboard. These critiques typically reflect knowledge not integral to the domain of the criticized process.

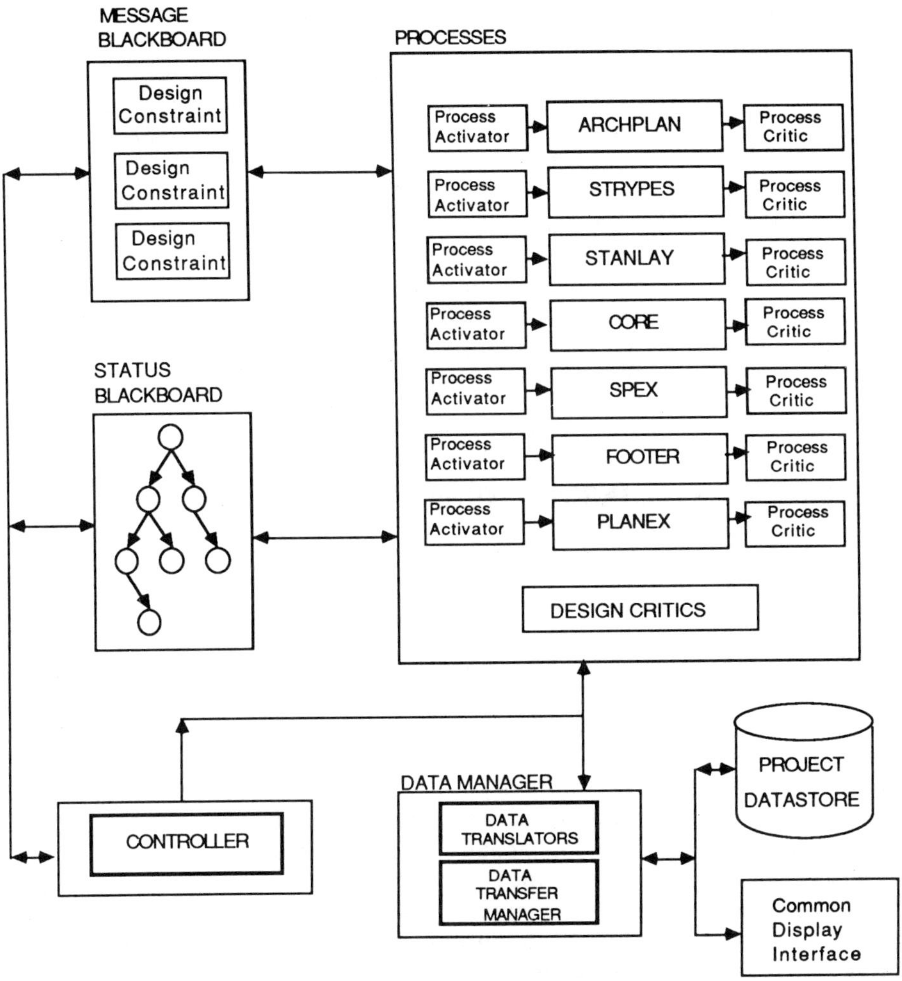

Figure 9.1. IBDE: overall architecture.

3. We are adding process-independent critics to critique the design using knowledge different from that used to generate the design. A prime example is a structural analysis critic that evaluates the performance of the structural systems and components using a formal analysis, while the design itself was generated using approximations and assumptions.

Performance evaluations thus provide strategically important knowledge to guide the overall process as well as its components. The architecture of the system allows each process to maintain its identity and follow a unique approach, while providing mechanisms for integration that steer these processes toward a common goal. The paper will elaborate on the multiple roles played by these evaluations and provide some details on the technical issues that arise.

A general description of the system can be found elsewhere [Fenves et al. 1990]. In the present paper we concentrate on the performance evaluations performed in the individual processes and in the environment as a whole. The knowledge-based processes make their decisions based on performance criteria or constraints that reflect the particular domain knowledge relevant for the type of problem handled by the process. The IBDE architecture allows each process to use its own knowledge representation and control strategy independently of the other processes. The actual mechanisms used are diverse; they reflect both the nature of the specific problems addressed and the preferences of the individual process designers. Taken together, they represent a collection of paradigms for the handling of performance knowledge. In the remainder of this paper we describe the representations and control mechanisms used and give examples of performance knowledge in the disciplines involved. We start with the knowledge-based processes, and then describe performance evaluation mechanisms for the environment as a whole.

PERFORMANCE EVALUATIONS WITHIN THE PROCESSES

ARCHPLAN

ARCHPLAN is a knowledge-based ARCHitectural PLANning expert system intended to assist in the development of a design concept. The input describes the site, the client's program and budget, and applicable geometric constraints. The output provides three-dimensional information on the overall geometry of the design, the distribution of functions within this geometry, and major circulation elements. The pilot version of ARCHPLAN was developed by G. Schmitt and is described in greater detail in [Schmitt 1988]. ARCHPLAN employs a form of *prototype refinement.* It is based on a geometric model of a typical office building and determines the parameters (such as number of floors or length and width of floors) that define a particular instantiation of the model and satisfy site- and problem-specific constraints.

Performance knowledge is represented in two forms: (1) as constraints, primarily in the form of upper or lower bounds, for example on the floor areas or total construction cost; and (2) as weights attached to constraints. ARCHPLAN computes a particular geometry so that the constraints are satisfied or conflicts are resolved based on the given weights. Constraints and weights can be varied interactively by the user through an interface that displays the current specifications, the resulting design, and (possibly) a previous version to which the current design can be compared; an example is shown in Figure 9.2.

CORE

CORE generates layouts of the elements in the service core of the building under consideration (elevators, elevator lobbies, rest rooms, emergency stairs, utility rooms, etc.). It is an adaptation of LOOS, a general system for the generation of layouts in various domains [Flemming et al. 1988]. LOOS places particular emphasis on the generation of layout *alternatives* with interesting *trade-offs.*

The major components of the LOOS architecture are a domain-independent *generator* able to generate layout alternatives, a domain-dependent *tester* able to evaluate

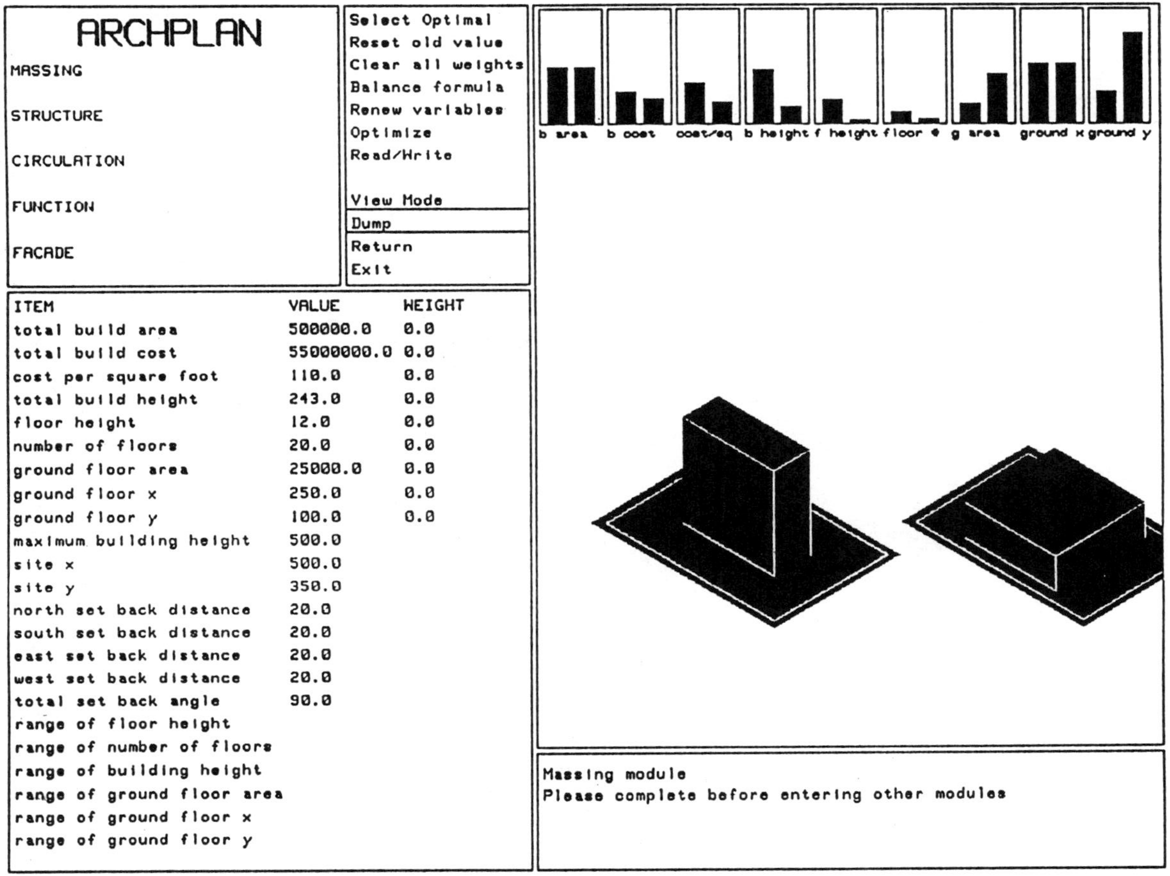

Figure 9.2. ARCHPLAN–user interface.

the layouts produced by the generator, and a *controller* that steers the generation process into promising directions based on the test results. The strategy employed is a form of *hierarchical generate-and-test* in which intermediate solutions are evaluated and, possibly, pruned from the search [Hayes-Roth et al. 1983, p. 72]. The generator starts from an initial state and recursively expands it by adding one object at a time in all geometrically possible ways, thus producing intermediate solutions that are immediately evaluated by the tester. The controller selects candidates for expansion based on these evaluations. (See Flemming et al. [1988] for a more detailed description.)

Performance knowledge is concentrated in the tester in the form of *test rules*, each of which evaluates a specific performance criterion. The tester for a particular domain is built up through the form of knowledge acquisition that proved effective particularly for diagnostic expert systems, where domain experts observe the system in operation and critique its performance. In the present case they are asked to check for unfavorable configurations of objects generated by the system. Each observation of this kind implies a test rule that checks for a condition to be avoided in a layout. This is an effective strategy for the extraction of domain-specific performance knowledge from experts and for its use in a generative system. (See Flemming [1990] for further elaborations.)

A subset of the conditions that the current version of CORE tries to avoid is shown in Table 9.1. They deal with a particular portion of the service core, the layout of elevator cars around a lobby, and are necessary and sufficient to produce all of the common arrangements. For example, the typical arrangements generated by CORE for a lobby with four cars are shown in Figure 9.3.

EDESYN-Based Processes

EDESYN [Maher 1987] is an expert-system shell for design synthesis in which the design knowledge base is represented as decompositions and constraints. The decomposition knowledge comprises a taxonomy of systems and components, both as templates for representing specific designs and as alternatives for generating multiple solutions. Its *hierarchical decomposition paradigm* generalizes decomposition schemes that have been proposed elsewhere for building design under names such as "morphological box" or "analysis of interconnected decision areas" [Luckman 1969]. Con-

TABLE 9.1 Subset of Conditions Checked by Current Test Rules in CORE

Lobby

1. Lobby is not accessible from main access direction.
2. If there exists only one bank, lobby is not accessible from opposite direction.
3. The number of cars on any side of lobby is greater than 4.
4. Car indices do not increase monotonically clockwise around the lobby, starting with the cars *above* the lobby.
5. If all cars in a bank have been allocated and are placed on opposite sides of lobby, the number of cars on one side differs by more than one from the number of cars on the other side.

Car (in an elevator bank)

6. Car is not adjacent to lobby serving the bank to which the car belongs.

straint knowledge represents combinations of alternatives that are not feasible. Performance evaluation can be represented within EDESYN through constraints.

In EDESYN, the synthesis process is a constraint-directed search through the taxonomy of systems and components in the knowledge base. Constraints are used to prune the search space at various levels of abstraction. As each subsystem or component is selected, it is checked for feasibility. An alternative is feasible if it is not eliminated by a constraint. Examples of constraints in STRYPES, STANLAY, and FOOTER are shown in Table 9.2.

SPEX

SPEX is responsible for the preliminary design of components of the structural system specified by STANLAY. It implements a heuristic *generate-and-test* strategy that applies three types of knowledge: knowledge contained in design standards; "textbook" knowledge of structural, material, and geometric relationships; and designer-dependent expertise [Garrett and Fenves 1986]. Design standards, such as those published by the American Concrete Institute (ACI) and the American Institute of Steel Construction (AISC), provide a set of performance requirements intended to achieve a satisfactory level of safety and serviceability for structures and their components. They consist of a set of provisions that define the limit states governing the behavior of the structure (e.g., component failure by yielding or excessive deflection) and the appropriate measure of acceptability for each potential limit state. These provisions are represented in SPEX as sets of constraints.

Design expertise is represented in SPEX by rules that express the designer's *intent* for the structural performance of the component. For example, a rule may specify conditions under which the designer intends a beam's performance to be controlled by the limit state of plastic yielding versus the limit state of lateral torsional buckling. This intent serves as the *focus* for retrieving the relevant provisions from the applicable design standard. SPEX generates a *trial* design by optimizing the structural component with respect to the constraints corresponding to the design focus. This trial design is then checked against *all* constraints applicable to the component. If the trial is unacceptable (i.e., it violates some constraint), a new focus is formed and the process repeated.

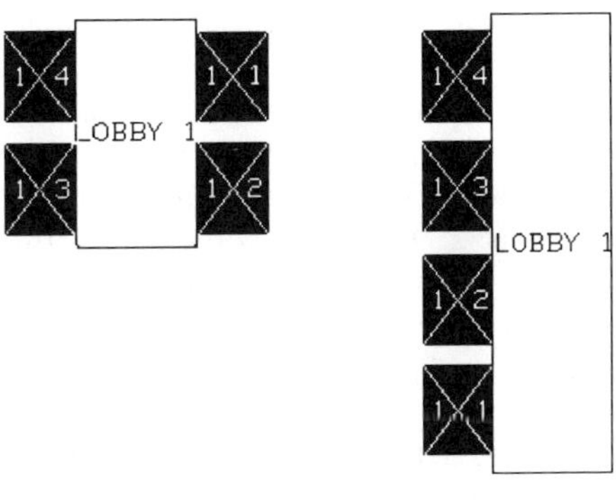

Figure 9.3. Alternative layouts generated by CORE for four elevator cars.

TABLE 9.2 Examples of Performance Constraints from STRYPES, STANLAY, and FOOTER

1. IF number of stores < 20
   AND 3D-lateral system = core
   THEN not feasible.
2. IF slab action = one-way
   AND slab ratio < 1.3
   THEN not feasible.
3. IF obstructions in soil = yes
   AND foundation-type = pile
   THEN not feasible.

Both SPEX and CORE employ a form of generate-and-test but with significant differences. SPEX aims at finding a single optimal design that satisfies the requirements and uses heterogeneous knowledge sources (textbook knowledge, specifications, heuristic rules), while CORE generates alternatives with interesting trade-offs based on a homogeneous knowledge base (a collection of test rules). The control strategies of the two processes vary accordingly. SPEX uses heuristic rules to generate a design focus for an optimal trial design; when it has to backtrack, the new design focus consists of the original focus and the violated constraint. CORE, on the other hand, develops alternatives incrementally by adding objects, one at a time, in all possible ways (without backtracking) and by pruning partial solutions based on intermediate evaluations. It finds the alternatives for adding an object by applying the highly context-sensitive rules of a formal grammar.

CONSTRUCTION PLANEX

CONSTRUCTION PLANEX estimates the basic attributes of cost, resource requirements, and time schedule for building construction [Hendrickson et al. 1987, Zozaya-Gorostiza et al. 1989]. These traditional measures are of considerable use in the evaluation of the overall design. Indeed, construction cost and time requirements can provide a basic critique of a design not in conformance with budget or time constraints. PLANEX is based primarily on the *nonlinear planning paradigm* developed in artificial intelligence [Sacerdoti 1973] in which elements of the initial partial plan are expanded, retracted, and reexpanded until the sequence of activities meets all precedence and resource requirements.

When forming the construction plan and computing resource estimates, PLANEX performs a significant amount of performance evaluation with respect to the constructibility of the proposed building. It first decomposes each structural or foundation component into element activities (e.g., formwork erection, reinforcement placement, concrete placement, curing, and formwork stripping for a concrete component) and aggregates element activities into project activities (e.g., formwork erection on floor 10). The number and kind of element and project activities are by themselves performance measures of the construction environment on the building site. PLANEX next selects construction technologies for each activity based on its knowledge about appropriate alternative technologies (e.g., place concrete by pumping versus lifting buckets by crane). Planex will reject design elements for which construction knowledge or necessary equipment are not available. Depending on the extent of the knowledge base, the construction performance can be fine-tuned (although not formally optimized) to a constructor's capabilities with respect to time and cost. Finally, PLANEX determines the resources required in terms of crews and equipment and produces the cost, resource requirements, and time schedule estimates.

The results of PLANEX (the project cost and duration estimates, the equipment and resource requirement profiles during the construction phase, and the element and project activities by number and kind) may be collectively viewed as critiques of the design produced by the other processes. If any of these measures exceed the owner's resources or estimates, constraints may be posted to reduce the appropriate measures.

PERFORMANCE EVALUATION AT THE SYSTEM LEVEL

The controller and critics are responsible for performance evaluation at the system level. The controller monitors the message blackboard and starts the appropriate processes. A process will initially be invoked only when its predecessor process(es) has been completed successfully. The status of a process is determined by the process itself. A process may also be reinvoked when constraints relevant to the process, posted by other processes or critics, warrant the initiation of a new version of the design. At present, the decision of initiating a new version is made by the user.

In the case of critics, performance knowledge is represented within the critic and communicated to the environment through constraints. Two classes of critics are incorporated in the system. To the first class belong *process activators* and *process critics* that extend the processes described in the preceding section. A process activator serves two purposes:

1. It evaluates the input data of the associated process. If the input includes requirements and constraints that fall outside the scope of the process, constraints are posted and the process is not activated. For example, ARCHPLAN may produce configurations with internal or external cores, but the current version of STRYPES can only handle internal cores; if its activator encounters an external core, it will post the constraint that the core cannot be external.
2. It reacts to critiques from other processes. If the blackboard lists a constraint relevant to a process, that constraint is directed to that process. When this process is reactivated, the constraint ensures that the conflict leading to the constraint does not arise again. In the example given above, ARCHPLAN will not propose an external core when it runs again.

A process critic also serves two purposes:

1. It eliminates causes of failure. If the process is unable to produce a feasible solution for its task, the critic poses constraints intended to eliminate the infeasibility.
2. It suggests possible improvements in performance. Even if the process produces a solution, the critic may provide critiques in the form of constraints aimed at improving the overall performance.

As an extreme example, assume that the technology selection component of PLANEX discovers that no crane of sufficient capacity is available to lift a beam designed by SPEX and posts the constraint "beam.weight $\leqslant$ some maximum capacity." The various processes may react to this critique as follows:

- SPEX may attempt to include this constraint directly, or its activator may pose a constraint to increase the grade of steel.
- STANLAY may attempt to use another layout involving shorter spans.
- STRYPES may choose a different structural system or material.
- ARCHPLAN may produce a different structural grid.

At the present, conflict resolution, that is, the decision where the processes should restart, is left to the user. We plan to introduce planning rules to guide the process automatically in such cases.

Critics in the second class are stand-alone components. They have been added to the IBDE to allow performance evaluations across domains or specializations within a domain. They are invoked to check the current state of the design using knowledge different from the knowledge that generated the design. They take as input the current state, process these data using their own knowledge bases, and produce one or more constraints. These constraints represent requirements that must be satisfied but are not satisfied by the current design.

A critic of this type that has been implemented is a structural analysis critic [Vora 1989]. It is invoked after the preliminary structural design has been completed; that is, STRYPES, STANLAY, and SPEX have been run successfully. The critic uses the current state of the datastore to produce a model and input for a structural analysis program. It communicates the results of the analysis as constraints on the parameters of over-stressed and grossly understressed (i.e., overdesigned) members. In the simplest case, SPEX would redesign the constrained components, but conceivably the constraints could propagate back to the other processes, as in the preceding example.

CONCLUSION

Performance evaluations provide strategically important knowledge to guide the design and planning of a building. The knowledge itself is diverse and can be used in different design and planning strategies. The Integrated Software Environment for Building Design and Construction Planning provides an architecture that allows processes dealing with specific performance aspects to maintain their autonomy and to use individual representation schemes and control strategies, while providing mechanisms for integration that steer these processes toward a common goal.

ACKNOWLEDGMENTS

The work described here has been supported by the Engineering Design Research Center, an NSF Engineering Research Center at Carnegie Mellon University. The contributions of G. Schmitt and many graduate students participating in the project are acknowledged.

REFERENCES

Fenves, S. J., U. Flemming, C. Hendrickson, M. L. Maher, and G. Schmitt (1990). "An Integrated Software Environment for Building Design and Construction," *Computer-Aided Design*, **22**(1): 27–36.

Flemming, U. (1990). "Knowledge Representation and Acquisition in the LOOS System," *Building and Environment*, **25**, pp. 209–219.

Flemming, U., R. Coyne, T. Glavin, and M. Rychener (1988). "A Generative Expert System for the Design of Building Layouts, Version 2," in *Artificial Intelligence in Engineering: Design*, J. S. Gero (ed.), Elsevier, New York, pp. 445–464.

Garrett, J. H., Jr., and S. J. Fenves (1986). *A Knowledge-Based Standards Processor for Structural Component Design*, Technical Report R-85-157, Department of Civil Engineering, Carnegie Mellon University, Pittsburgh, Pa., September.

Hayes-Roth, F., D. A. Waterman, and D. B. Lenat (eds.) (1983). *Building Expert Systems*, Addison-Wesley, Reading, Mass.

Hendrickson, C. T., C. A. Zozaya-Gorostiza, D. Rehak, E. G. Baracco-Miller, and P. S. Lim (1987). "An Expert System for Construction Planning," *ASCE Journal of Computing in Civil Engineering*, **113:** 253–269.

Luckman, J. (1969). "An Approach to the Management of Design," in G. Broadbent and A. Ward (eds.), *Architectural Association Paper 4: Design Methods in Architecture,* G. Wittenborn, New York, pp. 128–135.

Maher, M. L. (1987). "Engineering Design Synthesis: A Domain-Independent Representation," *Artificial Intelligence for Engineering Design, Analysis and Manufacturing*, **3:** 207–213.

Nii, H. P. (1986). "Blackboard Systems: The Blackboard Model of Problem Solving and the Evolution of Blackboard Architectures," *AI Magazine*, **7:** 38–53, Summer.

Sacerdoti, E. D. (1973). "Planning in a Hierarchy of Abstraction Spaces," in *Proceedings of the 3rd International Joint Conference on Artificial Intelligence*, Stanford, Calif., pp. 412–422.

Schmitt, G. (1988). "ARCHPLAN: An Architectural Planning Front End to Engineering Design Expert Systems," in *Expert Systems for Engineering Design*, M. D. Rychener (ed.), Academic Press, New York, pp. 257–278.

Vora, P. (1989). "A Structural Design Critic for the IBDE," Technical Report, Masters's thesis, Department of Civil Engineering, Carnegie Mellon University, Pittsburgh, Pa.

Zozaya-Gorostiza, C., C. Hendrickson, and D. Rehak (1989). *Knowledge-Based Process Planning for Construction and Manufacturing*, Academic Press, Orlando, Fla.

10

INTEGRATION OF PERFORMANCE EVALUATION IN COMPUTER-AIDED DESIGN

Avi Wiezel
Rachel Becker

Department of Civil Engineering
Technion–Israel Institute of Technology
Haifa, Israel

An integrated computerized system for the evaluation of the overall performance of a building was developed. The system exemplifies the capability of appropriate CAD techniques to upgrade the decision-making process and the design quality. In this paper we describe the specific problems arising from the integration of the performance evaluation into the existing CAAD process.

INTRODUCTION

An ideal, rational process of building design should be guided by the results of the performance evaluation of the generated solutions. During the design stages several solutions would be elaborated and detailed to achieve the goals stated initially. Although the solution generation depends mainly on the skills of the architect, the evaluation of the performance of the solution can be considered of technical concern and, consequently, shared by the computer. However, the complex input/output of specifically developed performance evaluator programs that are available today (i.e., thermal, acoustic, lighting, or fire safety) is a hindrance for their implementation in the design process, since they pertain to the experts more than to the architect. The evaluation tool that would be suitable for the designer has to derive all the necessary data from the proposed solution automatically and supply an overall and detailed performance evaluation. The capability of the integrated system to provide the evaluation results without interfering with the habitual design process is of major significance. This capability depends, furthermore, on the appropriateness of data structure to describe the building beyond the strict necessities of its graphic representation and to keep track of the design process itself.

THE PERFORMANCE EVALUATION

Performance requirements and criteria reflect, in engineering terms, the needs of human or other users, so that all the typical activities are duly accommodated and can be carried out without excessive disturbances. Levels of building performance are the degrees to which these needs are fulfilled. The performance approach regularly requires simultaneous consideration of the following attributes:

1. Serviceability limit state
2. Ultimate limit state
3. User's health and safety
4. Prevention of water penetration and moisture problems
5. Durability and maintenance
6. Fire safety
7. Internal climate
8. Acoustics
9. Lighting
10. Spatial characteristics

The numerous criteria to be considered, combined with the complexity of the relationships between the different parts of the building, usually lead to a deterministic approach to design problems and impede the use of methods based on a comparison of alternatives. Nowadays the spatial characteristics (attribute 10) are fulfilled extensively by the architect, serviceability and ultimate limit states are under the competence of the structural engineer, and attributes 3 to 5 are settled by proper design detailing procedures. Problems concerning fire safety, internal climate, acoustics, and lighting are in some cases covered by codes of practice and solved by domain specialists.

Although the aim of the design is to achieve the best fulfillment of all the performance criteria, the contradictory influences of some factors on building performance requires certain trade-offs within the proposed solution. The suitability of a solution can be measured only by the improvement it effects on all of the attributes listed above. For this purpose an overall performance grade can be used. It can be defined as a weighted average of the grades received by the solution for each of its qualities. The performance grade, Q_k, of the kth alternative is obtained by

$$Q_k = \sum_{n=1}^{N} W_n q_{n,k} \tag{10.1}$$

where $q_{n,k}$ denotes the nth subgrade of the kth alternative and W_n is the weighting factor. Since a percentile system is used,

$$\sum_{n=1}^{N} W_n = 100 \tag{10.2}$$

The overall cost divided by this grade may express a judgment about the optimal solution (i.e., "minimum cost for maximum quality").

The criteria that were considered for the actual state of the present computerized system are those for which there are recognized conventional algorithmic methods for performance evaluation [IMCH 1981]. These include internal climate, acoustics, and fire safety. The weighting factors, as they result from a study concerning a larger set of 26 criteria conducted on a typical urban population sample [Becker 1985], are given in Table 10.1.

The methods for the estimation of the performance grade can be roughly divided into two classes:

1. **Threshold value type.** The possible grades are only 0 or 100%, as the criterion does or does not reach a certain value
2. **Continuous function type.** When the possible grades rank between 0 and 100%, as the criterion ranges between certain boundary values

TABLE 10.1 Building Parameters and Weighting Factors W_n in Descending Order

Building Parameter	Enhancing Criterion (in Terms of "Prevention of")	W_n (%)
1. Acoustic insulation of the external envelope	Noise penetration from the outside	8.5
2. Thermal insulation and resistance to fungi of internal coverings	Mold growth due to condensation in the dwelling	8.5
3. Acoustic insulation of interdwelling partitions and floors	Noise penetration from neighboring dwellings	7.6
4. Balance between thermal insulation and thermal time constant (TTC) of envelope	Excessive expenditures for cooling during summer	7.4
5. Fire endurance of external envelope	Fire penetration into the dwelling from outside	7.4
6. Medium TTC of external envelope	Excessive heat at noon during summer	7.4
7. Low fire spread of interior coverings	Fire spread in the dwelling	7.2
8. Location of thermal insulation close to the inside	Excessive cold in the evening during winter	7.2
9. Low TTC of the envelope	Excessive heat in the evening during summer	7.0
10. Thermal insulation of external envelope	Excessive cold at night during winter	7.0
11. Thermal insulation, and high TTC of envelope	Excessive expenditures for heating during winter	6.6
12. Location of thermal insulation close to the inside	Excessive cold in the morning during winter	6.3
13. Acoustic insulation of intradwelling partitions	Noise penetration between rooms within dwellings	6.1
14. Acoustic insulation of intradwelling partitions	Being heard in other rooms (privacy)	5.9

A particular case is the minimum value prescribed by obligatory standards, when grade 0 coincides with the solution that just fulfills this standard. Negative grades represent nonacceptable solutions. For each criterion a specific algorithm for estimating the performance grade is thus applied.

THE COMPUTERIZED SYSTEM

Theoretically, CAD systems can assist designers in two ways [Kalay 1985]:

1. They can fully model the artifact that is being designed and automatically maintain the semantic integrity of the model.
2. They can guide the designer through the planning process and provide him or her with informative feedback on design decisions.

While the first request is accomplished extensively by the numerous available CAAD packages, practically none of them today offers the necessary feedback on design decisions. This leads to the necessity for either attaching an evaluation tool to existing CAAD package or for integrating the package into a more comprehensive design tool. The second solution was found to be more appropriate since it permits a better modeling of the design process and a more efficient management of the involved databases.

In the system proposed, a manager program conducts the process through the design stages and it is responsible for activation of the other two main components of the system: the graphic package and the performance evaluator (Figure 10.1). Both components have access to most of the specialized databases that complete the system. Sharing the process control between the two components, the manager also enables to overcoming the trouble caused by the single-tasking limitations of the chosen operating system (IBM DOS 3.0 or later version) (Figure 10.2).

The design process is influenced by multiple factors concerning building owner and user requirements, codes, regulations and standards, regional and local site conditions, properties of materials, manufactured products, past project experience, and so on. These factors, representing a prime set of information to be gathered, can be classified into two interactive classes: declarative and/or procedural. The declarative information activates the proper procedures dealing with the generation or gathering of new declarative information. This information is stored in the appropriate database and accessed whenever needed by any of the system components described above. The procedural information, related to the performance evaluation techniques, is embedded in the evaluator programs. At present the system is concerned only with domains for which the performance evaluation can be accomplished by means of recognized conventional (algorithmic) methods. The development of the system is still in progress, and while the databases are already prepared to receive information regarding thermal behavior, acoustics, fire safety, and lighting, the procedural information has been completed only for the first two aspects mentioned (Figure 10.3).

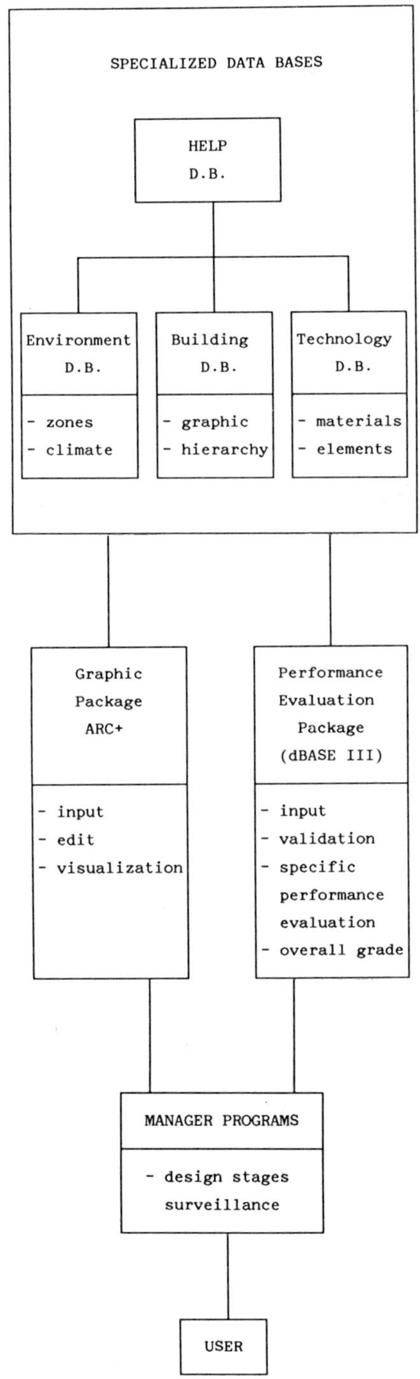

Figure 10.1. Conceptual structure of the system.

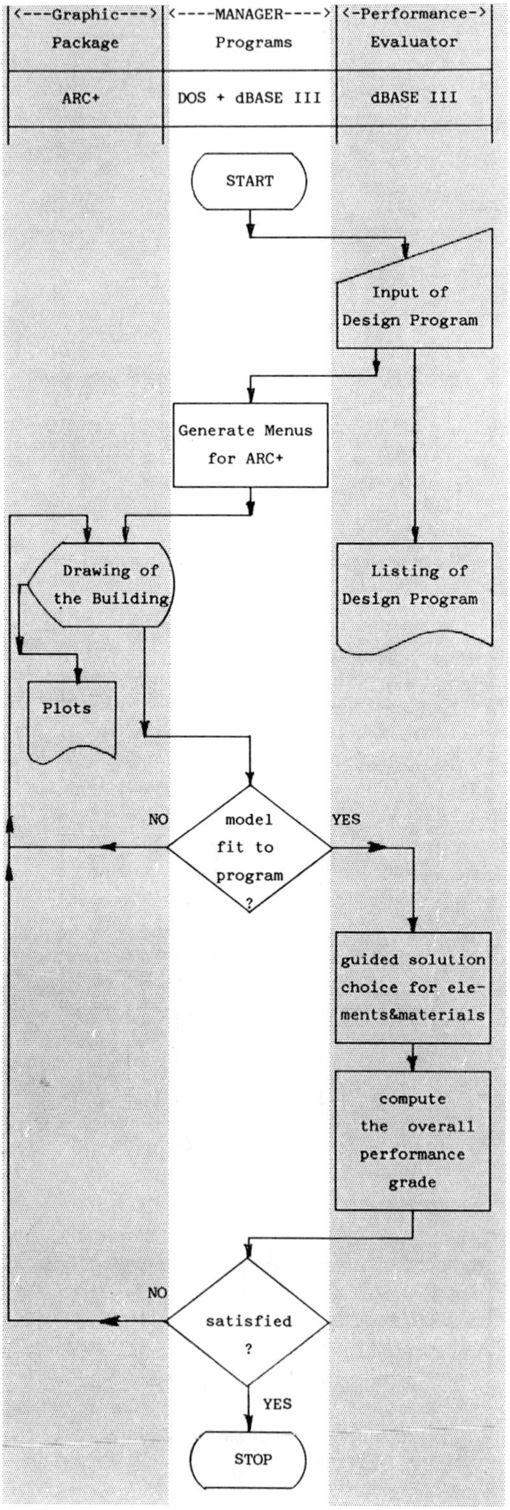

Figure 10.2. General flowchart of the system.

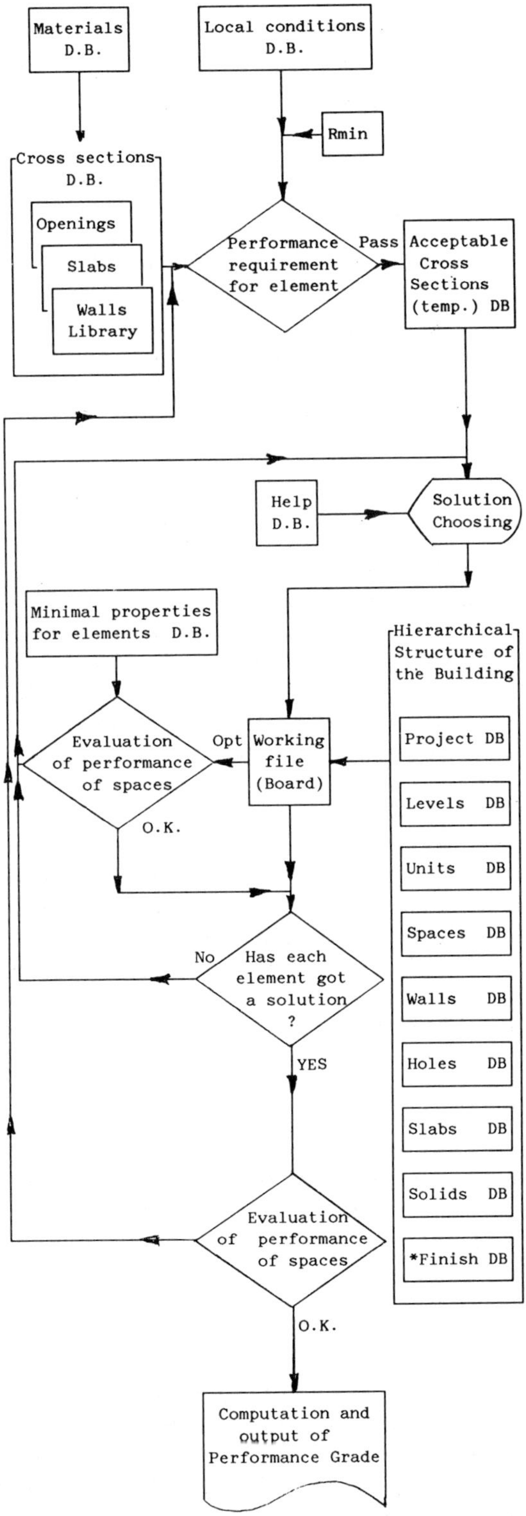

Figure 10.3. Performance evaluation process.

THE BUILDING DESCRIPTION DATA STRUCTURE

The CAAD package selected, ARC+ (a trademark of ACA, Architecture Computer Aids Ltd., Israel), which is the drawing component of the system, produces a database to represent the geometry of the design solution. This database, as in most of the available commercial CAAD packages, does not recognize all the building components, nor does it deal with the relationships between them. This fact induces the necessity to create a new data structure, parallel to the geometric one, to describe the relevant building components, their properties, and their connections, and to point back to the geometric data. This structure, which follows the hierarchical model of the building (Figure 10.4), is based on the functional definition of each represented part of building, starting with the site and the environment at the top, and ending with pointers to the physical properties of the building materials from which each element is to be made.

While the definition of entire building parts, such as wing, floor, or room, seems to be very close to the conventional perception, the partition elements created some special problems related to the definition of the performance functions. The mathematical model that supports the data structure requires the performance function to be single valued, which means that partition elements (both horizontal and vertical) may be defined between two, and only two, (adjacent) spaces. This definition imposes a rigorous delimitation of the building elements in the CAD model. In the case of vertical partition elements (walls and openings) the delimitation problem reduces to the condition that no entity continues beyond its intersection with another. This intersection problem, otherwise not present, has to be overcome either by means of suitable design procedures or by automation (which has not yet been fully implemented), within the selected CAAD package. Columns are also treated as a special case of wall, in which a column passing through more than two rooms is apportioned into more logical parts (Figure 10.5). The height of walls and columns has to be delimited by the horizontal elements (slabs or beams) they join. Although walls are recognized as building elements by ARC+, the horizontal elements are not. A special topological algorithm was developed to deal with the problems arising from slab design. The algorithm deals with an appropriate definition of the floor whenever it was not explicitly included in the three-dimensional model, or in such special cases as vertically nonrepetitive floors. The algorithm is restricted, however, to rectangular spaces, having walls only on the main directions (X and Y).

Special procedures dealing with the adjacency of the spaces enable the automatic identification and classification of the elements as internal or external ones and establish the relevant performance requirements. (The environment outside the building is also considered as a space by these procedures.)

THE HUMAN INTERFACE

The main purpose of the system is to liberate the user from boring evaluation procedures and to grant him or her useful information during the design process. This endeavor requires a user-friendly interface that minimizes the input of alphanumeric data (from the keyboard) by use of a mouse or digitizer. The solution that was preferred is that of a window-based screen (Figure 10.6), in which, whenever possible, any available relevant

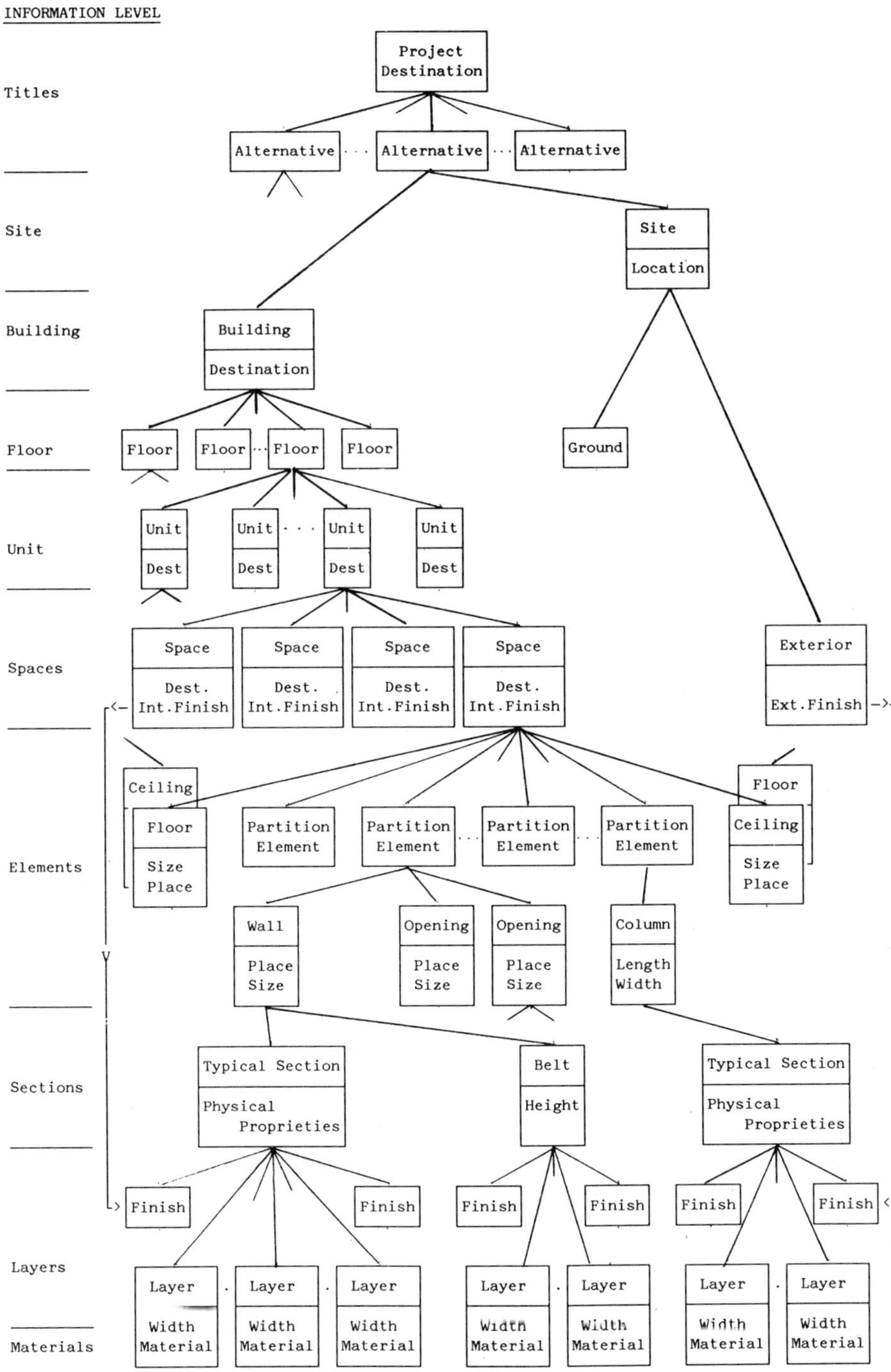

Figure 10.4. Hierarchical structure of a building.

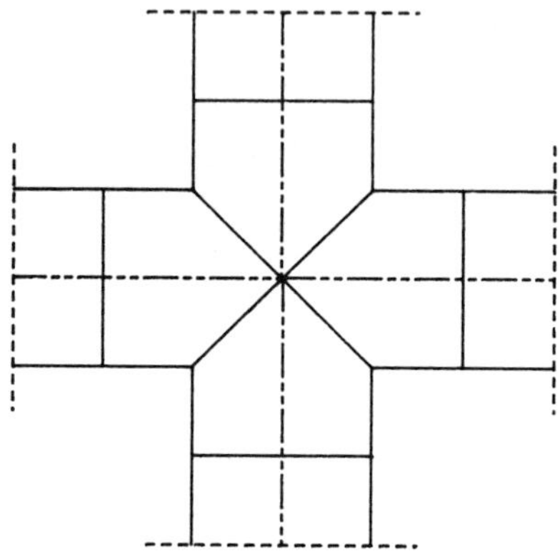

Figure 10.5. Logical apportionment of a column.

```
                        NEW BUILDING
                       DWELLING HAIFA
                  DEFINE FLOOR: TYPICAL
            DEFINE THE SPACES FOR : TYPICAL
                  UNIT: 2 BEDROOMS APT.

TYPE:BALCONY -CLOSED NAME: BALC      ROOMS/UNIT: 1 AREA BETW.:  1 AND 20
TYPE:BATHROOM        NAME: BATHROOM  ROOMS/UNIT: 1 AREA BETW.:  2 AND  6
TYPE:BEDROOM         NAME: BEDROOM   ROOMS/UNIT: 2 AREA BETW.:  8 AND 12
TYPE:KITCHEN         NAME: KITCHEN   ROOMS/UNIT: 1 AREA BETW.:  5 AND 12
TYPE:LIVING ROOM     NAME: LIVING    ROOMS/UNIT: 1 AREA BETW.: 24 AND 28
TYPE:TOILET          NAME: TOILET    ROOMS/UNIT: 1 AREA BETW.:  1 AND  4
TYPE:           CHOOSE

                1- BALCONY -CLOSED
                2- BALCONY -OPEN
                3- BATHROOM
                4- BEDROOM
                5- DINING ROOM
                N-NONE OF THESE
                H-HELP
                X-EXIT to previous menu
                ENTER YOUR CHOICE ===> X
```

Figure 10.6. Design program definition screen.

information emerges as a list of choices and users have to pick their choices. Furthermore, the menu of the graphic package (ARC+) is also dynamically adapted to the conditions deriving from the earlier design stages (Figure 10.7). When the geometrical design is completed the program presents the minimal performance requirements for each type of element and helps to choose one of the existing solutions for the elements, or, alternatively, to define new ones. Solutions with negative grades are not permitted by the system. In the present form the system can handle up to two exceptions in the typical cross section of each element (e.g., concrete belts in brick walls). Finally, the system attaches the chosen solution to each element and displays a list of all the elements and the selected solutions. It also presents a grade for each computed performance and a final overall grade, which is based on a weighted average for the whole building.

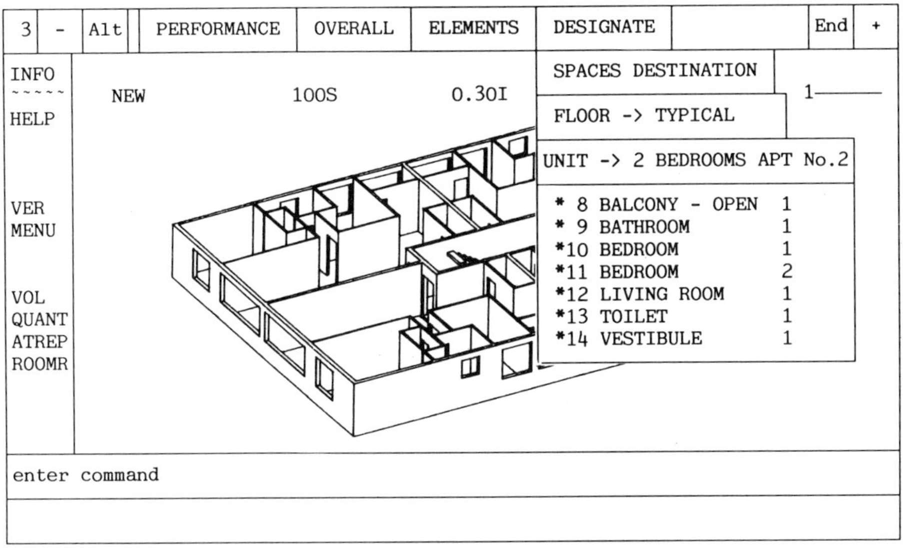

Figure 10.7. Screen display of a stage of consultation.

CONCLUSION

The efficiency of the design can be increased by providing a proper method for choosing the solution and an evaluation of the solution chosen. In the detailed design stages this can be achieved by means of appropriate queries applied to specific solution libraries. The automatic generation of suitable queriers relies on the ability of the system to recognize and describe the relationships between the represented elements and to keep track of the design stages. This ability is conditioned by the structure of the data that describes the design artifact and by facilities to access these data. In the present research the feasibility of this approach was demonstrated. The prototype of a decision support tool, integrated into the late stages of existing CAAD process, was developed. Further research, concerning the decision support in early design stages, is in progress.

REFERENCES

Becker, R. (1985). "A Method for the Generation of Weight Factors for Performance Evaluation Systems," *Building and Environment*, **20**(4): 195–200.

IMCH (1981). *Performance Criteria for New Building Systems*, Design Guides Series No. 5, Israeli Ministry of Construction and Housing, Jerusalem, December.

Kalay, Y. E. (1985). "Redefining the Role of Computers in Architecture: From Drafting/Modelling Tools to Knowledge-Based Design Assistants," *Computer-Aided Design*, **17**(7): 319–328.

11

ARCHITECTURE AND SOFTWARE FOR INTERACTIVE LEARNING ABOUT TOTAL BUILDING PERFORMANCE: EXPERIENCE-BASED OR OCCUPANCY EXPERT SYSTEMS FROM OCCUPANTS TO FACILITIES MANAGERS AND DESIGNERS

Volker Hartkopf
Vivian Loftness
Peter Mill
Melvin Siegel

Center for Building Performance and Diagnostics
Department of Architecture
Carnegie Mellon University
Pittsburgh, Pennsylvania

For evaluating and predicting design performance, CAD should contribute far more than the simulation of existing mathematical models in singular performance agendas. Building experience-based CAD systems of overall performance in the occupied setting can create new models for building use, management, and design. In this paper we outline the criteria and sources for multivariate expert systems for building design and control, based on a collective expertise from building occupants.

INTRODUCTION

Computers offer a tremendous capability for synthesizing large multivariate databases into tools for design decision making. To date, performance evaluations in buildings have been focused on the mathematical simulation of single performance criteria, such as energy and lighting, in the design development stages. There are major opportunities, however, in deductive analysis of multiple performance criteria in the occupied setting. Tying CAD systems to occupancy action records and to diagnostic sensors would allow the creation of an occupancy expert system or an experience-based system that could inform not only building management and future design but also the individual occupant.

There are several fundamental conditions that are needed in the pursuit of this objective. First, there must be a definition of the minimum range of interdependent performance criteria to be analyzed. This team has been actively involved in a definition of total building performance to include thermal, visual, acoustic, spatial, and air quality, as well as long-term building integrity. This definition builds on professional precedents but focuses on the simultaneous sensing of the human occupant. There must also be a definition of the components of the integrated system to be recorded, those causing variations in sensor readouts and occupant control actions. Finally, to ensure compatibility and dialogue there is the need for hardware and software development both in the sensor/controller area and in CAD hardware/software.

PERFORMANCE ANALYSIS AND THREE-DIMENSIONAL MODELS

There is still a significant division between the rapid development of three-dimensional and geometric modeling programs for architectural design and the two-dimensiosnal, often algebraic programs for building performance assessment. This division reduces the value of computer program developments in both areas, since light and visual quality must be assessed three dimensionally, and three-dimensional manipulation must be assessed in terms of its impact on visual, acoustic, thermal, and air quality, as wsell as spatial quality. Among "performance" professionals there is also a division of responsibility, with simulation of visual quality developed independently of acoustics, which in turn is developed independently of thermal quality.

The divisions between three-dimensional model development and performance modeling might easily be attributed to the size and complexity of the task, but this is an inadequate excuse. Three-dimensional space manipulation cannot be further promoted without some performance accountability, a task that the computer can effectively undertake.

FULL DEFINITION OF BUILDING PERFORMANCE

There are two critical definitions needed for the task of merging three-dimensional and performance modeling: the definition of total building performance or overall performance, and the definition of the integrated system comprising the occupied building. We have written extensively on total building performance [Hartkopf and Loftness 1986,

Loftness and Hartkopf 1989], building on the work of national and international organizations [NBS 1972, CIB 1982, ASTM 1982]. It is our contention that a minimum of six performance areas are needed to describe effectively the performance of the built environment for building occupants. First, there has been a fundamental mandate over the centuries for building integrity—protection of the buildings visible, mechanical, and physical properties from environmental degradation through moisture, temperature, air movement, radiation, chemical and biological attack, and environmental disasters. Established by concerns for health, safety, welfare, resource management, and image, the requirements for building integrity are set within limits of "acceptable" degradation, ranging from slight decay, to debilitation in the ability to provide weather tightness, to total devastation or destruction. Second, there are a series of mandates relating to the interior occupancy requirements (human, animal, artifact, machine) and the elemental parameters of comfort—thermal comfort, acoustic comfort, visual comfort, air quality, and spatial comfort—dependent on physiological, psychological, sociological, and economic limits of acceptability (see Table 11.1).

The key conditions for developing this list of performance mandates are that the list be limited in number (fewer than seven), be mutually exclusive, and deal wholistically with the interdependent human senses. The performance requirements in each of the six categories cannot be understood completely in isolation from the other. First of all, they are related in a complex manner through their physiological, psychological, sociological, and economic/resource subsets. For instance, the concept of privacy has acoustic, spatial, visual, and air quality dimensions. Second, in trying to fulfill the requirements of one performance mandate, side effects are created which impinge on the fulfillment of another (Table 11.2). The neeed for acceptable air quality, for example, may lead to a ventilation rate which will affect both thermal and acoustic comfort. Third, performance requirements are related to each other through the multiple effects of building component choice. The selection of a ceiling light fixture has implications in terms of light as well as heat, noise, and radiation. Although the building component may provide adequate performance in one dimension, it may fail in other dimensions, due to specification, context, or maintenance. To evaluate and eventually deliver a project that is acceptable in all performance areas, therefore, conflicts must be resolved between performance mandates and limits, and priorities must be set based on the building or space function. Starting, then, with a discrete list of performance mandates, the two- and three-dimensional computer simulation of buildings could evaluate the impact of spatial manipulation on visual quality, thermal/energy quality, acoustic quality (in a limited fashion), and even air quality, given critical system specifications.

DEFINITION OF THE INTEGRATED SYSTEMS

This demand for specification identifies a second critical factor in the task of integrating spatial and performance simulations: the definition of the integrated system comprising the occupied building. Many computer programs effectively model structure and interior systems spatially, while a few independent programs model mechanical, transport, and telecommunication systems spatially. They do not, however, model the performance specifications of these systems, nor the interdependency of one system on another in delivering spatial quality, much less thermal, visual, or acoustic quality.

TABLE 11.1 Minimum Definition of Building Performance

I. FUNCTIONAL/SPATIAL QUALITY = satisfactory:
Based on knowledge of the building occupancies, occupancy functions, and organizational structures
- A. Individual space layout quality
  Usable space, furnishings, layout efficiency, acces, anthropometrics, ergonomics, image, flexibility/growth, occupancy controls
- B. Aggregated space layout quality
  Proximities, access, compartmentalization, usable space, layout efficiency, image, amenities, flexibility/growth
- C. Building siting layout quality
  Access, public interface/image, indoor–outdoor relationships, outdoor space layout, flexibility/growth
- D. Quality of conveniences and services
  Sanitary, fire safety, security, transportation, electrical, telephone, information technology; flexibility/growth

II. THERMAL QUALITY = satisfactory:
- A. Air temperature
- B. Mean radiant temperature
- C. Humidity
- D. Air speed
- E. Occupancy factors and controls

III. AIR QUALITY = satisfactory
- A. Fresh air
- B. Fresh air distribution
- C. Restriction of mass pollution: gases, vapors, microorganisms, fumes, smokes, dusts
- D. Restriction of energy pollution: ionizing radiation, microwaves radio waves, light waves, infrared
- E. Occupancy factors and controls

IV. ACOUSTIC QUALITY = satisfactory
- A. Sound source: sound pressure levels and frequency
- B. Sound source: background noise
- C. Sound path: noise isolation (air and structure borne)
- D. Sound path: sound distribution; absorption, reflection, uniformity, reverberation
- E. Occupancy factors and controls

V. VISUAL QUALITY = satisfactory:
- A. Ambient light levels: artificial and daylight
- B. Task light levels: artificial and daylight
- C. Contrast and brightness ratios
- D. Color rendition
- E. View: visual information

VI. BUILDING INTEGRITY = satisfactory:
Based on knowledge of loads, moisture conditions, temperature shifts, air movement, radiation conditions, biological attack, human-made and natural disasters
- A. Quality of mechanical/structural properties
  Compression, tension, shear, abuse
- B. Quality of physical/chemical properties
  Watertightness, airtightness; transmission, reflection, absorption of heat, light and sound energy; fire safety
- C. Visible properties
  Color, texture, finish, form, durability, maintainability

TABLE 11.2 Interrelated Nature of Performance Variables through Building System Decision

Decisions made for:	Leading to failures in:
Acoustical Performance	**Thermal Performance**
freestanding acoustical partitions flush to the floor	poor air circulation at the working plane
soft, sound-absorbing surfaces on floor, walls, ceiling	rapid heating-up with sunshine or running equipment, due to poor thermal capacitance
isolation of noise-generating equipment	large heat buildup, unexpected air conditioning needs
drywall partitioning into grid in ceiling plane	possible cutoff of supply air diffuser
operating supply air or fans continuously as masking sound	possible thermal discomfort, draftiness
subdivision for security, space definition, and privacy	possible misplaced thermostats, control in adjacent spaces
Acoustical Performance	**Visual Performance**
freestanding acoustical panels, especially if dark colored	poor light levels at the working plane
high density freestanding partitions	often askew with lighting grid in ceiling plane, inadequate light supply within workstation
subdivision for security, space definition, and privacy	possible misplaced light switches, control in adjacent spaces
Acoustical Performance	**Spatial Performance**
partitions from floor to ceiling slab for acoustical privacy	poor flexibility for rapid change
isolation of noise-generating equipment	poor accessibility
acoustically isolated typing stations	poor supervision
Acoustical Performance	**Air Quality**
isolation of noise-generating equipment	concentration of potentially polluting equipment
increased use of sound-absorbing materials	possible outgassing of binders, release of particulates
Acoustical Performance	**Building Integrity**
soft, open-pore interior surfaces for acoustical absorption	easily damaged, torn, punctured, and soiled

Table 11.3 illustrates that the integration of two systems is as important or more important to the spatial and environmental quality in a building as the selection and manipulation of independent systems. For this reason it is necessary to fully describe the integrated building in the computer, to enable quick manipulation with assurance of positive impact on spatial, thermal, visual, acoustic, and air quality.

In *The Building Systems Integration Handbook* [Rush 1987], we began the definition of critical systems input by exploring the choices and performance implications of integration between structural, enclosure, mechanical, and interior systems (Table 11.4). A fuller definition of the integrated system in a CAD database would enable meaningful diagnosis of the critical factors affecting air quality, thermal, visual, acoustic, and spatial performance in buildings.

GROWTH IN OCCUPANCY EXPERT SYSTEMS TO ACCOMPANY ELITE EXPERT SYSTEMS

In the computer-aided design field, there is an amazing reliance on the expertise of independent building designers—architects, engineers, and interior planners—a reliance that is being slowly transformed into expert decision-making or design tools. Sur-

TABLE 11.3 Impact of Systems Integration on Building Performance Beyond Individual Disciplinary Decision Making

	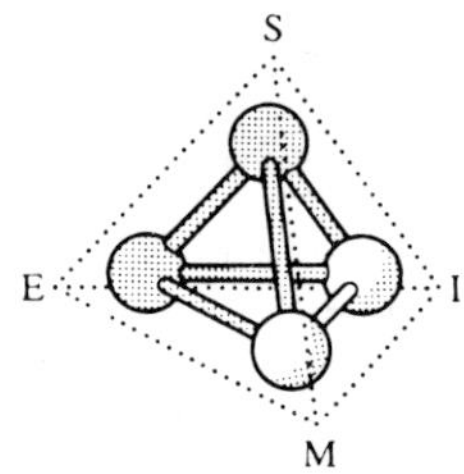 Integrate Mechanical and Interior Systems for:
Spatial	• service module for present use and flexibility • space volume and ceiling height, functional identity • individual space layout, compartmentalization & flexibility • suitable servicing to individual space and aggregated space • expansion and change potential • individual control/space management
Thermal	• balanced mean radiant temperatures • air distribution, heating and cooling effectiveness, comfort • humidity impact on interior components • volume to be heated, stratification • mechanical efficiency from source to task
Air Quality	• fresh air distribution effectiveness • flushing vs. outgassing • protection from radiant pollution • individual control, compartmentalization vs. pollution migration
Acoustical	• background noise coordinated with appropriate room absorption • minimization of mechanical noise disturbance • minimization of vibration disturbance • effective acoustical compartmentalization around mechanical components
Visual	• adequate lighting distribution, balance • mechanical components for lighting dispersion, reflection • integration for space/function identity • lighting efficiency from source to task (lens) • flexibility potential • individual control, compartmentalization
Building Integrity	• humidity and temperature compartmentalization for artifact/machine/human "comfort" • staining and dirt buildup from mechanical on interior • vibration disturbance/destruction • mechanical accessibility, maintenance and repair • fire safety

TABLE 11.4 Minimum Definition of the Integrated System for Evaluating Performance

PHYSICAL DECISIONS IN MECHANICAL SYSTEM DESIGN

- HVAC

Service Generators
Size, Volume
Form, Configuration
Expansion Capability
Material, Ornament
Service Conduits
Thickness, Volume of Service
Form, Shape
Configuration, Distance, rise/run
Interface/Expansion Capability
Material, Ornament
Connection to Other Systems
Access
Service Terminals
Planning Module
Number, Size, Capacity
Form, Material, Ornament
Interface/Expansion Capability
Relocation Capability
Connection to Other Mechanical Systems
Control Systems
Central Management Systems
Local Management—Automatic/Manual

- Lighting

Service Generator—Size, Capacity
Service Conduit
Thickness, Volume of Service
Interface, Expansion—Capability
Material, Ornament
Access
Service Terminals
Planning Module
Size, Capacity
Form, Material, Ornament
Interface, Expansion Capability
Relocation Capability
Connection to Other Mechanical Systems
Control Systems
Central Management
Local Management, Automatic/Manual

- Telecommunications, Power, and Security

Service Generator—Size, Capacity
Service Conduit
Thickness, Volume of Service
Interface/Expansion Capability
Material, Ornament
Access
Service Terminals
Planning Module
Number, Size, Capacity
Form, Material, Ornament
Interface/Expansion Capability
Relocation Capability

PHYSICAL DECISIONS IN STRUCTURAL SYSTEM DESIGN

General System Type:
Frame, Diaphragm, Tensile, Inflated
System Materials and Properties:
Steel, Concrete, Wood, Plastics
Span, Bay Sizes, Column Spacing
Floor to Floor Height
Cross Section of Structural Elements (Height, Width, Depth)
Building Form: Plan, Section
Expansion Capabilities (Vert., Horiz.)
Connections to Other Systems

PHYSICAL DECISIONS IN ENVELOPE SYSTEM DESIGN

Exterior Wall/Roof/Floor
Exterior Surface, Material Properties
Composite Materials, Thickness
Interior Surface
Form: Planar, Curved
Slope, Orientation
Module Size, Shape
Connection of Envelope Components
Windows/Openings
Material Properties
Size, Shape, Spacing
Orientation
Control Systems, Sunshading
Control Systems, Heat Loss
Control Systems, Security/Privacy
Frame, Connections, Plan/Sec/Elev.
Access, Visual and Physical
Expansion Potential, Horiz. & Vert
Change Potential for Access & Image
Color, Texture, Ornament

THERMAL PERFORMANCE

To Achieve Appropriate:
1. Air Temperature
2. Air Movement
3. Relative Humidity
4. Median Radiant Temperature
5. Occupancy Factors

To improve thermal performance and to foster design innovation, examine any two ***bold-faced*** *systems decisions.*

PHYSICAL DECISIONS IN INTERIOR SYSTEM DESIGN

Ceiling
Ceiling Height, Shape, Form
Ceiling Depth
Ceiling Surface, Material Properties
Planning Module
Connection to Other Interior Systems
Floor
Floor Depth
Floor Form/Slope/Steps
Floor Surface, Material Properties
Floor Module
Connection to Other Interior Systems
Wall
Wall Form, Depth
Openings (Interior)
Wall Surface, Material Properties
Planning Module
Connection to Other Interior Systems
Furnishings
Work Surface: Material, Height, Depth
Sit/Sleep Surface: Material, Ergonomics
Storage
Vegetation/Ornamental Objects
Connection to Other Interior Systems
Office Equipment and Appliances
Access/Circulation
Security Configuration
Expansion Potential, Horiz. & Vertical
Change Potential for Access & Image
Color, Texture, Ornament

From *Building Systems Integration Handbook*,

prisingly, there is little development of expert systems based on the vast occupancy expertise in buildings. Yet occupants are multisensory, able to capture all senses simultaneously, and have built their expertise within the fully integrated and occupied setting. The development of occupancy expert systems or experience based systems would enable the vast database management capabilities of the computer to build understandings of spatial, thermal, visual, acoustic, and air quality simultaneously, in a diversity of integrated settings, testing for critical conditions and flexibilities. The lack of performance knowledge transfer from the occupants and facility managers to the building designers and constructors is a far more serious void than the ability to manipulate space three dimensionally, an effort that is receiving so much expert attention today.

FROM WHERE WOULD THE INPUT COME?

The "mass" expertise of building occupants can be gleaned through a number of avenues: through interviews and questionnaires, through smart sensor/controllers, through complaint and health records, and through behavioral patterns and physical traces [Preiser 1989, Bechtel et al. 1987, Mill et al. 1986). There are even a few examples of developments in sensors tied to mainframes for interactive learning and control of building performance. In Osaka, Japan, the Umeda Center has both local sensors and controllers for temperature, air (on/off with a minimum), and lights (on/off), tied to the central computer database (with local PC dial-up for occupant readout and control). The computer software enables the building managers and design team to assess the variations in thermal, visual, and air quality desired in an evolving intelligent office. In another Japanese project, Ark Hills, in Tokyo, a mobile diagnostic unit measuring air temperature, CO_2, humidity, and light levels is pushed systematically through the project to ascertain the actual environmental quality in the changing workplace, and the performance of local sensors and controllers. These data could be tied to a three-dimensional CAD program incorporating both spatial and system specifications for causal analysis. In Lloyd's of London, individual environmental control of lights (on/off), air (on/off with minimum), and heat (on/off) is controlled and monitored through a four-port cable network (with power, telephone, and data management). The port can be programmed locally or at the central computer to manage as many or as few lights, floor air diffusers, and perimeter heat pumps as may be in a large conference room or a single workstation, to allow interactive learning about the density of environmental control needed in various work settings. In the United States, Johnson Controls has also begun the development of an occupancy expert system for interactive learning about total building performance in the occupied setting. The Personal Environments Module enables the monitoring of individual modifications in light (dimming), heat (dimming), air temperature (dimming), air speed (dimming), and even white noise (dimming). The workstation harness that brings conditioned air and metered power to each desk enables the control and monitoring of multiple performance variables over time of day and season (Figure 11.1).

In addition to metered data through sensors and controllers, the expert database for building design and management can also be effectively built or supported by statistically valid questionnaires and interviews, complaint records, as well as observations of behavioral patterns (putting on sweaters, drawing blinds) and physical traces (added heaters, fans, light shields over the computers). The database would yield a wealth of design information that would eventually feed into more performance-based three-dimensional CAD programs. However, the fusion of multiple information sources will not be trivial, requiring iterations for robustness and clarity, as well as eliminating conflicts and establishing priorities.

TO WHERE WOULD THE OUTPUT LEAD?

The benefits of experience-based/occupancy expert systems range from verification of building or design effectiveness, to user preferences, control strategies, and directions for design innovation in building products and integrations. Experience-based CAD

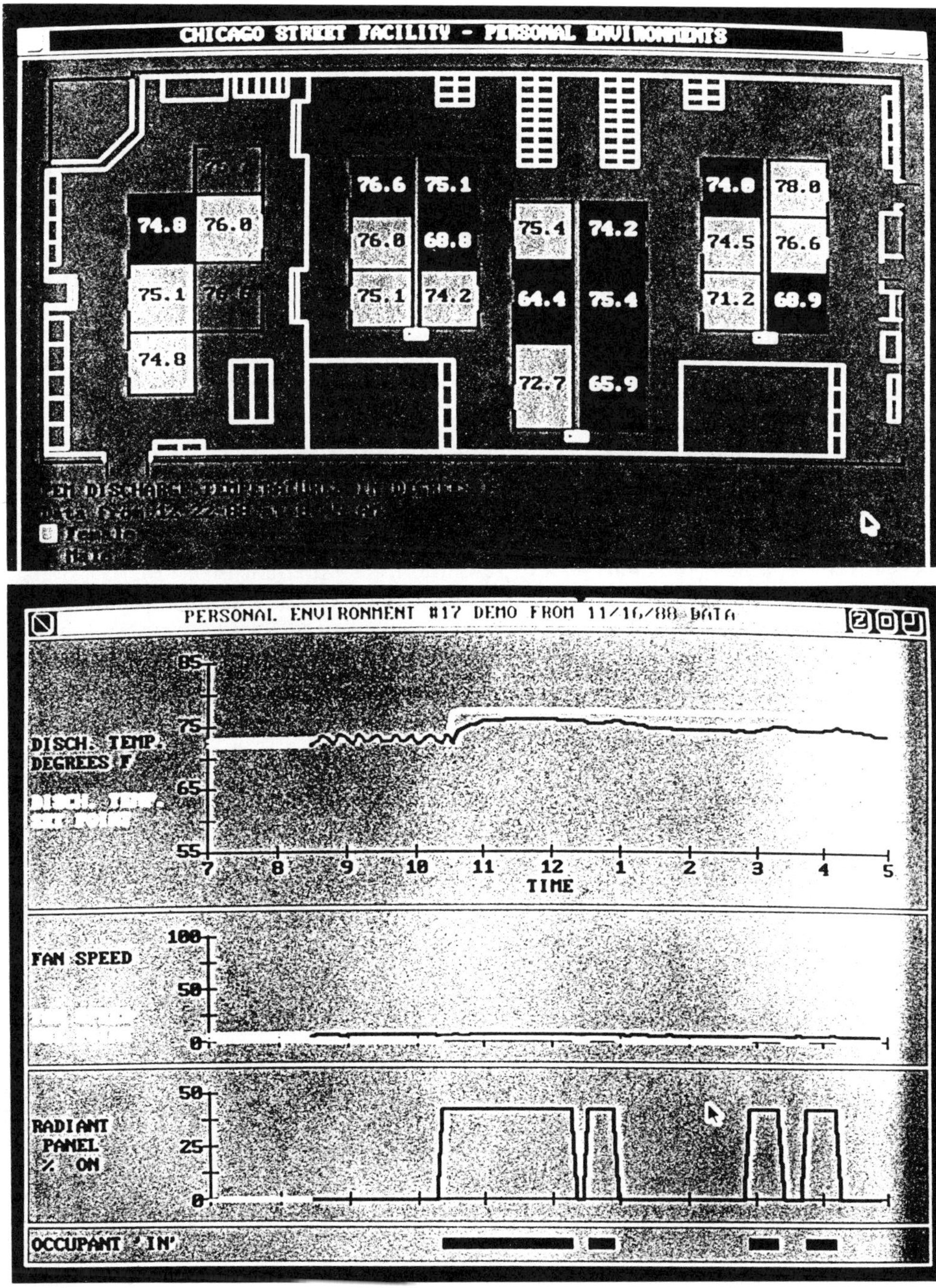

Figure 11.1. Computer monitoring of occupant modification of multiple performance variables over time of day and season at Johnson Controls.

systems would inform other performance evaluations of integrated systems, completing assessments—by comparison and by interpolation—of acoustic quality, thermal quality, air quality, visual quality, and long-term integrity. Repetitive errors found in today's buildings would be avoided (e.g., flanking sound transmission, inadequate diffuser densities or problematic heat islands, pollution migration paths, excessive brightness contrasts). The need for new products or new integrations of existing products would be highlighted in a pursuit of better building performance for the occupants.

In the conventional building, the existing task-ambient systems for heating, air conditioning, lighting, and even masking sound can be optimized through feedback from user modification or critique. In the buildings of the future, the shift away from blanket systems to local control systems will be facilitated, enabling the wide-band ambient system to be maximized (as broadly as possible) and the range of task controls to be established. The systems (or assembled products) intended to provide thermal comfort, acoustic comfort, spatial comfort, air quality, and/or visual comfort will also receive badly needed expert input from the body of building occupants and facility managers, enabling innovative designs to emerge (Table 11.5).

A critical area of application for these experience-based/occupancy expert systems is for monitoring building health and illness. Since physiological inputs (visual, air quality, thermal, acoustic) and psychological inputs into building sickness cannot be effectively separated for analysis and resolution, an experience-based CAD system with both multiple performance and integrated system inputs is critical. Clearly, these mass expert systems will contribute to the single-person expert systems under development today.

TABLE 11.5 CAD Software Tying Spatial Exploration to Thermal, Visual, Air Quality and Acoustic Performance

Base Building Givens as input:
Structural Grid
Enclosure Grid
Ceiling Grid
Acoustic Budget
Floor Grid
Mechanical Grid (Diffuser Layout and Distribution Pattern)
Mechanical (Cooling) Budget
Lighting Grid (Layout and Distribution Pattern)
Telecommunications Grid (Power, Data, Telephone)
Limits and Budgets as input:
Furniture Modules of Choice (Plan and Section)
Furniture Components of Choice
Furniture Group/Templates of Choice
Furniture Specifications of Choice (Light, Acoustics,...)
Sizes of Aisles
Sizes of Workstations
Limits of "Humanity" (Density)]
Daylight and View Access
Telecommunications Density (Connectivity and Heat Generation)
Occupancy/ Performance Rules in the Software:
Workstation Sizes and Configurations
Free Address, Hierarchical, Village and Campus, Open/Closed Ratios
Distance/Layout for Windows, Light and View (perp. and parallel planning)
Density/Layout to Mechanical Grid
Density/Layout to Lighting Grid
Density/Layout to Telecommunication Connections
Acoustic Absorption Capacitance
Air Conditioning Capacitance
Ergonomic/Anthropometric Rules
Component Assembly Rules
Output:
Free Design Decisionmaking with Rule Based Evaluations
Rule Based Design Decisionmaking — Limited (or Ranked) Multiple Options
Cost and Performance Benefits

CONCLUSION

Computers can make major contributions to many stages in the design process. In the early design stages, simulation can enhance spatial exploration through three-dimensional modeling. More critically, simulation can enable spatial exploration to be tied to performance implications, through appropriate subroutines in visual, thermal/energy, acoustic, and air quality analysis. In the design development and working drawing stages, performance specifications can be tied to the three-dimensional geometries, and major building performance failures can be avoided through expert assessment. In the commissioning and occupancy stages, computers tied to sensors, controllers, questionnaires, complaint records, and even mobile diagnostic units can manage large multivariate databases to inform engineering and facilities managers, as well as occupants, regarding correction and optimization. Most critically, the development of occupancy expert or experience-based systems can greatly inform the design process, filling gaps that have not even today been resolved through expert designers and design processes. It must be understood that human beings are complex sensors responding to a fully integrated and occupied physical setting that will be difficult to approach through any software architecture that does not include their expertise.

REFERENCES

ASTM (1982). *3rd Symposium Proceedings, Performance Concept in Building*, ASTM/CIB/RILEM, Lisbon, Portugal.

Bechtel, R., R. Marans, and W. Michelson (eds.) (1987). *Methods in Environmental and Behavioral Research*, Van Nostrand Reinhold, New York.

CIB (1982). *Working with the Performance Approach in Building*, CIB Report Publication 64, Center Internationale du Batiment, Paris, January, pp. 5, 10, 17, 20.

Hartkopf, V., and V. Loftness (1986). "The Concept of Total Building Performance and Building Diagnostics," in *Building Performance: Function, Preservation, and Rehabilitation*, ASTM Special Technical Publication STP 901, presented October 1983, published May 1986.

Loftness, V., and V. Hartkopf (1989). "The Impact of Building Design and Use on Air Quality," in *Occupational Medicine: Building-Associated Illness*, Hodgson (ed.), Hanley & Belfus, Philadelphia.

Mill, P., V. Hartkopf, and V. Loftness (1986). "Evaluating the Quality of the Workplace," in *The Ergonomic Payoff: Designing the Electronic Office*, R. Lueder (ed.), Nichols, New York.

NBS (1972). *Proceedings, Performance Concept in Buildings*, NBS 361, National Bureau of Standards, Gaithersburg, Md.; e.g., G. Blanchere, "The Notion of Performance in Building: Building Requirements," and "What Are the Natures of Performance and Evaluation for the Three Levels: Building, Component, Materials?"

Preiser, W. (ed.) (1989). *Building Evaluation*, Plenum, New York; e.g., V. Hartkopf, V. Loftness, and P. Mill, "A Critical Framework for Building Evaluation: Total Building Performance, Systems Integration, and Levels of Measurement and Assessment."

Rush, R. (ed.) (1985). *The Building Systems Integration Handbook*, Wiley, New York; e.g., V. Hartkopf, V. Loftness, and P. Mill, "Integration for Performance."

12

A BUILDING DESIGN AND EVALUATION SYSTEM

Eek J. H. Hacfoort
K. Jan Veldhuisen

Design Methods Group
Faculty of Architecture, Building and Planning
Eindhoven University of Technology
Eindhoven, The Netherlands

Within the field of architectural design there is a growing awareness of imbalance among the professionalism, the experience, and the creativity of the designer's response to the up-to-date requirements of all parties interested in the design process. The building design and evaluating system COSMOS makes it possible for various participants to work within their own domain, so that separated but coordinated work can be done. This system is meant to organize the initial stage of the design process, where user-defined functions, geometry, type of construction, and building materials are decided. It offers a tool to design a building to calculate a number of effects and for managing the information necessary to evaluate the design decisions.

The system is provided with data and sets of parameters for describing the conditions, along with their properties, of the main building functions of a selection of well-known building types. The architectural design is conceptualized as being a hierarchy of spatial units, ranking from building blocks down to specific rooms or spaces. The concept of zoning is used as a means of calculating and directly evaluating the structure of the design without working out the details. A distinction is being made between internal and external calculation and evaluation during the initial design process. During design on screen, an estimation can be recorded of building costs, energy costs, acoustics, lighting, construction, and utility. Furthermore, the design can be exported to a design application program, in this case AutoCAD, to make and show drawings in more detail. Through the medium of a database, external calculation and evaluation of building costs, life-cycle costs, energy costs, interior climate, acoustics, lighting, construction, and utility are possible in much more advanced application programs.

INTRODUCTION

The design process is traditionally seen to be exclusively in the hands of individual designers and is therefore less open to suggestions from other interested parties. As a result, the quality of the end product, the design, is essentially limited by the amount of information that can be processed by one person. Moreover, this does not facilitate adjustment to changes in contemporary socioeconomical and technological requirements, not to mention the rapidly increasing amount of technical information available. Hence within the field of architectural design in the Netherlands there is a growing awareness of imbalance among the professionalism, the experience, and the creativity of the designer's response to the up-to-date requirements of all parties interested in the designing process.

To overcome some of these difficulties, a design methodology has been developed that takes into account the following:

- The growing differentiation in the building design profession
- The growth of specialized knowledge
- The interests of the various parties and their willingness to take responsibility for their part in the end product

This methodology opens the possibility for different participants to work within their own domain, so that separated but coordinated work can be done. These aims can be so arranged as to make computer-assisted building design and evaluation possible. Thus within any phase of the design process professionals and user groups can make clear decisions on spatially defined levels.

The design methodology is based on the concept of domain and zoning. A *domain* is one of three functional fields: (1) usability (usage), (2) durability (construction and installation), and (3) production (technology) [Bax 1990]. A *zone* is a characteristic linear part of a spatial unit which gives direction to a further geometric definition. For the zoning of a spatial unit, zones and margins are differentiated. The zones are used for positioning the spaces, the margins are used to position the materials [Habraken et al. 1976].

COSMOS is a system developed to meet these requirements and the ideas presented in earlier paragraphs. COSMOS is a Dutch acronym meaning "computer-assisted methodological architectural design system." This system is meant to organize the initial stage of the design process, where user-defined functions, geometry, type of construction, and materials are decided. In addition, the system simultaneously calculates and evaluates a number of different aspects of the design. These range from building costs, energy consumption, acoustics, lighting, and construction, through to the utilities that users may possibly derive from some of the properties of the design.

COSMOS began by offering the user the opportunity of rapidly calculating a small number of effects by means of separate programs for which interface files have been developed. However, it soon became apparent that the means of evaluating (variants of) designs could not adequately deal with the growing amount of effects. Moreover, the computer programs themselves tended to become more complicated and much more difficult to operate.

There were two steps for solving this problem. The first lay in incorporating a number of effect calculations (in a simplified form) into a second version of COSMOS. The second step was to construct a database for using the various programs for calculating effects. This database had to be consistent with the ISO-STEP standard (STandard for the Exchange of Product model data [Altemueller 1988]). These various external programs can include all computer programs that accept a STEP file or an interface file based on a STEP file.

Earlier we mentioned the increasing number of effects that need to be taken into account while designing. Their combination, together with the possibility of generating a large number of alternative solutions for the same design problem, creates a great deal of data, which this system can easily handle. These data now constitute a data matrix that can be analyzed by multi-criteria or single-criterion analysis. The system is based on the assumption that trade-offs between aspects must be possible at any stage of the design process. Bivariate analysis is proposed for managing the amount of information contained in the data set.

The main advantages of this system can be summarized as follows. First, the effects of a design decision on different aspects of a building can be calculated directly, while the "default values" of the necessary materials and other parameters are prespecified according to the type of building. Moreover, in any phase of the design process the designer can substitute his or her own values, at will, for the default values. Second, the effects are calculated for spatial zones, spatial elements that can be regarded as clusters of rooms and spaces. These zones, which are generated automatically by the system, function as units of analysis for evaluating the design so far. In addition, the zones work out the design in further detail. Third, bivariate analysis means that the effects for the different aspects are neither combined into one criterion nor treated as being altogether separate aspects. On the contrary, the various effects are combined into 2 by 2 tables or diagrams for facilitating trade-offs. Finally, there is a clear distinction made between the internal calculation by COSMOS itself and the external calculation of effects by means of specialized computer programs. The internal calculations of COSMOS are essentially simplified, whereas the external calculations are very much more advanced. The standardized file for these external calculations has already been developed.

We set out to develop a system that can make use of standard programs to perform such tasks as the presention of tables or three-dimensional projections of architectural designs. In the following sections we describe the methodology of the design system, the calculation of effects outside and inside COSMOS, and the methodology of the evaluation system.

METHODOLOGY OF THE DESIGN SYSTEM

Basic Spatial Units of the Design

The design is conceptualized as being a hierarchy of spatial units going from building blocks down to specific rooms and spaces. During the design process a number of attributes will be increasingly defined for each spatial unit. One of the main attributes is the spatial unit's user function. For example, the regime for energy and lighting conditions is determined by the function of each relevant spatial unit, while another set of attributes determines what materials are necessary to make the spatial units. The spatial units and their main attributes are:

	Functions	Construction Type	Materials
1. Building block	1 of 10	1 of 5 types	Type dependent
2. Building layer	Identical	Identical	Identical
3. Section	Section functions	Identical	Specified
4. Zones	Function types	Identical	Identical
5. Rooms/spaces	Room functions	Identical	Specified

(The specification of materials can involve changing type-dependent materials into other materials on request.)

Three screens are used for defining the geometry of the design:

1	2	3
1. Building block	1. Building block	1. Section/layer
2. Building layer	2. Roof	2. Zone
3. Section		3. Rooms/spaces

In principle, only modules of the type section, roof, or room/spaces are defined:

1. Sections are defined by means of geometrical forms.
2. Roof volumes are defined by means of their specific shapes.
3. Rooms/spaces are defined by dividing up the zones.

Building Blocks and Layers

Part of the basic data for building blocks and layers is a set of functions ranging from housing to storage. There is a predefined set of parameters for describing the interior climate, light, and acoustic conditions of the main activities for each building function. In a similar way, the building construction types are defined a priori, and this includes, for example, the use of load-bearing walls and various prefab systems. Since the type of construction and the choice of materials are usually related, there is a predefined default set of materials for every construction type. The materials include those for external and internal walls, floors, and roofing. The properties of the materials from which the elements of construction are composed are used to calculate the parameters for heat, lighting, and sound transmission. Additionally, calculation is based on the cost of the materials for manufacturing the construction elements.

The building blocks and layers of sections are composed horizontally and vertically. For definition of the sections, only two diagonally placed points on the section base, together with the height, are necessary. Two diagonally placed points on the roof base plus some points of the roof tree are sufficient for determining the roof volumes. These sections, together with the roof volumes, are automatically translated into a set of panels. The properties of these panels, such as type of material, glazing, and type of glass, are set on defaults according to the construction type chosen. If some changes in material types are desired, a set of questions is shown on the screen. Building blocks composed of building layers can be repeated, whereas for those composed by sections in the same horizontal plane, it is automatically assumed that the original adjoining external walls are replaced by internal walls.

Building Sections

The building's function primarily defines the function of each individual section. Whenever the designer elects not to specify a particular function for all of the sections or only for some of them, this serves as a "default." Again, for each function a choice can be made from a limited set of section functions. For example, the designer can specify a "living" or a "sleeping" section for the "housing" function, and for other building functions similar sets have been specified. The main section function is used to combine a number of separate functions since the number of sections may be quite extensive; for instance, the "sleeping" section may contain bedrooms, bathrooms, and corridors. The parameters for interior climate, lighting, and acoustic conditions are identical to those describing the building function, since the main function refers to the main activities. A different set of parameters is defined for the other functions.

The materials for the construction elements are also prespecified since it is assumed that the function of the sections is identical with that of the building, unless the designer decides otherwise. Apart from choosing the prespecified materials the designer may also assemble his or her own particular set of construction elements from basic materials (brick, concrete, wood, insulation material, etc.). A set of building material parameters for calculating the properties of heat or sound transmission of construction elements is available in the system. Figure 12.1 shows the process of defining sections and zones.

Zoning

Daylight and nondaylight, or alpha and beta, are in principle two distinct types of zone. A third type, the gamma zone, represents a mixture of both, while a fourth zone, delta, defines the relationship with open space [Habraken et al. 1976]. The chosen section function, the proportion of glass in the exterior walls of the section, the dimensions, and the form (either linear or concentric) initially determine the dimensions of the zones and their functions. The zoning can be changed or discarded if the user of COSMOS so desires. In the latter case the section designed is regarded as one zone.

Within COSMOS the zones perform two functions. In accordance with the method developed by Habraken et al. [1976], the first is meant for designing the place and position of various rooms and spaces within dwellings. The method, within the context of COSMOS, is used to define clusters of user functions with roughly comparable attributes, for example, interior climate and other conditions. The second function of zones is as basic units for statistical analysis of the properties of the design. The functions of the zones are defined initially as a section function, with one of the function types already described, consistent with the system of default values. The final function of the zones per section can be chosen from a list. If necessary, the limits of the zones may be changed in the same process.

COSMOS offers a basic layout for zoning according to the dimensions of the sections. The function type of the zones is defined by default. If changes are desired, a new function can be chosen from a list presented on the screen. The zoning is meant to facili-

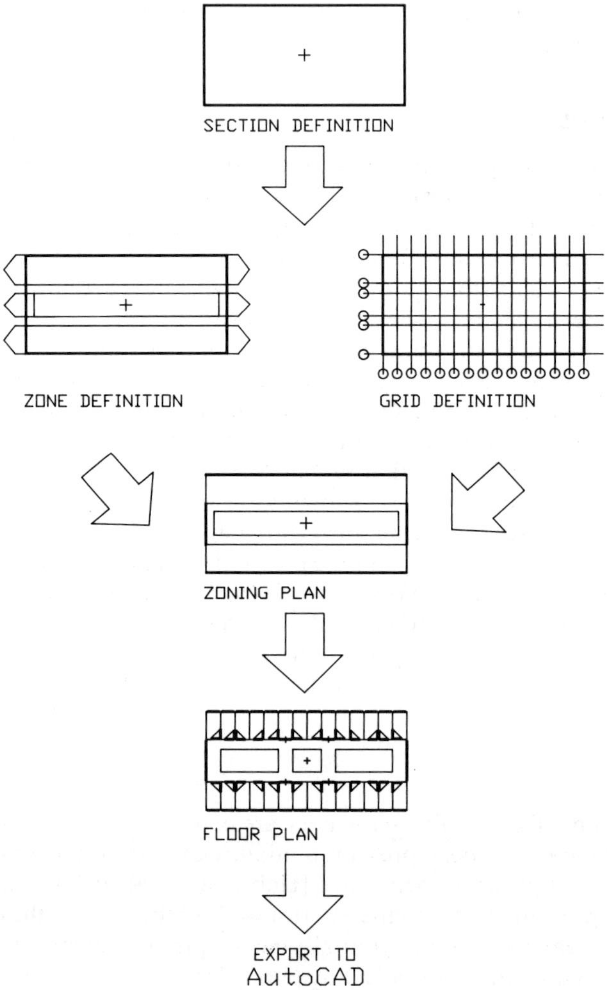

Figure 12.1. Definition of sections and zoning.

tate the positioning and dimensioning of the rooms and spaces in general. Apart from positioning internal walls, sites for the windows and doors can be chosen. Figure 12.2 shows the definition of zones in COSMOS.

Rooms and Spaces

Rooms or spaces form the basic spatial level. For interior climate and lighting conditions, the possible functions of the building blocks, the sections, the zones, and the rooms are related to a number of parameters. These parameters become increasingly specific as the spatial level is reduced. Figure 12.3 shows the definition of a floorplan.

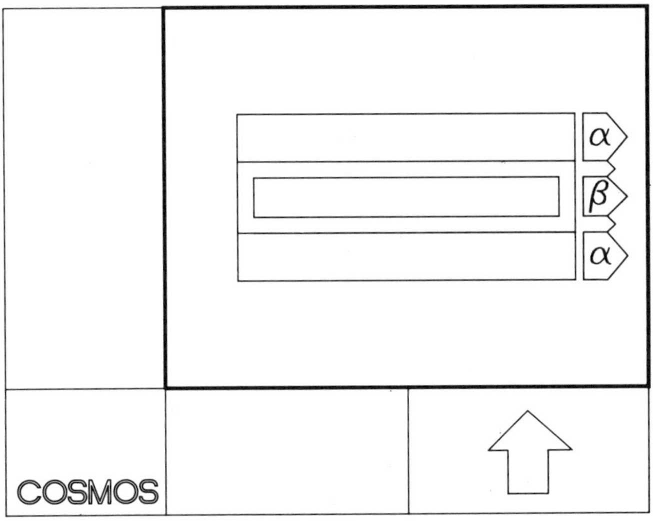

Figure 12.2. Zoning.

CALCULATING EFFECTS OUTSIDE AND INSIDE COSMOS

One of the problems already mentioned lies in the fact that the range and complexity of specialized computer programs are rapidly developing for a number of aspects. At the same time, if the designer's solutions are to be adequate, he or she needs an equally rapid feedback. Therefore, a difference is made between internal calculations, which are part of COSMOS, and external programs for giving direction to the design. This makes it possible for the designer to apply a set of easy-to-use extensions of the rule-of-thumb measures, and at the same time, allows the designer to apply specialized calculations. An external relational database is produced by COSMOS using SQL statements for the use of external programs (independent applications), to calculate effects and eventually to evaluate the various solutions for a design. The form of the database is in accordance with STEP. The supposition is that major programs for calculation effects will, in the not too distant future, require a database according to this standard. In the meantime, interface files based on the STEP format are being used. Building designs made by COSMOS or by any other design system can then be applied by these programs. Figure 12.4 shows a diagram of the design and evaluation system. In the remainder of this section each of the various aspects is treated.

Building Costs

A commercially developed program, IBIS-calc (Tricht and Brink 1989), is used for external calculation of the building costs. This program is built around the concept of a "price hierarchy." Each price hierarchy consists of a building element with a set price at the top and a number of elements at the base which determine the calculation of the price at the top. The "ultimate top" represents the complete costing of the project. Intermediary

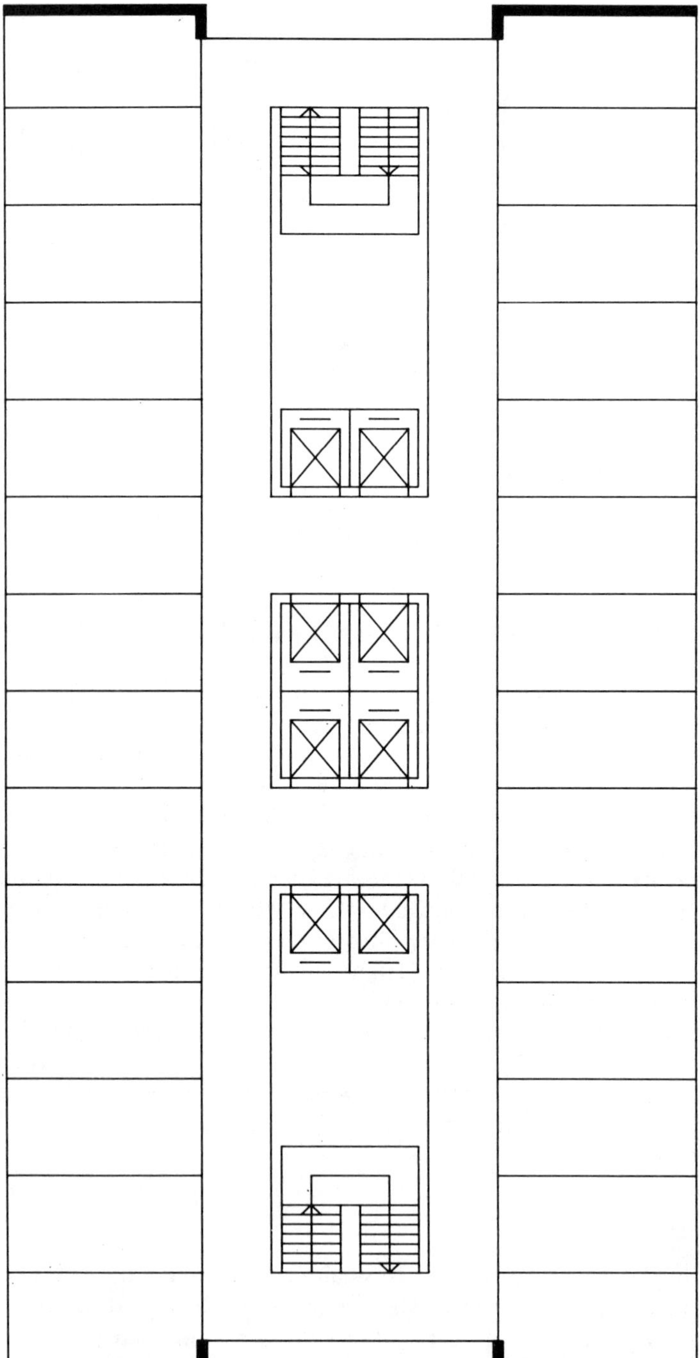

Figure 12.3. Floor plan.

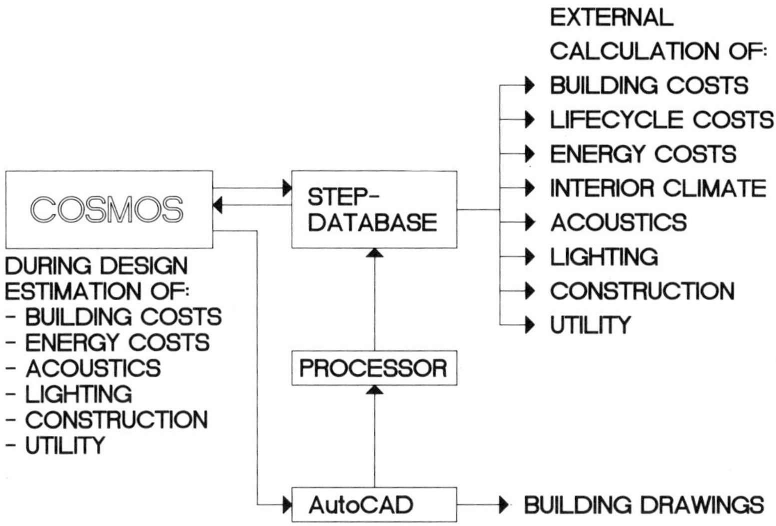

Figure 12.4. Diagram of the design and evaluation system.

"bases" of the hierarchy may consist of comparable elements, ranging from parts of buildings to bricks or mortar. Hierarchies that occur more than once in a recipe can be stored in a database. Together with IBIS-calc a set of standard recipes and materials and manufacturing costs are supplied on the lowest base level. Calculation of the project price follows from a number of specific recipes and standard recipes.

To estimate the internal costs during the first run of COSMOS, a number of recipes are assembled to find the cost per meter and per square meter for the panels, openings, doors, and windows. When the first run is ended and the STEP database is produced, an estimation can be made on the basis of specific and standard recipes while the recipes for running COSMOS are updated.

Dynamic Life-Cycle Costs

Apart from calculating building costs, a model (IRRG) is formulated to calculate the dynamic life-cycle costs which also uses the IBIS-calc hierarchies [Prins 1989]. The flexibility of a building depends on the composition of its building parts and the possibility for upgrading and replacing these in time, according to changes in demand in the market for building services. As a consequence, each building part has to be defined by its own cost data and life span. In this model the depreciations of the building parts appear as periodic costs. The calculation is carried out by a spreadsheet program for which COSMOS provides a database. As this subject is relatively new, no provision has yet been made for internal calculation in COSMOS.

Energy Costs and the Interior Climate

The externally calculated energy costs and the climate conditions are produced by the program ELAN [de Wit and Driessen 1988], based on an ELectric ANalogon model of heat flows. ELAN is a product of the building physics group on our faculty and is aimed at the dynamic calculation of energy costs for a specific period. These costs are calculated on an hourly basis for heating or cooling in a reference year. Also for the same period the number of hours a specified temperature is maintained is recorded.

The internal simplified calculation is twofold. First, a thermal insulation index is established using the weighted insulation index for all the outside panels. This index is subject to Dutch government regulations for buildings and more specifically for dwellings. Second, to shorten the processing time, the reference year is simplified into a selected number of days. Actually this option is feasible only for a fairly fast computer, such as a VAX or comparable system.

Acoustics

The externally calculated sound transmission is established on a model based on experiments formulated and carried out by the institute of TPD-TNO [Gerretsen 1979, 1983] in cooperation with our building physics group. This model is formulated to calculate the frequency-dependent sound transmission between two rooms by partitions and by flanking structures. The most important data are the sound reduction indexes for direct transmission of various structures, and the vibration level differences across junctions. This computer program has been developed by our group. The internally calculated noise reduction is further simplified by only estimating the noise reduction on the outside structures independent of their frequencies.

Lighting

The externally calculated lighting conditions are evaluated by the VISTA programs [Stearn 1985] and more recently by DIM (a co-production of ABACUS, LIDEC, and Calibre [Maver 1989]). The interior lighting conditions can be inspected visually, in color and with at least five light sources, by taking into account the position of the furniture. Apart from this a program for calculating both the daylight factor and the isoline map is presently being developed (see [de Wit 1990]). For the application of VISTA or DIM, no simplification is possible. A simple lighting program is used for a rapid, but only approximate, internal calculation of the daylight factor for a number of points.

Construction

For the external construction calculations, the programs GTSTRUDL [1989] and DIANA [1989] are available, but can not yet be applied to standard input files. A separate ASCII input file has been produced for each building block in both programs. A check is carried out for the possible spans during the input of the design, since no effect can be calculated for the construction afterward.

Utility

The calculation of the utilities that users may derive from the attributes of buildings in general, and of dwellings in particular, is based on a model developed by the Architecture and Urbanistics group of our faculty [Veldhuisen 1988, Veldhuisen et al. 1987]. The utility that users may derive from or attach to one of the design properties has to be established at the very outset. The relationship between the theoretical values that properties may acquire in the course of the design process and their utility values can be expressed as a mathematical function. These functions are in turn used for predicting responses to design decisions. COSMOS can deal with externally established utility functions for subjects that are of interest to the designer. No difference is made between the external and internal calculated utility.

Spatial Units and the Calculations

The relationships between the programs for internal and external use and the phases in the design process are summarized in Table 12.1.

Use of AutoCAD

One of the main principles of the system developed is that standard programs will be employed for output or other purposes whenever possible. For a number of reasons, in the present situation AutoCAD is used. The main reason lies in the open character of AutoCAD, which enables communication both to and from AutoCAD and the use of LISP routines for drawing problems. For the same reason, other organizations have adopted it to develop drawing aids: for instance, the Association of Computing Architects [ACAD 1989] in the Netherlands for producing scaled building drawings according to the Dutch (NEN) standard. Apart from that, a large number of program libraries are available for the standardized drawing of building elements and utilities. AutoCAD is used specifically for:

1. Making perspectives
2. Working out details
3. Making the scaled drawings

TABLE 12.1

	Section Zones (Int./Ext.)	Room Spaces (Int./Ext.)	Rooms—Furnished (Int./Ext.)
Building costs	COSMOS/IBIS	COSMOS/IBIS	—/—
Life-cycle costs	—/IRRG	—/IRRG	—/—
Energy costs	COSMOS/ELAN	COSMOS/ELAN	—/—
Acoustics	COSMOS/—	COSMOS/prog.	—/—
Lighting	COSMOS/—	COSMOS/prog.	—/DIM[a]
Construction	—/GTSTRUDL[a]	—/—	—/—
Utility	COSMOS/prog.	COSMOS/prog.	—/—

[a] For GTSTR, DIANA can act as a substitute; for DIM, VISTA.

Figure 12.5 shows a perspective of the designed building, using AutoCAD.

METHODOLOGY OF THE EVALUATION SYSTEM

Zones are spatial units essential for the evaluation system. The differences in zonal effects are used as a basis for evaluating the building as a whole. At least this means that in principle it is not necessary to work out different solutions for each design problem. Various design solutions are probably very useful, for comparison, but the method is aimed at making possible an evaluation of the differences between the zones. Furthermore, most of the decisions in the designing process involve the making of trade-offs. The question that now arises is how to reduce the number of existing data on the values of different effects for each zone. On the one hand, values for the various aspects for each zone may be combined into one characterizing measurement using multicriteria analysis, but on the other hand, we could use these values directly. Neither prospect is very attractive.

In its simplest form, MCA can be regarded as a method to derive, for each unit of analysis, a single evaluation score from a set of different effects and a set of priorities. In a slightly more complex context the priorities can be based on several different visions or viewpoints, so that a matrix can be calculated from the various evaluation scores. Different levels of measurement of the original effect matrix lead to different kinds of evaluation scores [Voogd 1983]. Some of the subjects to which it may be applied can be

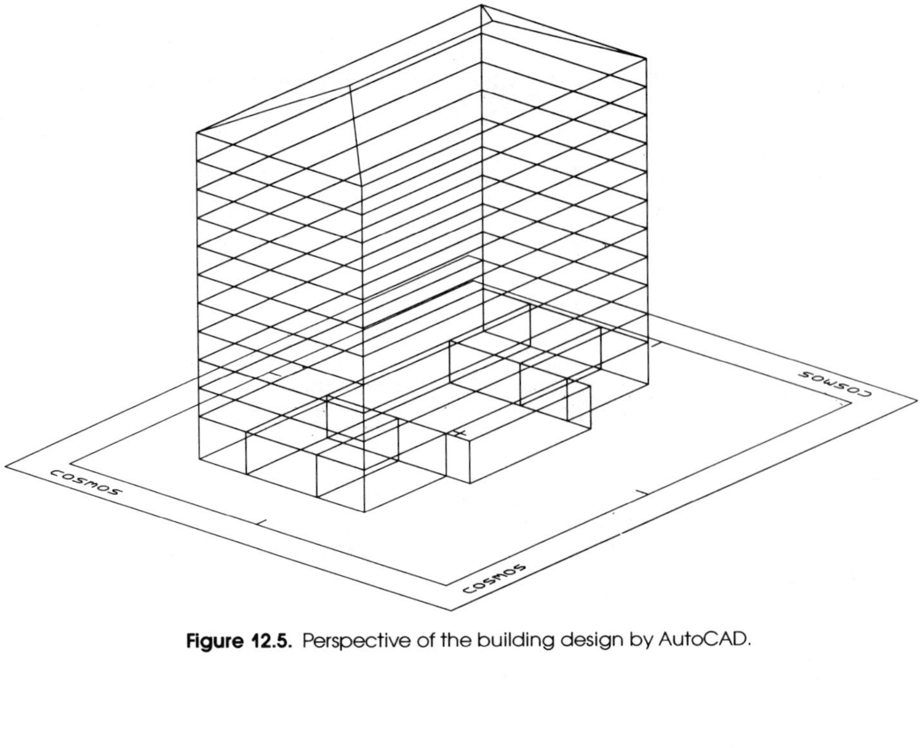

Figure 12.5. Perspective of the building design by AutoCAD.

found in the field of spatial planning. There it could be used to find the best location for a new residential area or the most suitable location for an industrial site.

The major advantage of the application of MCA lies in the fact that it forces participants in the planning process to reexamine their own biases while discussing their particular priorities. As a further attractive application, it can be used to express the value of a possible solution in one figure. However, establishing a set of priorities or weights necessary for calculating single evaluation scores proves to be rather complex, while information about individual effects is lost in the process. It should be noted that the calculated evaluation scores may have two values, and these may be similar, although based on calculations from conflicting sets of effects. This means that the evaluation scores cannot form the basis for making trade-offs.

For evaluating zones, we are interested primarily in maintaining most, preferably all, the information contained in the effect matrix without calculating evaluation scores, since we are not primarily interested in finding the "best" zone. An alternative would be to use the effect matrix directly, but this does not work at all well. The way we propose to reduce the number of data is to assemble the data in 2 by 2 scattergrams or tables. Since we have dealt only with a limited set of effects, only 15 combinations have to be calculated and shown. The main advantage of this is that the effect matrix is available at all times for inspection and so can be used as a basis for eventual trade-offs. However, as the number of effects increases, so does the number of possible combinations: the greater the number of combinations, the more one has to use the correlation matrix.

Effect Matrices

To describe the evaluation system, some points have to be made concerning:

1. Calculated effects and the evaluation of various design solutions
2. Rooms, zones, and the various design solutions
3. Directly calculated and normalized effects (e.g., costs per cubic meter)
4. Levels of measurement of the effects (qualitative versus quantitative)

A distinction has been made between the calculations and the resulting evaluation, which means that an analysis is to be a necessary step to making the evaluation. The concept of zones is used as a means for calculating and directly evaluating the structure of the design concept without actually working out the details—designing the rooms and spaces on a lower spatial level. In fact, the zones can be considered as a not yet existing cluster of spaces with roughly the same attributes or properties. However, the evaluation of the effects for simple designs can be based directly on the rooms or spaces, while for complex designs a number of effects can easily be based on the zoning. The proposed analysis takes little account of the difference in scale between rooms or spaces and zones.

Not all the effects can be expressed relatively. Here the qualitative effects are treated as a possible option, although we prefer the utility model, where the reactions of (groups of) persons can be established as interval values, as a result of trade offs.

The analyses of the effects concern primarily the internal calculations made by COSMOS for each zone of a particular design. The first data matrix consists of a combination of the main design variables and the effect variables for each zone. Preferably, all these variables should be normalized in meters and square meters, so that, for example, per zone external walls are related to the volume in cubic meters. However, this is not possible for material combinations which have to be indexed.

	Design and effect variables				
	1	·	·	·	m
Zone 1					
·					
·					
·					
n					
Design 1					

The design per column forms an aggregate of the zonal variables, eventually combined with values for unique elements.

The second data matrix consists of a number of normalized variables for each design:

	Design and effect variables				
	1	·	·	·	m
Design 1					
·					
·					
·					
n					

The main purpose of the analysis of these data is to find optimal solutions and, if necessary, make a number of trade-offs.

Analysis

The first step in the proposed analysis is to establish the relationship between each pair of variables. In principle, two forms of expressing these relationships are possible, thus allowing for the different levels of measurement for possible variables (ratio, interval, ordinal, or nominal). The first form requires a linear least-squares equation to describe the relationship, the second form a contingency table (see, e.g., Blalock [1960] and Siegel [1956]).

	Ratio	Interval	Ordinal	Nominal
Ratio	ls	ls	ct	ct
Interval	ls	ls	ct	ct
Ordinal	ct	ct	ct	ct
Nominal	ct	ct	ct	ct

For example, the relationship between costs per cubic meter and the mean thermal insulation, which is expressed in a least-squares model, and the relationship between the quality of the view and sound in dB(A), which is expressed in a contingency table, is shown here. In both cases the relationship is considered to be symmetrical, since it cannot be causal. Figure 12.6 shows the use of scattergrams and contingency tables.

The relationship between the mean thermal insulation (dimensionless) and the cost per cubic meter is shown in the scattergram. Although the least-squares line in the scattergram shows a linear relationship, the zones with a higher MTI and lower costs are of specific interest. Inspecting data matrix 1 may give some insight into the causes for the deviation.

The quality of the view may be expressed by four possible classes, while the sound levels on the outside of external walls are given in dB(A). The view may be classified according to the characteristics of the particular site for which the design has been made. The classes could be defined by the amount of information available, the amount of traffic, the amount of open (green) space, or by possible combinations of all three. Depending on how the the analysis is to be applied, the view may be either ordinal or nominal. The table shows the relationship between the views and the sound levels irrespective of the level of measurement. Once the quality of the views is known, the sound effects may be expressed in a number of zones showing more or less dB(A).

Utility

It is possible to replace the views with their respective utility values by replacing the table with the least squares model. In general, the utility of a particular property can be established by allowing would-be users to make trade-offs between this and a number of other properties. The analysis of the data set can be carried out either for individuals or for groups (see, e.g., Veldhuisen [1988] and Veldhuisen et al. [1987]). The results of the measurements and the analysis can be expressed in two forms. If the property in question is continuous, for instance the size of the living room in a dwelling, the relationship between size and utility is shown in the diagram on the left-hand side of Figure 12.7. If

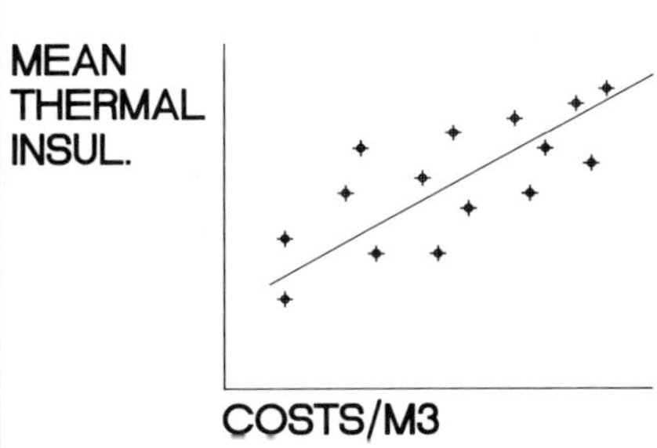

	VIEWS				
Db(A)	1	2	3	4	T
30-39		4		1	5
40-49	3			2	5
50-59			7		7
60+			10		10
TOTAL	3	4	17	3	27

Figure 12.6. Scattergram (left) and contingency table (right).

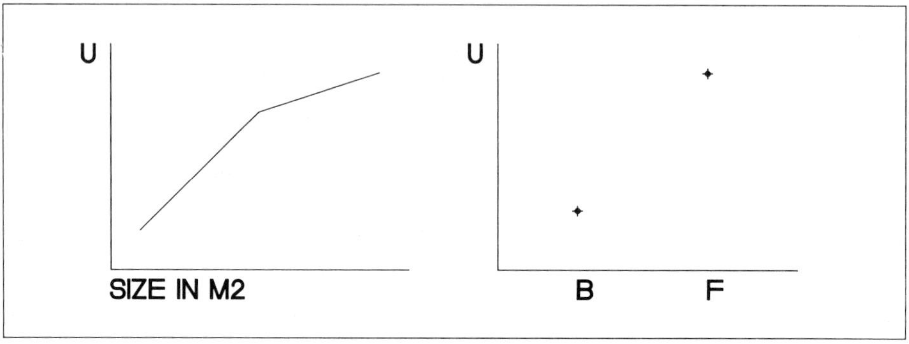

Figure 12.7. Two forms of the results of utility measurements.

the independent variable is discrete, for instance the place of the kitchen at the front or the back of a dwelling, the diagram on the right-hand side of Figure 12.7 will result. The utility values can be considered as special types of effects variables, which may be used to express subjective judgments on certain design variables. COSMOS accommodates the inclusion of a number of these variables, assuming that the necessary relationships have been quantified.

Correlation Matrix

Appropriate correlation measures exist for all combinations of the levels of measurement and effect variables for the design. However, most of the relationships are between ratio or interval variables. The results of the analysis can be reduced to a correlation matrix in the form of an overview.

			Variable		
	1	.	.	.	*m*
Variable 1					
.					
.					
.					
m					

If the number of effects can be kept within manageable limits, however, the evaluation can be carried out easily by inspecting the most salient scattergrams and, if necessary, the contingency tables. The relationships can be used to change the design.

CONCLUSION

This design and evaluation system offers a means for designing a building, calculating a number of effects, and for managing the information necessary for evaluating the design decisions. The design can be defined on the level of building blocks (sections with zones), or in more detail on the level of rooms or spaces, at the initial stage of the design process. The effects for both levels can be based on either integrated (simplified) or independent (extensive) applications. The concept of zones is used to describe as yet nonexisting clusters of spaces with similar attributes or properties, in order to calculate effects realistically on the basis of building blocks.

By subdividing building blocks into zones, the designer is given the opportunity of analyzing the design on the basis of the relationships between partial effects. The differences between zones should give insight into where the solution for problems in the design as a whole may be found. Apart from the analyses based on zones, complete designs can be evaluated along the same lines.

REFERENCES

ACAD Users-Group (1989). *VCA Tablet Menu*, Rijswijk, The Netherlands.

Altemueller, J. (1988). *The STEP File Structure*, ISO TC184/SC4/WG1 Draft N279.

Bax, Th. M. F. (1990). "Design of a Taxonomy for Architectural Education," *Proceedings of the SEFI Conference*, Dublin.

Blalock, H. M., Jr. (1960). *Social Statistics*, McGraw-Hill, New-York.

de Wit, M. H. (1990). "Light Transfer Models," in *European Reference Book on Daylighting*, CEC, Brussels.

de Wit, M. H., and H. H. Driessen (1988). "ELAN: A Computer Model for Building Energy Design," *Building and Environment*, **23**(4): 285–289.

DIANA (1989). *TNO-IBBC: Numeric Mechanics*, Rijswijk, The Netherlands.

Gerretsen, E. (1979). "Calculation of the Sound Transmission Between Dwellings by Partitions and Flanking Structures," in *Applied Acoustics*, Applied Science Publishers, Barking, Essex, England.

Gerretsen, E. (1983). *Berekening van luchtgeluid en contactgeluidisolatie*, Rapport 007.101/2, TPD-TUE.

GTSTRUDL (1988). GTICES Systems Laboratory, Atlanta, Ga.

Habraken, N. J., J. T. Boekholt, A. P. Thijssen, and P. J. M. Dinjens (1976). *Variations: The Systematic Design of Supports*, MIT Laboratory of Architecture and Planning, Cambridge, Mass.

Maver, T. (1989). *DIM: Dynamic Illumination Model* [co-product ABACUS, LIDEC (Philips), and Calibre (TUE)], University of Strathclyde, Scotland.

Prins, M. (1989). "Decision Making and Flexibility: An Integral Design Decision Making Model," *Proceedings of the CARDO Symposium on Quality in the Built Environment*, Newcastle upon Tyne, England.

Siegel, S. (1956). *Nonparametric Statistics*, McGraw-Hill, New York.

Stearn, D. D. (1985). *VISTA: An Introduction*, ABACUS, University of Strathclyde, Scotland.

Tricht+Brink (1989). *IBIS-calc*, Voorschoten, The Netherlands.

Veldhuisen, K. J. (1988). "Some Empirical Tests of the Predictive Strength of Three Multi-attribute Preference Models," in *Sociometric Research*, W. E. Saris and I. N. Gallhofer (eds.), Macmillan, London.

Veldhuisen, K. J., A. P. Thijssen, and H. J. P. Timmermans (1987). "Flexibility in Architectural Design: An Experimental Assessment," *Journal of Architectural and Planning Research*, **4**: 343–348.

Voogd, H. (1983). *Multicriteria Evaluation for Urban and Regional Planning*, Pion, London.

PART FOUR

MULTILEVEL EVALUATION

13

A PRELIMINARY TO DEVELOPMENT OF EXPERT SYSTEMS FOR TOTAL DESIGN OF ENTIRE BUILDINGS

Peter Manning*

Technical University of Nova Scotia
Halifax, Nova Scotia, Canada

Samir Mattar

United Nations Centre for Human Settlements
Nairobi, Kenya

Objectives of This Paper. This paper has two primary objectives. The first is to represent the practicability of making the design of entire buildings a conscious, craftsmanlike, activity conducted in the clear, without the mystery that tends, because of designers' usual "black box" methods, to surround it. To this end, a design strategy and some tactics for resolving decisions at critical stages in the design process, which we have described elsewhere, are recapitulated to show how total design of buildings can be pursued in a generic manner. This done, the way is opened for the second objective: to make the large and important field of work that is building design amenable to computerization.

The form that pursuit of this second objective is taking is being influenced greatly by growing interest in expert systems, which for everyday professional building design appears a more useful development than previous CAD emphases on drafting and graphics. Application of our design methods to a series of expert systems for the total design of entire buildings is therefore indicated.

For such a vast project—the formulation of bases for design assistance and expert systems that can be integrated and used as a generic method for the total design of entire

**Present address:* 1470 Summer Street No. 1601, Halifax, Nova Scotia, Canada B3H 3A3

buildings, so that the results are more certain and successful than the outcome of the generality of present-day building design—the most that can be attempted within the limits of a single paper is a set of examples of some of the stages in the process. Nevertheless, since the design method described begins at the "large end" of the process, where the most consequential decisions are made, it is hoped that the major thrusts and the essential CAD activities will be evident. All design is substantially iterative, and provided that the major iterations are intelligible, there should be no need for this demonstration to labor over the lesser ones.

INTRODUCTION

The Nature of Building Design

The characteristics of building design that, for the purposes of this paper, need to be regarded as self-evident are as follows:

1. Design is an "intellectual activity that produces artifacts" [Simon 1969]. It certainly is not drafting, which is a tool for communication and for aiding comprehension that a designer might *choose* to use (and of course most, in fact, do).

2. Design is a decision-making process.

3. Design requires several very different kinds of cerebral activity, principally those kinds that result in analytical, creative, evaluative, communicative, and executive skills and modes of action. Because of this complexity and the broad range of design subsystems within the overall design system:

4. Building design is a multiprofessional activity, with design subsystems the responsibilities of several professions with quite different traditions, attitudes, and interests: architects, engineers of several kinds, quantity surveyors, frequently landscape architects, and often others.

5. Design work in the various subsystems required of each type of design professional involves besides considerations of choice and arrangement of hardware, decisions about human factors, performance, economics, aesthetics, and other and different matters.

6. Design is an integrative activity. At every level of design there must be consistency, harmony, and compatibility evident in the manner by which spaces, materials, assemblies, and components are integrated, one with another. But design also requires the marriage of all those decisions with the types of mental activities and modes of action, subsystems, professional representations, and so on, that have already been mentioned. This characteristic is of crucial significance in the approach presented in this paper. Throughout such integration, a particular difficulty, innate to building design, becomes apparent: the task of effecting integration of totally different things that have great varieties of measure and no common base.

7. Design is conducted in the world of practical realities, where it is necessary for designers, armed as they will be with imperfect resources (of money, information, etc.), to make the best decisions possible. Design is the blending of knowledge with experience, a synthesis of theory and practice, and the translation of theories and ideas into courses of action and three-dimensional reality. (A. Bell, personal communication, 1988).

8. Design for a projected building [or group or part(s) of buildings] is a process that begins with complete uncertainty and amorphous, nebulous ideas or concepts and proceeds through stages, in each of which the original idea is enlarged and successively given more and more definite shape until, finally, the smallest parts have been decided. An analogy can be drawn with mathematicians' "successive approximations."

9. Dominant characteristics of design activity are that (a) many things need doing concurrently (in practice this, or much of it, is entirely practicable, because most design is undertaken by multiprofessional teams, whose members work separately, in their separate professional offices; of course, such working generates substantial need for effective management and coordination of the separate contributions); and (b) decisions taken are illuminated by later decisions, so there is a continuing need to recycle the decision process, when decisions already taken may be reviewed in the light of decisions taken subsequently, and revised, or former decisions canceled and entirely new decisions made. A simplification of the process, to which recycling and feedback loops must be added, is shown in Figures 13.1 (for one element: constructional components) and 13.3 (for one cycle of design of entire buildings).

10. Design is a matter of compromises and of optimizing conflicting subsystems: it is virtually never possible to achieve an integration of "ideals" or "bests" (to take a simplistic example, it is clearly impossible to achieve maximum space at least cost). Each decision has to represent, for its particular time and context, the best compromise that can be effected.

11. Because the human, physical, economic, and other performance criteria of buildings, and the design options that are responses to those criteria, are measured in units that are generally incommensurable, most design practitioners have considered it impossible to design using rational, logical methods, except for particular parts or units that lend themselves to those approaches. Indeed, virtually everywhere, the design of entire buildings is still pursued in a largely unsystematic (intuitive) and unquantified manner.

Although in the design process considered in this paper, concerns for form, external appearance, and most other design goals habitually pursued by designers are intrinsic, the search is focused on what has been termed *evolutionary* design: designs are *not* sought that are original and totally different from previous solutions to the same or similar problems. Rather, the search is for developments from earlier solutions to building design problems that keep or improve their successful characteristics and change, for the better, those characteristics that have been less successful.

The Practice of Building Design Today

The everyday process of contemporary design of entire buildings embraces the activities of *programming* (or *briefing*, as it is called in places other than North America), generation of solutions via *schematics* (otherwise *sketch designs*), instruction and coordination of consultants, design devleopment, filling in the framework of the schematic to make a final design, and communicating that conclusion to those who will approve, construct, supervise, pay for, (and so on) the building.

> The outline concept of a design is usually produced by an experienced architect, after personal contact with the client. . . . more detailed decisions are usually made by less ex-

perienced staff . . . who . . . cope with the more detailed technical aspects which often require reference to written information. . . . little information is kept to record the reasons for decisions taken. . . . Use of written data . . . is limited [this being] seen as a time-consuming activity. . . . Designers tend to refer to a few standard publications only. . . . [A] fairly consistent pattern . . . [of the design process consists of] . . . an initial concept for the building plan, form and general construction [being] developed rapidly, using little information other than the client's brief, the site constraints and the designer's own experience. In nearly all cases studied the initial concept formed the general basis of the final design, only undergoing minor changes. Time . . . seemed to restrict . . . alternatives. . . . requirements of statutory authorities were generally responsible for the few major changes to the designs [Mackinder and Marvin 1982].

When buildings are designed as individual cases rather than as examples of types, every single design issue and decision tends to be determined from new. A consequence of such nonroutine attention being given a vast number of what could be routine matters is that so much time is consumed that many such issues are decided arbitrarily and few opportunities are allowed for those design features that could to benefit from the results of recent building research. (Indeed, there is much evidence to show that research completed half a century ago is still not used.)

Because buildings are made individually, few opportunities are found for the advantages of mass production to be incorporated in their design, except in the use of manufactured components. At the same time, it often seems that for a number of reasons (cost, speed, low standards of available craftsmanship, etc.) few opportunities are found for them to benefit from the advantages and special qualities that usually are associated with custom-made artifacts.

Systematic design (design methods) enjoyed some vogue during the 1960s and early 1970s. Interest in that subject still exists, but some of that period's most active and creative practitioners of systematic design appear to have had second thoughts (e.g., Jones [1980b]). Most of today's building designers, like building designers of other ages, are not directly interested in the methodology of design. Beyond their natural professional concerns to satisfy their clients' expectations and bring their projects in on time and budget, they seem most interested in matters of style (e.g., presently, in Great Britain, in combating the Prince of Wales' urgings that they adopt a set of fashions different from their own).

The State of Building CAD Today

An unavoidable first step toward computer-aided design of buildings must be a greater employment of rationality and system in the field of building design, which traditionally has been more generally associated with artistry and individuality. When computers are used in architects' offices, it is usually for word processing, scheduling, or drafting. It is only exceptionally that they are used for *design*—and then it is only for the design of particular facets, not for the design of entire buildings.

Paterson [1980] has been featured in one literally visible case (in a movie, "A New School for Eastergate") of CAD being used for the design of whole buildings: the drawing of plans of a school, integration of heating installations, concurrent calculation of materials requirements and costs, and much more. However, in this case the design projects were within the SCOLA school building system, a prefabricated system for school construction developed by a consortium of county education authorities in Great Britain, which incorporated complete, predesigned, constructional and servicing systems. Moreover, the buildings were part of an extensive program of many new schools designed within a common philosophy and constructed, year after year, by the consortium. The development was impressive and it is a matter for regret that as Reynolds [1987] has explained, design "went back to manual methods in 1974 after it was decided that the reduced use of the SCOLA system did not justify the cost of using the computer." Although admirable, this use was actually more akin to choice and combination of parts from a sophisticated Meccano set to build a model of already existing things than the original creation, traditionally, but on a computer, of entire buildings. Despite this, SCOLA's CAD showed, as few other large-scale devleopments have yet shown, how *generic* design of entire, non-system-built buildings might evolve.

Many practitioners of conventional building design, as well as some CAD practitioners, believe that CAD of entire buildings is attainable, but not immediately: that for the moment it is necessary to proceed cautiously, and step by step, from detail upward. The more optimistic possibility—indeed, the *probability*—is that it is more desirable to proceed from concept downward, if for no other reason than because this is the obvious and most direct way of reducing, from the beginning, and then progressively, the sheer scale of the design problem and the ranges of options in "the space." It is on this basis that this paper proceeds.

EXPERT SYSTEMS

Expert systems apply inference techniques to a knowledge base in a narrow domain in order to achieve high levels of performance in that domain and simulate the application of human expertise. Feigenbaum and McCorduck [1983] suggested that the "knowledge necessary to perform at such a level, plus the inference procedures used, can be thought of as a model of the expertise of the best practitioners of the field." Unlike conventional programming, the knowledge base that contains the facts, rules, and heuristics (i.e., the acquired knowledge in the domain) is separate from the inference methods used to process the knowledge; it can grow incrementally as additional knowledge is acquired. The inference engine comprises general problem-solving methods that draw conclusions based on the knowledge base. The third part of the expert system, the input/output system, is essentially a human–computer interface that handles facts relevant to the current problem: facts input by the user relating to the specific problem and facts ascertained by the expert system as it goes about solving that problem.

It seems clear to us that for work in design methods to be implemented, that work should advisedly be translated to expert systems in design. So far, expert systems have seemed to work best for specific constrained problems. It is likely, then, that computer-aided design of entire buildings will best be served by not one but several (or even many) expert systems arranged in sequence (Figure 13.1).

PREVIOUS WORK CONTRIBUTING TO THIS PAPER'S OBJECTIVE OF A GENERIC APPROACH TO BUILDING DESIGN OF ENTIRE BUILDINGS

In 1965, J. C. Jones's paper on "Systematic design methods and the building design process" presented to the third CIB Congress [Jones 1965] suggested a systematic phasing of

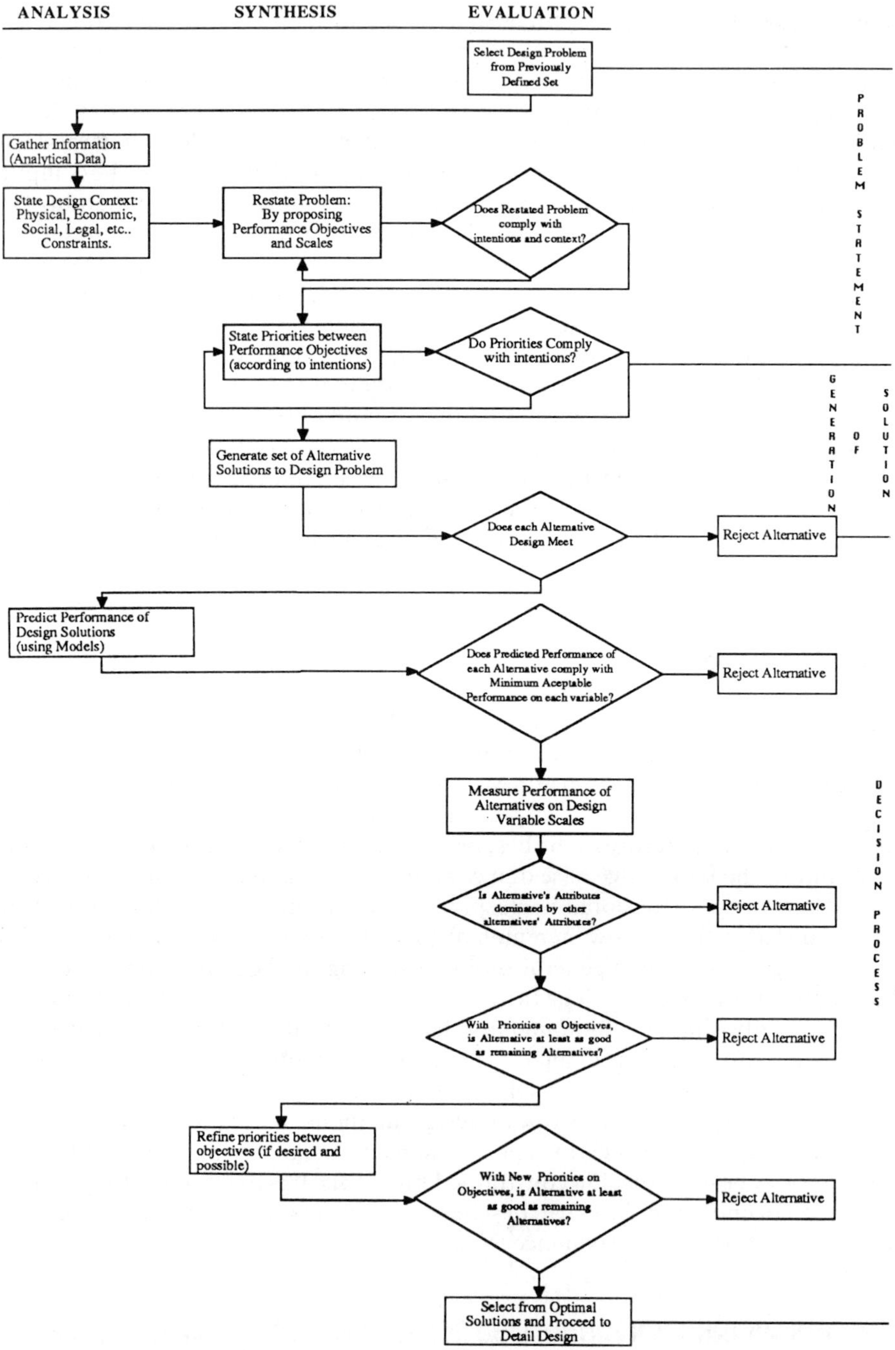

Figure 13.1. Operational framework of design sequence (in this case, for components of building fabric). (Reprinted with permission from S. Mattar et al. *Building and Environment*, **13:** 201–216, copyright 1978, Pergamon Press PLC.)

building design and construction processes that, among other things, allowed more ready application of design methods. Another effect was to make the design process more public ("externalized") and therefore more apparent to and communicable among the participants. The paper indicated some of the more critical design decisions that need to be made during each phase.

Influenced by Jones's paper and other work presented at a working conference at the Hochschule für Gestaltung, Ulm, in 1966 (including Jones and Thornley 1963), Manning developed a *design process flowchart* and *design framework* [Manning 1968]. These provided a structure for the overall design of entire buildings. That structure also accommodated a variety of programming requirements and technical, economic, and psychological techniques for evaluating different features of buildings' performance that had been devised or used, mostly by Liverpool University's Pilkington Research Unit, in studies of the design and internal environments of farm buildings, factories, offices, schools, mental hospitals, and other building types. The design framework was elaborated, by means of schedules, to illustrate the "outline design" of entire buildings. These methods were later used to coordinate the professional design of offices and a large plant for printing a regional newspaper [Anon. 1975a]. This use was successful; in fact, the building won a national design commendation [Anon. 1975b].

In a variety of later studies and related activities this work was used to provide a structure for design teaching [Manning 1974]. It influenced the arrangement of guides written for design of industrial production space [Manning 1966a], thermal environments in buildings [Manning et al. 1968], daylighting in factories, [Manning 1982a], and the formulation of projective management case studies for the identification, preparation, and appraisal of large-scale developments in the Third World [e.g., Manning and Biskaps 1981]. It has been used to suggest a strategy for lighting design for health science buildings [Manning 1981], for ensuring energy-conserving building design [Manning 1982b, 1983], for presentation of climatic information needed during building design [Manning and Hassan 1982], and for other purposes. An updated account was published as a strategy for multiprofessional design of buildings in 1984 [Manning 1984].

Building design must usually be considered an ill-structured problem. The general strategy mentioned above permits structuring of the design of an entire building along three continua: (1) from first things to last, from abstraction and concept to concrete and detail; (2) from awareness of need through program (or brief) and design to communication in preparation for realization; and (3) among the different building design subsystems (Figure 13.2) and their respective design professions.

An adjunct to this strategy whose availability is crucial to a computer-aided design method is an objective way of making choices between alternatives that have many and different attributes. These usually involve various kinds of scale (nominal, ordinal, interval) and such incommensurable units as dollars, square meters, gigajoules, and lux. Apart from subjective judgments and the technique of comparing design options after they have been translated to some common currency (such as annual costs, amount of labor or material use, or energy commitment), no ways of making reasoned choices appeared to exist until a solution to the problem was established by Mattar in 1976 [Mattar 1976, Matter et al. 1978a,b]. This has become known as the *decision by exclusion rule*. The essence of this rule is that alternative designs are plotted against axes representing the design attributes (structural strength, fire resistance, thermal insulation, capital

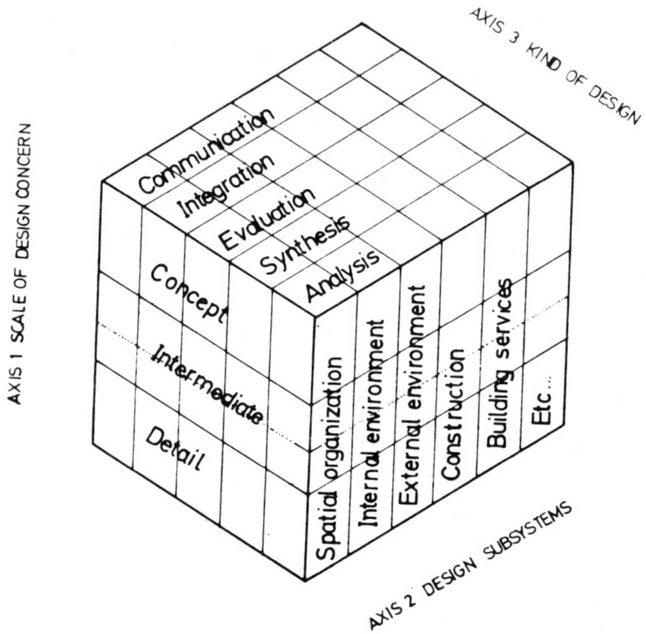

Figure 13.2. Three-dimensional matrix of the design process.

cost, sociopreferential choice, etc.) with the least desirable values at the origin, the most desirable values at the extremities of the axes. Values used are the most precise available: physical for physical attributes, economic for economic attributes, subjective for human behavioral and experiential attributes, and so on. It will readily be seen that in a two-attribute comparison, the preferred alternatives will be positioned within the area farthest away from the origin. The designer's (or building owner's or others') preferences and leanings toward the attributes can be implemented by deliberate choice toward one or the other axis within the "preferred" zone. In a multiattribute comparison, preferences and leanings are implemented similarly from among the alternatives plotted in multidimensional space.

The decision by exclusion rule originated as a computer-driven building design concept and is readily applicable to the overall design processes described in this paper. By its use subjective attributes such as appearance, sociability, and ambiance, as well as objective attributes, may be evaluated one against another so that design alternatives can be ranked by performance overall.

ASSEMBLING THE MATERIAL: A METHOD FOR THE TOTAL DESIGN OF ENTIRE BUILDINGS SUMMARIZED AND ILLUSTRATED

The form of generic, computer-aided design process summarized here from earlier papers can be represented as a series of steps, each of which consists of a sequence of considerations, analyses, and decisions. Individual actions required of the designers are presented as prompts (memory checks). In most cases, any one required action can be matched by a range of possible responses from which a designer makes a choice. Choices open to designers may be regarded as stages in continua, of which there will be many (or very many) to each design subsystem. These choices (or options) can be presented as menus. Where objective, entirely rational decisions are precluded (because of the presence of uncertainty, due, for example, to the subjective nature of the issues, or the absence of specific data, or similarly imprecise conditions), decisions are made (e.g., choices are made from the menus) on the basis of judgment. Where knowledge exists and precise responses are possible, decisions are made appropriately.

The main characteristics of this design strategy, with concurrent activities and feedback loops, can be illustrated as a complex sequence in any of the forms shown in the figures and in other ways: as a critical path network, as the logic for a computer program, as a succession of frames constituting a series of prompts for traditional design, and so on. The process can be summarized as follows:

1. Decompose the entirety of the design task into discrete parts that may be managed individually, using as axes for the decomposition (a) the scale and stage of the design concern (e.g., early stages and conceptual formulation, later stages, and detail definition); (b) the design subsystems and the design professions mainly responsible for the design of those subsystems (e.g., spatial organization and architects, external environment and landscape architects, constructional design and architects and structural engineers, building services and mechanical and electrical engineers); (c) type of design activity (e.g., analysis or problem statement or programming or briefing, synthesis or problem solution or selection among options, evaluation or optimization and justification of solution or option). See Figure 13.2 for a simplified model of such decomposition; see Table 13.1 for an illustration of the principal parts of such a decomposition with respect to five design subsystems against the axis of design scale.

At the conceptual level:

2. For each design subsystem at this level, determine the program (or brief). See line P.2 in Figure 13.3 for illustration.
3. Review the collectivity of programs and ensure appropriateness, completeness, consistency, and so on. See line P.3 in Figure 13.3.
4. For each design subsystem at this level devise ranges of alternative actions and building characteristics that will lead to answers to the program (options) for that subsystem. See line D.1 in Figure 13.3 for a general illustration of this set of actions. See Table 13.2 for an illustration by schedule of environmental design decisions needed to be made for the design subsystems Siting, Planning, Construction, and Services at the conceptual level (intermediate and detailed levels also shown). Tables 13.3 and 13.4 provide examples of more detailed considerations involved in particular design decisions and illustrate the need for compromises in design solutions to optimize attempts to satisfy complex requirements (that may often conflict with one another).
5. Review each design solution for each design subsystem at this level to ensure that it is compatible with at least one design solution in every other design subsystem. Discard every design solution that does not achieve such minimum compatibility. See line E.1 in Figure 13.3.

TABLE 13.1 Principal Parts of Five Selected Subsystems for a Complete Design for a Building

	Construction Design	Construction Management	Internal Environment	Building Services	Building Economics
Conceptual Level	Constructional systems, e.g., load bearing, frame; timber, steel, concrete, etc., enclosure etc., systems.	Project management system including type of contract, method of construction	Main character and quality of internal climate and ambiance	Kind and sources of services systems and means of provision	Main economic characteristics of project including quality, cost standards; desired ROI; overall cost, energy budgets.
Intermediate Level	Constructional elements (foundations, walls, roofs, floors, windows, etc.) including loadings, weather resistance thermal insulation.	Site organization and management system, including safety	Kinds and qualities of lighting, thermal environment, and acoustical conditions.	Building services including selection and location of main equipment and layout of main servicing routes including heating, ventilating and air conditioning, drainage including rainwater.	Budgets for capital, operating costs; energy consumption.
Detailed Level	Constructional details including materials, internal joints, fittings, finishes, colours	Project planning and scheduling	Specific values of heat light, and sound for particular locations; environmental fittings colours, finishes; textures, colours; furnishings.	Details of building services, including minor equipment, materials and finishes, and layout of servicing routes to fittings.	Detailed budgets for capital expenditures, operating costs; energy consumption.

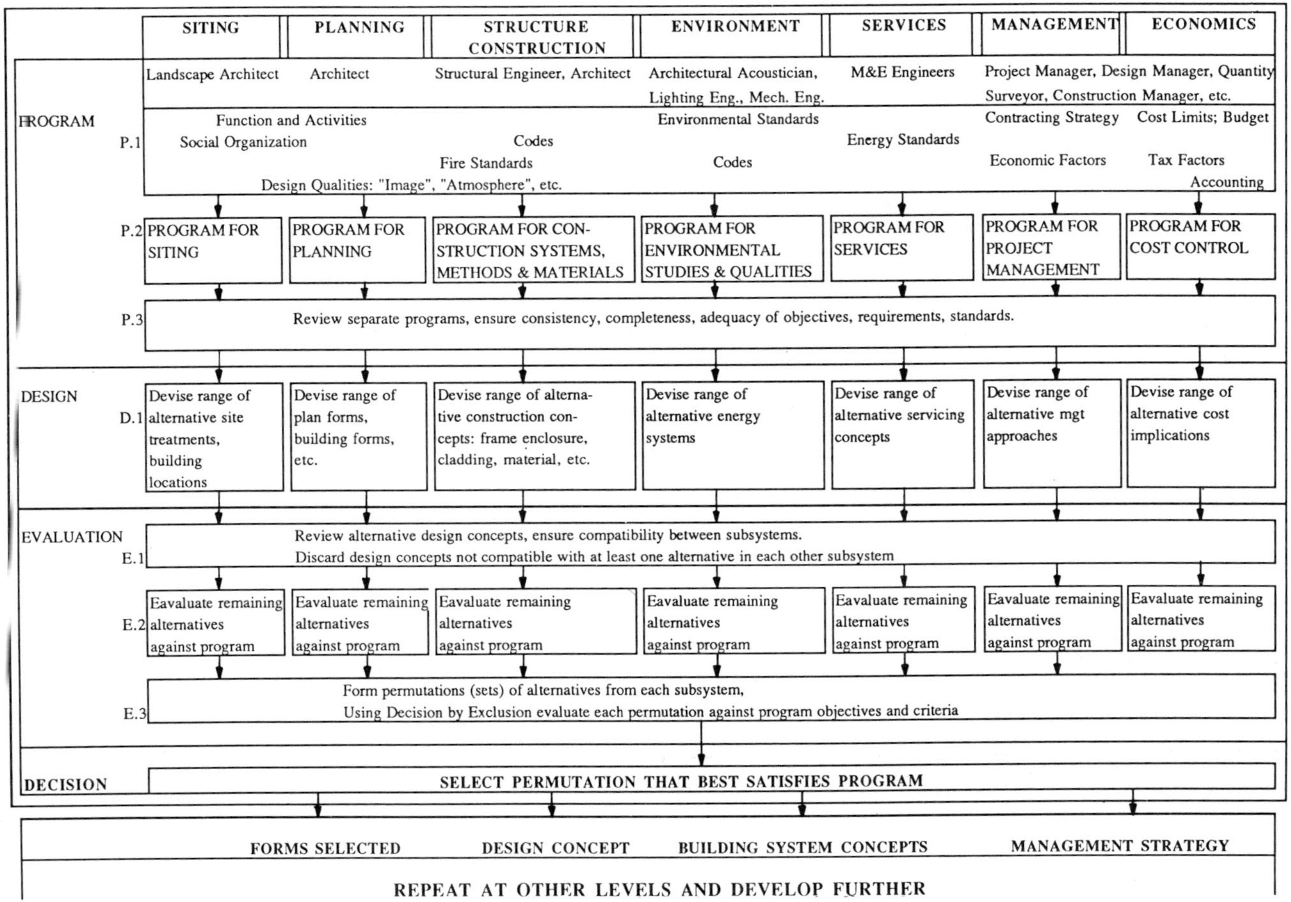

Figure 13.3. Key: design process at conceptual level.

TABLE 13.2 Summary of Design Decisions by Design Subsystems for Internal Environmental Design

SCHEDULE: ENVIRONMENTAL DESIGN
Summary of Design Decisions by Design Sub-System
(Work on horizontal levels undertaken approximately concurrently)

SITING	PLANNING	CONSTRUCTION	SERVICES
Identify alternative possible sites. Locate the building. **Conceptual Level**	Decide general character and kind of space and environment. Determine plan type and overall building form. Determine principal separations of planning subdivisions of the building. Plan major spatial units. Establish general massing	Select constructional system(s) (in, combination of structural framework and building fabric). Determine principal separations of structural subdivisions of the building.	Select building services systems. Select energy/fuel type(s) to be used.
Modify the site and landscape; earthmoving, planting, water, site construction, etc. **Intermediate Level**	Determine building volumes, shapes, orientations, number and height of floors. Determine main circulation routes for people and goods. Locate building's entrances, exits for people, materials, etc. Determine general arrangement of fenestration (solids and voids).	Select main constructional materials. Determine duct dimensions, construction, interior finishes, etc. Determine enclosure system, including walls, roofs and fenestration.	Select major mechanical, electrical, etc., plant and determine its location. Determine main circulation routes for services. Plan general distribution of mechanical, electrical, etc., services. Plan routes for all environmental services from mains to rooms and other spaces. Determine dimensions of service mains.
Detailed Level	Design individual room spaces in detail	Design other constructional elements: partitions, floors, ceilings, windows, sunshades, doors, etc. Determine constructional details: window and door openings, windows, etc.	Design organization of services within ducts. Determine environmental services control systems. Select secondary environmental services equipment and determine its location. Select and locate switches, controls, stats, etc. Determine detailed layouts of servicing routes from point of entry to rooms to terminals, outlets, etc.

TABLE 13.3 Example of Part of One Design Decision: Location of Building on Site with Respect to Environmental Factors

	Thermal	Visual	Aural
3.04 *Identify preferred location within site for the building:* assess site's characteristics to maximize advantages and minimize any inherent disadvantages, e.g.:			
Dimensions, area, size	A large site provides freedom to select or avoid shading, give clear access for solar, direct or obstruct wind patterns, etc.	A large site permits planned views within the site and freedom to select daylight exposure	A large site permits distance from noise sources (such as traffic, a school, etc.) and may give freedom to plan "back" of building to noise
Level, elevation	Shelter from or expose to sun and wind	Obtain views	Building cannot be placed "above" noise
Orientation, aspect	Receive/avoid solar radiation; minimize/maximize effect of cold/cooling winds/breezes; face building to maximize winter heat gain/summer shade	For views; minimize obstruction of daylight and sunsets and sunrises	Avoid direct paths from noise sources
Topography	Obtain shelter from, or expose to, sun and wind (and rain, frost, snow, etc.)	Secure views; avoid obstruction of daylight; obtain shade	Obtain screening from noise sources

Source: From Manning [in preparation].

TABLE 13.4 Example of Part of Another Design Decision: Selection of Constructional System with Respect to Environmental Factors

	Thermal	Visual	Aural
3.06 *Select constructional system(s)* (e.g., combination of structural framework and building fabric) to achieve economic, aesthetic, etc. objectives, to ensure appropriate strength and stability, fire resistance, durability, weather exclusion, etc.			
	. . . and desired thermal conditions. Note that "heavy" constructions (e.g., concrete-encased steel frames and floors, etc.) act as "thermal flywheels" and smooth effects inside buildings of diurnal temperature variations. Heavy constructions usually predicate continuous heating/cooling. "Light" constructions (e.g., uncased steel, timber frames, panelized cladding) respond more rapidly than "heavy" to temperature variations and so are often associated with intermittent heating/cooling. Note possible effects of choice of structural system (e.g., solid versus lattice beams) upon enclosed volume to be heated/cooled	. . . and desired visual conditions: e.g., consider appearance, possible wish for daylighting/windowless conditions	. . . and desired aural conditions. Some constructional systems and structural forms (e.g., domes) have profound aural consequences

Source: From Manning [in preparation].

6. For each design subsystem evaluate each remaining design solution against the requirements established in the program (or brief) for that design subsystem. Retain all design solutions that at least minimally satisfy their programs. See line E.2 in Figure 13.3.
7. Form alternative permutations of retained design solutions, each permutation containing one design solution from each design subsystem. See line E.3 in Figure 13.3.
8. Using the decision by exclusion technique, plot each permutation in multidimensional space and evaluate against program objectives and criteria. See line E.3 in Figure 13.3.
9 Select the permutation that best satisfies the collective program.

At one or more intermediate levels:

10. Using as basis the permutation selected at step 9, repeat the process described in steps 2 to 9.

At a final, detailed level:

11. Using as basis the permutation selected in the final stage of step 10, repeat the process described in steps 2 to 9.
12. The permutation selected from this final, detailed level should be (or approximate) the final, entire design.

Table 13.5 provides, in the form of a schedule, an abbreviated illustration of this process and the sequence of considerations, design decisions, and other design actions involved in the total design of entire buildings.

The types of design actions employed within the general strategy may extend throughout the *design spectrum* articulated by Broadbent [1966]: algorithm, ratio, deduction, induction, analogy, metaphor, chance. Different design methods can provide systematic means for determining needs, formulate series of design alternatives for each stage and subsystem, predict design performance, evaluate options methodically, and provide justification for selection or rejection. Design methods used for the determination of specific parts of buildings' design include:

1. Design methods (usually, but not always, concerned with *product*) that are derived from the building sciences (and in which there has already occurred a good deal of CAD activity): for example, calculation of framed structures; levels of daylight and electric lighting; heat loss, needed amounts of sound absorption, etc.)
2. Design methods (usually, but not always, concerned with *process*) that are substantially derived from the (design) professions: for example, activity data methods [Ministry of Public Buildings and Works 1966], brainstorming [Jones 1980a], environmental evaluation [Manning 1987], synectics [Gordon 1961], and so on. (A valuable toolbox of these and other design methods is contained in Jones [1980a].)

It is at the stage of choosing between alternative permutations of options from the several subsystems (e.g., at stage E.3 in Figure 13.3) that, traditionally, the greatest requirements for judgment have arisen. Hitherto it has not been possible to reduce the

TABLE 13.5 Outline (and Incomplete) Program for Sequencing Building Design Decisions

1000	*Assemble building program at conceptual level: requirements, constraints, etc.*
1100	State program requirements, constraints, etc. for:
1110	Function and activities
1115	Social organization
1120	Fire standards, means of escape
1125	Desired "image," building character; context within which to relate, etc.
1130	Intended building life: physical
1135	Maintenance standards
1140	Replacement cycles
1145	Desired environmental character(s)
1150	Environmental standards
1155	Energy standards
1160	Project management strategies: design; contracting; commissioning; etc.
1165	Intended building life: economic
1170	Limits of building cost: capital
1175	Limits of building cost: operational
1180	Expected ROI for building
1185	Legal constraints
	Etc.
1200	*Formulate programs for design subsystems at conceptual level*
	Translate program requirements, constraints, etc. into conditions/criteria for design subsystems
1210	*Siting*
1212	Access
1214	Provision for function/activities
1216	Environment: External
1218	Internal
1220	*Planning*
1230	*Structure and construction, methods, and materials*
1240	*Building, including environmental, services*
1250	*Project management*
1260	*Cost standards*
1300	*Review separate programs*
1310	for: Consistency:
1312	Objectives
1315	Requirements
1318	Standards
1320	Completeness
1322	Objectives
1325	Requirements
1328	Standards
1330	Adequacy:
1332	Objectives
1335	Requirements
1338	Standards
2000	*Devise range of alternative design concepts at conceptual level*
2100	*Site layouts*
2110	Alternative possible sites
2120	Select particular sites
2130	Alternative locations for the building, defining:
2131	Dimensions, area, size
2132	Level, elevation
2133	Orientation, aspect
2134	Topography
2135	Existing buildings
2136	Trees

TABLE 13.5 (*Continued*)

2137	Ground surfaces
	Pavings, other hard surfaces
	Grass, plants
2138	Water surfaces
2139	Atmospheric pollution
2140	Noise climate
2200	*Building plans/forms*
2210	General character/type of space/environment
2220	Overall building form/establish general massing
2222	High/low
2224	Regular/irregular
	Etc.
2230	Plan type
2231	Deep/shallow
2232	Compact/extensive
2233	Open/cellular
2234	Symmetrical/asymmetrical
2235	Rectangular/curved
	Etc.
2240	Principal separations of subdivisions of the building:
2241	Planning:
	Grouping parts of the building intended for particular
	Spatial/internal visual treatment
	External visual treatment
	Forming subdivisions: for different:
	Thermal conditions
	Thermal engineering services
	Lighting engineering services
	Creating acoustic isolation
	Etc.
2242	Structural
	to minimize problems of:
	Thermal movement
	Moisture movement
2250	Plan major spatial units
	Etc.
2300	*Construction systems*
2310	Foundations
2320	Structural frameworks
2330	Floors
2340	Enclosure system; fenestration system
2350	Internal space divisions
	Etc.
2400	*Building servicing systems*
2410	Energy supplies
2420	Water
2430	Communications
2440	Plumbing
2450	HVAC
2460	Electric lighting
2470	Garbage disposal
	Etc.
2500	*Project management systems*
2600	*Cost plans*
2610	Capital costs—by subsystems and elements
	Siteworks

Continued

TABLE 13.5 (*Continued*)

2600	*Cost plans (Continued)*
	Constructional:
	Structural frameworks
	Enclosure system; fenestration systems
	Floors
	Partitions
	Finishes
	Etc.
2620	Annual provision for replacement costs
	Finishes: Floor, roof
	Light fittings
	Electrical installation
	Fans, pumps
	Boilers
2630	Operational costs
	Building maintenance
	Services operating:
	Heating
	Lighting
	Services maintenance
	Building cleaning
	Etc.
3000	*Review design alternatives at conceptual level*
3100	Ensure compatibility with design alternatives of other subsystems at conceptual level
3200	Discard every design concept at conceptual level that does not have such compatibility with design alternatives of other subsystems at this level
3300	*Evaluate remaining design concepts at conceptual level against program*
3310	*Site layout*
3320	*Building plans/forms*
3330	*Construction systems*
3340	*Building services systems*
3350	*Project management systems*
3360	*Cost plans*
3400	Discard every design concept that does not meet objectives
3500	Discard every design concept that does not meet criteria
3600	*Form sets (permutations) of design concepts*, each set comprising one alternative from each design subsystem
3700	*Evaluate each permutation* using decision by exclusion technique: for performance:
	Against program objectives
	Against program criteria
4000	*Select permutation that best satisfies program objectives and criteria* as revealed by decision by exclusion
4100	Clarify characteristics of selected permutation:
4200	Describe, within limits of selected permutation:
4210	Siting
4220	Plan and form
4230	Structural system, constructional system, fenestration system, etc.
4240	Internal environment
4250	Building and environmental systems
4260	Project management
4270	Cost plan
	Etc.
5000	*Elaborate building program at intermediate level*
5100	Restate program requirements, constraints, etc. in appropriately greater detail
5200	Translate program requirements, constraints into conditions/criteria for design subsystems at intermediate level

TABLE 13.5 (*Continued*)

5210	*Siting*			
5220	*Planning*			
5230	*Structure and construction, methods, and materials*			
5240	*Building, including environmental, services*			
5250	*Project management*			
5260	*Cost standards*			
5300	*Review separate programs*			
5310	for:	Consistency		
5320		Completeness		
5330		Adequacy		
6000	*Devise range of alternative design concepts at intermediate level*			
6100	*Site layout*			
6200	*Building plans/forms*			
6210	Determine building	Volumes		
6212		Shapes		
6214		Orientations		
6216		Number of floors		
6218		Height of floors		
6220	Determine main circulation routes:	for:	People	
6225			Goods	
6230	Locate building's entrances/exits:	for:	People:	Occupants
6232				Visitors
6234				Others
6236			Materials	
6238			Goods	
6240	Determine general arrangement/location of fenestration (solids and voids)			
6300	*Construction system*			
6310	Select main constructional materials:			
6311		Structural framework		
6312		External walls/cladding		
6314		Roof		
6316		Floors		
6318		Other main elements:	internal	
6320	Determine duct:	Dimensions		
6323		Construction		
6326		Internal finishes		
6330	Determine enclosure system:	External walls/cladding		
6333		Roof		
6336		Fenestration		
6400	*Servicing systems*			
6410	Select major plant and determine its location:			
6412		Mechanical		
6414		Electrical		
6416		Other		
6420	Determine main circulation routes for services:			
6421		Mechanical		
6423		Electrical		
6426		Other		
6430	Plan general distribution of services:			
6431		Mechanical		
6433		Electrical		
6436		Other		
6440	Plan routes from mains to rooms			
6450	Determine dimensions of mains			
6500	*Project management systems*			
6600	*Cost plans*			

Continued

TABLE 13.5 ***(Continued)***

7000	*Review design alternatives at intermediate level*
7100	Ensure compatibility with design alternatives of other subsystems at intermediate level
7300	Discard every design alternative at intermediate level that does not have such compatibility with design alternatives of other subsystems at this level
7300	*Evaluate remaining design alternatives at intermediate level*
7310	*Site layout*
7313	Against program objectivs
7316	Against program criteria
7320	*Building plans/forms*
7323	Against program objectives
7326	Against program criteria
7330	*Construction systems*
7333	Against program objectives
7336	Against program criteria
7340	*Building services*
7343	Against program objectives
7346	Against program criteria
	Etc.

And continue with this (and possibly additional) intermediate level(s) and final detailed level

level of judgment and increase the degree of confidence and certainty because the different components of the permutations have been characterized by their incommensurable units and incompatible scales. As mentioned earlier, this problem is now resolved by use of the decision by exclusion rule.

DISCUSSION

Campion [1968] expected the readers of his book on computers in architectural design to be disappointed ". . . with regard to the amount of light which it sheds on architectural applications of computers. . . . it will be realized both how little work has so far been carried out in this field and how much lies ahead." Twenty-odd years later much work has indeed been undertaken into computer-aided design, yet from the point of view of a designer wanting to design buildings with the assistance of computers, the position is, to all intents and purposes, unchanged. This may be attributed to a variety of reasons, but there should be no doubt of the significance of two particular factors: (1) that the problem has generally been tackled from the wrong end (i.e., from concern for detail or parts rather than from a design's overall conception and entirety) and (2) that the significance of computer graphics has been greatly overrated. The elemental role in design played by chains of decisions, and the importance to such decision chains of "menus" of alternatives from which choices are made—a model both of design processes used by experienced designers and processes that are basic to computer programming and applications—is made more applicable to CAD by recent developments in ES.

We are in substantial agreement with Kalay's [1985] belief that computers should be used to augment designers' decision-making capabilities and in seeking solutions from within "spaces of alternative solutions" that satisfy specified design criteria. The authors' previously published works in the general fields of design methods and design evaluation provide practical bases for the establishment of criteria, the generation of

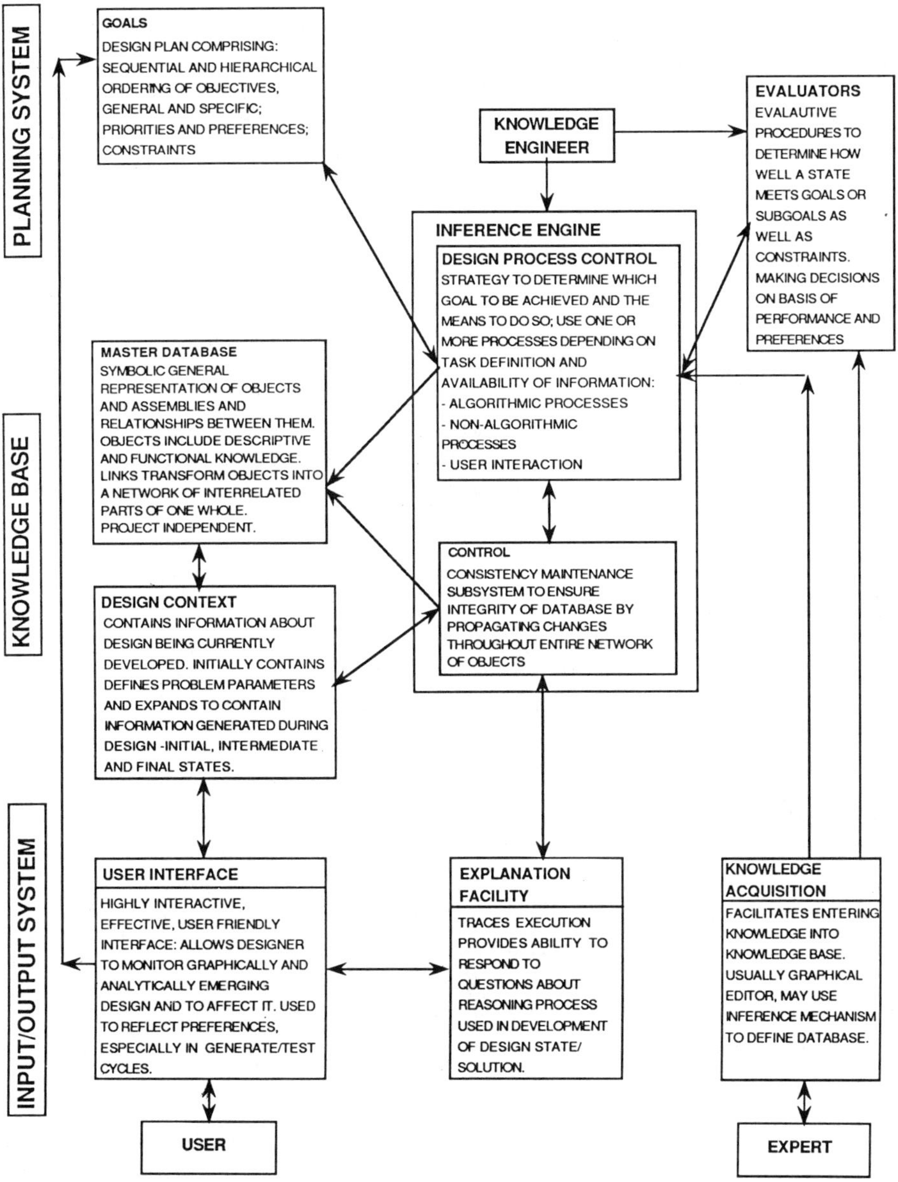

[DEVELOPED FROM KALAY'S FRAMEWORK 1985]

Figure 13.4. Knowledge-based computer-assisted design system: a conceptual model.

alternatives, and most critically (because otherwise, without a method, the process must be arbitrary and not *by design*), the evaluation of alternatives on multicriteria bases where there have been no common measures.

The authors' current work in this field concentrates on the identification of design menus and their evaluation (i.e., an extension of the activities illustrated in Tables 13.3 and 13.4) and on the incorporation of such design considerations in expert systems (for which Figure 13.4 illustrates a conceptual structure).

CONCLUSION

The task of formulating a generic means for design of entire buildings is an enormous one that some might think has characteristics not unlike development of systems for perpetual motion. The design task, though, has not yet been ruled physically impossible, so though it still appears unsolved, those who labor at it do so in the belief that, one day, all will be accomplished: the materials of the solution exist; some labor is required to organize those materials into working shape. Meanwhile each step taken is a step toward that goal. This paper has been prepared in that belief.

ACKNOWLEDGMENTS

To Pat Kelly and Don Westin, both of the Faculty of Architecture at TUNS, who helped one of the authors begin to comprehend and use what previously had been totally incomprehensible.

REFERENCES

Anon. (1975a). "Bristol United Press Building," *Architects' Journal*, February 26, pp. 445–458.

Anon. (1975b). *Royal Institute of British Architects Journal*, July–August, p. 18.

Broadbent, G. (1966). "Design Method in Architecture," *Architects' Journal*, September 14, pp. 679–685.

Campion, D. (1968). *Computers in Architectural Design*, Elsevier, Amsterdam.

Feigenbaum, E. A., and P. McCorduck (1983). *The Fifth Generation*, Addison-Wesley, Reading, Mass.

Gordon, W. J. J. (1961). *Synectics*, Harper & Row, New York.

Jones, J. C. (1965). "Systematic Design Methods and the Building Design Process," in *Proceedings of the 3rd CIB Congress*, Copenhagen; republished in *Architects' Journal*, September 22, 1965, pp. 685–687.

Jones, J. C. (1980a). *Design Methods: Seeds of Human Futures*, 2nd ed., Wiley, New York.

Jones, J. C. (1980b). "Opus One, Number Two," *Design Studies*, **1**(6): 373–377.

Jones, J. C., and D. G. Thornley (1963). *Conference on Design Methods*, Pergamon Press, London.

Kalay, Y. E. (1985). "Redefining the Role of Computers in Architecture: From Drafting/Modelling Tools to Knowledge-Based Assistants," *Computer-Aided Design*, **17**(7): 318–328.

Mackinder, M., and M. Marvin (1982). *Design Decision Making in Architectural Practice*, Research Paper 19, University of York, Institute of Advanced Architectural Studies.

Manning, P. (1966a). "Briefing Guide: Industrial Spaces, Fixtures and Equipment," *Architects' Journal*, April 13, pp. 953–984.

Manning, P. (1966b). In *Programme, Group 4, Ulm, 1966*, pp. 60–62. Waltham Forest Technical College.

Manning, P. (1968). "Appraisals of Building Performance: Their Use in the Design Process, *Architects' Journal*, October 9, pp. 793–800.

Manning, P. (1974). "Research and Education for Building Design," *Canadian Architect*, June.

Manning, P. (1981). "A Strategy for Lighting Design for Health Services Buildings," in *Proceedings of the Health and Welfare Canada Conference on Light, Health and Design*, Ottawa, May.

Manning, P. (1982a). "Daylight in Factories," Chapter 6 in *Developments in Lighting, 2: Industrial*, D. Pritchard (ed.), Applied Science Publishers, London, pp. 141–157.

Manning, P. (1982b). "Energy Conservation Through Use of Systematic Design Methods by Multi-professional Building Design Teams," in *Proceedings of the CIB 3rd International Symposium on Energy Conservation in the Built Environment*, Dublin, pp. 3.10–3.22.

Manning, P. (1983). "The Generic Design of Major Building Types for Energy Efficiency," in *Proceedings of the 2nd International Conference on Building Energy Management*, Ames, Iowa, pp. 11.3–11.12.

Manning, P. (1984). "A Strategy for Multi-professional Design of Buildings," *Canadian Journal of Civil Engineering*, **11:** 25–34.

Manning, P. (1987). "Environmental Evaluation," *Building and Environment*, **22**(3): 201–208.

Manning, P. (in preparation). *Environmental Design.*

Manning, P., and O. Biskaps (1981). "A Method for Projective Case Studies Used to Foster Inter-professional Skills in the Identification, Preparation and Appraisal of Large-Scale Development Proposals Involving the Construction of Facilities, in *Proceedings of the CIB 3rd International Symposium on Organization and Management of Construction*, Dublin, Vol. 2 pp. 101–120.

Manning, P., and S. Hassan (1982). "Presenting Climatic Information in Relation to the Building Design Process: A Tentative Approach," in *Proceedings of the CIB W71 International Symposium on Building Climatology*, Moscow, pp. 336–350.

Manning, P. et al. (1968). "AJ Design Guide: Building Environment," Section 4, "Thermal Properties," *Architects' Journal*, December 18 and 25, pp. 1483–1500.

Mattar, S. (1976). "Decision by Exclusion: On the Design of the Building Enclosure," D. Eng. thesis, Concordia University, Montreal, Quebec.

Mattar, S., W. Bitterlich, P. Manning, and P. Fazio (1978a). "A Decision Model for the Design of Building Enclosures," *Building and Environment*, **13:** 201–216.

Mattar, S., P. Manning, P. Fazio, and W. Bitterlich (1978b). "Application of the Decision by Exclusion Rule: The Design of External Walls," *Building and Environment*, **13:** 217–232.

Ministry of Public Building and Works (1966). *Activity Data Method*, HMSO, London.

Paterson, J. (1980). *Architecture and the Microprocessor*, Wiley, New York.

Reynolds, R. A. (1987). *Computing for Architects.* Butterworth, London.

Simon, H. A. (1969). *The Sciences of the Artificial*, MIT Press, Cambridge, Mass.

14

ON MEASURING PROJECT PERFORMANCE: SOME PROBLEMS OF AGGREGATION

R. A. Mohsini

School of Architecture and Planning
State University of New York at Buffalo
Buffalo, New York

Performance evaluation of buildings—by means of postconstruction evaluation—has traditionally focused on the performance measurement of building subsystems. This approach, based on the horizontal view of the building process, suffers from two important shortcomings. First, its underlying assumption that a completed building's performance can be arrived at simply by aggregating the individual performances of its parts is not valid. Like most complex systems, a completed building is much more than the sum of its parts. Second, the focus of traditional appraisal on subsystems, with its feedback information on individual products and services, actively encourages suboptimization of tasks. It is a condition that is contrary to the requirements imposed by the multiorganizational nature of the building team, and the resulting interdependence of tasks, where useful feedback information is that which relates to the project as a whole. To overcome these shortcomings of the traditional building appraisal model, an alternative model based on the vertical view of the building process is proposed.

INTRODUCTION

Building performance is usually appraised once the building is in use; "postoccupancy evaluation," or better, "postconstruction evaluation" compares the performance of the building, *considered as an object*, with the expectations of the users, *expressed in terms of activities*. Postconstruction evaluation bears on "the stuff of which the building is made, the bricks-and-mortar, components, service installations and so on" [Building Performance Research Unit 1972]; various procedures, based on a comparison between performance *criteria* and performance *characteristics*—the former of concern to the users and the later inherent in the building—attempt to relate the *building* to the *activities* it houses [Camus 1972].

Postconstruction evaluation carries with it an inherent problem that shows up in two ways:

1. Inability to deal with changes of user activities occurring after the evaluation
2. Inability to draw conclusions where activities *at the time of evaluation* differ from those that were anticipated in the design brief

In other words, postoccupancy evaluation can cope with the perfectly natural evolution of user patterns only to the extent that the building itself can accommodate it, which is usually very little.

An additional problem arises when relating performance *as evaluated* to costs *as accounted for.* A number of documents on performance include costs but often under headings such as "economic performance", as if they were of the same nature as "thermal performance" or "performance in the presence of fire" [CIB 1972]; however, costs should really be considered as if they were "the opposite side of the coin," as a measure of the "sacrifice" that the users are called upon to make in order to attain the satisfaction being evaluated [Parsons 1972]. In other words, costs are a direct and quantifiable indicator of the *processes* used (1) to build and (2) to oeprate/maintain/adapt the building, with its inherent levels of performance and with the satisfactions they provide at any moment in time. In this article we address the problem of evaluating the first part of these processes—the entire sequence of events that leads from the first inception of a project to its final completion.

The building process, it should be stressed, has two distinct structures: one that can be called "horizontal" and one that is "vertical." The *horizontal* structure comprises the iterative and cyclical process of decision making and production that occur again and again at each of the different phases of design and construction of each of the many different projects being built over any given period of time. For example, the architects' office may be carrying out working drawings in its production department, following almost unvarying routines, or the plumbing subcontractor may be gearing up to install, and installing, sanitary facilities within his or her region over a period of years; in all cases, costs are incurred, which although related to different building projects, form the basis for the continuous cash-flow picture of the individual enterprises that make up the building industry.

The *vertical* structure of the building process comprises the sequence of phases by which the outcomes of *one* of each of these phases of the horizontal structure is integrated into the work required to produce a single completed building. For example, one set of working drawings will be produced, and one plumbing subcontractor will install the appropriate sanitary facilities; all the partial costs add up to the total project cost, in other words, add up to fix the level of sacrifice the user has, sooner or later, to pay.

For all practical purposes, performance evaluation of buildings is limited to the evaluation of their various parts (i.e., to the evaluation of the products used in them) even if these parts are, in practice, lumped together in a way that better reflects the users' activities. The justification for this approach lies in the fact that since the overall building process consists of a series of subprocesses, each of which produces a part of the final product, and since a building is in fact a coordinated assembly of selected products, its performance can also be assessed by aggregating the individual performances of these products. Furthermore, as these products are generated by successive, iterative processes (which characterize the "horizontal" structure), the results of building parts evaluation can be fed back to the appropriate offices and firms, which can improve the performance of their services or their products "next time around."

Up to now, there seems to be general agreement in the literature on performance appraisal that evaluation of the building process (understood in the sense of the vertical structure mentioned previously) is of no great use because the results of the evaluation cannot be used to improve the performance of the building itself. This argument, it should be explained, reflects a linear view of the sequence of events involved in building: design–construct–occupy; by the time a building is occupied (and postconstruction evaluation can begin) it is too late to improve the building process, even if better processes might have produced a better building. Useful feedback is no longer possible.

BUILDING PROCESS APPRAISAL

Focusing appraisal on subproducts—and thus on subprocesses (the "horizontal" view)—isolates the interactive phases of the building process from each other and leads to suboptimization. It is a highly significant consideration, particularly when one realizes to what extent the building costs and quality are influenced by the interactions between the aspects of the total building and between the phases of the total building *process*. For example, while the appraisal data obtained from the evaluation of the HVAC design and performance of a building certainly enables its mechanical engineering designers to build up their own experience and improve later design, it is of little use to a building owner whose prime concern is with the comfort levels, as a whole, of the building that he or she procured.

Incidentally, the model underlying appraisal at the subprocess level is that of a single organization, such as a manufacturer of doors, producing and marketing a standardized product. Supposing that the entire process at such a door manufacturer consists of four phases—design, prototype testing, manufacturing, and sales/use—then under normal industrial conditions, the following iterative appraisal process becomes effective:

Step 1: During the prototype testing phase, the design of the door is scrutinized and the appropriate feedback is sent to the designers, who, in turn, improve the design.

Step 2: During the manufacturing phase, the design is once again evaluated, this time from the manufacturing point of view, and this new feedback is directed at the designers.

Step 3: Finally, the sales/use phase further evaluates the design of the door and these data come back to the designers of the door.

It is evident that the three iteration loops created here can lead to a continuous modification of the product toward one that would perform more or less optimally. In fact, this model is almost standard for most manufacturing situations, and its use in the building process may also be justified where strong analogical relationship between building and manufacturing exist (in the prefabrication industry, for example). For most conventional building activity, where projects differ and the central concern of performance appraisal is focused at the building project level, it can be argued that placing emphasis on the performance evaluation of parts and subprocesses is actually counterproductive to the central objective of performance appraisal, that is, that it leads to better buildings (Mohsini 1989). Two unique characteristics of the building process support this proposition:

1. The interdependent nature of decision making in a sequential process [Cherns and Bryant 1984, Hillebrandt 1984]
2. The temporary multiorganizational nature of the "building team" [Mohsini and Davidson 1986, Davidson and Mohsini 1987]

BUILDING PROCESS AND ITS ORGANIZATION

In organizational terms, the building process is markedly different from other manufacturing processes—and not only for the oft-mentioned reason that its product is location specific (i.e., fixed to a particular site and inserted into a particular socioeconomic environment). Generally, the building process is handled by a number of independent "task organizations" (architectural or engineering offices, general and subcontracting companies, product manufacturers, etc.), which establish links with each other when the need arises. The relationships between the different organizations involved in any building project are therefore temporary—and short-lived when compared to the life spans of each organization individually.

Furthermore, the formation of the building team is progressive, from the time the building client (or future building owner) makes an initial contact—directly or through a representative—with the first party whose services he or she requires (usually, the architect) until the last specialist subcontract is signed. Thus although a large number of task organizations are involved at some time or other, or for some period or other, while the building project is being carried out, not all are involved in the process simultaneously.

It is important to remember, however, that while the task organizations appear at different points in time and appear to act independently, the decisions they make are highly interdependent. The task organizations not only begin their involvement with an initial list of task-specific requirements, but also with the outputs of preceding task organizations as influential constraints for them. Furthermore, this high degree of interdependence works two ways. On the one hand, the organization appearing upstream in the building process passes on its "product" as a constraint to the organization appearing downstream, while on the other hand, the ultimate performance of the upstream task organization's "product" may indeed be determined by the downstream task organization's performance of its tasks. For example, a perfectly balanced heating, ventilating, and air-conditioning (HVAC) system may go haywire if the design or installation of windows turns out to be substandard and the resulting air leakage exceeds the values anticipated by the HVAC designers.

The pivotal point is that the performance of a built facility cannot be arrived at simply by aggregating the discrete performances of its parts. Yet what emerges from the focus on the horizontal structure of the building process is precisely that—an underlying assumption that the completed building's performance can be determined simply by adding the individual performances of its subsystems.

TEMPORARY MULTIORGANIZATIONS

While temporariness is a common characteristic of any project organization (by its nature), the multiorganizational characteristic is specific to building project organizations. This refers to the fact that each organization participating in the building team is itself an independent organization, with its own objectives, long-term plans, and so on [Mohsini and Davidson 1986, Davidson 1988].

A temporary multiorganization thus has to cope with disparities between two levels of organizational objectives:

1. The temporary objectives of the organization responsible for the project
2. The "permanent" objectives of each participating task organization

The first level of objectives in building projects is generally defined by the client system (client/user/community). These are the requirements and the constraints that are imposed upon the project by the purchaser and by the environment within which he or she operates and within which he or she must build; the project organization is obligated to meet them.

The second level of objectives includes those that are typical of all permanent organizations: namely, survival in the marketplace, enhancement of their domains and their position in it, and so on. The architectural firms, the consulting engineers, and the general and specialty contractors are all permanent organizations in their own right, and as such also have their own permanent organizational objectives. However, once they become part of the project organization, the independent organizations have to undertake to subsume, to a certain extent, the achievement of their independent objectives in order to respect the project's system of objectives. Under these conditions, the degree to which these second-level objectives are met depends largely on the relative bargaining power of the firms within the project organization, and on the extent to which they dare to use what bargaining powers they possess. Indeed, a situation could develop where one organization gains advantages at another's expense to such an extent that the disadvantaged organization may not be able to meet its survival objective, thus abandoning its role in the project organization and jeopardizing the first-level objectives of the temporary organization as a whole.

Looked at from the point of view of the building project, this need for compromise has interesting consequences. A project organization that is mandated to serve the interests of the client/user/community set, and hence committed to fulfilling the first-level objectives, must create an environment that minimizes the risk of individual participants failing to meet their own objectives adequately, in other words, that controls the disparity between the project-level objectives and the participant-level objectives. This requirement, however, is inconsistent with the outcomes of evaluation and suboptimization of discrete products and subprocesses. It is important to note that by focusing on the performance evaluation of individual task organizations, their processes, and their products, the horizontal-structural approach gives rise to an organizational environment where:

1. A task organization's rewards depend almost entirely on the extent to which it is able to maximize its individual task performance.
2. The absence of a project-level mechanism to constrain the individual maximization demands progressively reduces incentives to integrate, coordinate, and cooperate.

As a result, a situation that requires joint decision making is transformed into a kind of contest to claim greater bargaining powers. And all this despite the abundance of precedence, both in practice as well as in literature, which suggests that in situations where interdependence exists, the combined returns of all task organizations that seek to maximize their separate return functions is less than the joint returns in situations where the task organizations are willing to coordinate their actions under a superordinate objective. It leads to the conclusion, then, that the evaluation of discrete subsystems and subprocesses (reflecting the "horizontal" view of the building process) not only generates competitive pressures among the participant task organizations, but also leads to a final product that is less than it could be.

AN ALTERNATIVE APPROACH: THE VERTICAL MODEL

The central problem concerning performance appraisal is one of setting up the most useful feedback loop, so that evaluation findings can influence subsequent building projects as beneficially as possible. In the case of the building projects, however, the strategic decision making rests with the client/user/community set—whose decisions are concerned with the project as a whole and not with its phases or parts in isolation, however important they may be. Therefore, to be useful, performance appraisal must satisfy two primary conditions:

1. It must take place at the level of the project
2. Its results must be expressed in a format that helps the client system to optimize its strategic decision-making function.

But what are the strategic decisions that must be made by the client/user/community set and which have so much influence on the project as a whole? Whereas in the past it was assumed that there was only one process by which buildings could be designed and built, several approaches are now available for systematic selection. The former approach consisted in hiring a professional designer, *then* calling for bids on complete documents, *then* constructing with the lowest bidding company; called traditional procurement, the *design* function was effectively separated from the *construct* function (with its *supply* and *erect* tasks further separated). Nowadays, a number of generic classes of arrangements governing designing and building have been described by various authors [Glover 1976, Haviland 1976, Hillebrandt 1984].

These procurement strategies establish the nature of the relationships between the various participants in the project team, together with the timing and duration of their involvements. In due course, the procurement strategy decided upon will lead—without further high-level decisions needed—to a set of contracts legally binding the participants to fill a determined role. Given the multiorganizational nature of the project organization, these different procurement strategies correspond to the many possible forms that the relationships between the different independent participating organizations can adopt—and describe them more or less exactly.

Selecting one particular strategy implies establishing in advance the types of relationships that can be formed between the independent organizations, the extent of the bargaining powers allowed to each of them, and by setting the timing, the degree of uncertainty within which they must operate. Thus, selecting a procurement strategy is, in fact, equivalent to deciding how the building process is to be structured (vertical view) and is therefore the most fundamental decision to be made by the client/user/community set. To guide in making this decision, two types of information must be forthcoming from the appraisal activity:

1. Information about how different task organizations fare under different procurement strategies
2. Information as to how the aggregated performance of all individual task organizations, under a particular procurement strategy, compares with the project-level expectations

THE AGGREGATING FRAMEWORK

Studies dealing with measurements at both of these levels have been reported in the literature [Ibbs and Ashley 1987, Nahapiet and Nahapiet 1985, Sidwell 1983]; what is needed, however, is a new aggregating framework to replace the simple summation of performance of subsystems and subprocesses with one that is compatible with the objectives of evaluation reflecting the vertical view of the building process. As the overall performance of building projects is dependent on the extent to which *all* participating task organizations are able to achieve their objectives without anyone gaining undue advantage at the expense of another, the strategic decision maker must design the project organization such that both project's as well as task organizations' objectives are achieved satisfactorily. The new aggregating principle must therefore be sensitive to these new imperatives.

Most of the multiattribute utility models can easily accomplish these objectives [Cohon 1978, Edwards and Newman 1982, Keeney and Raiffa 1976]. For example, the aggregation of task organizations' performances may be achieved by using the following simple form:

$$Uj = \sum_{j=1}^{n} w_i U_{ij}$$

where U_j is the overall performance for the jth objective of the project organization, w_i is the normalized weight (i.e., importance factor or the bargaining power) assigned to the ith task organization, and U_{ij} is the performance of the jth objective by the ith task organization. The symbol $\sum$ means to sum the weighted performance over all the task organizations from the first (1) to the last (n).

AN EXAMPLE

Consider a hypothetical situation where three task organizations—an architectural firm, an engineering company, and a general contractor—have teamed up to deliver a small project. The application of multiattribute utility technique to evaluate the project performance requires the following steps:

Step 1: Identification of the function that the evaluation is intended to perform. In the present example, of course, this is to compare the "delivered" project performance with the "expected" project performance.

Step 2: Identification of the stakeholders: Essentially, these are the individuals or groups who have an interest in the decision. In the present case, the stakeholders are identified as the client/user/community set.

Step 3: Identification of stakeholders' value attributes. This step may produce a structure as follows:

[A] Architectural firm	[AA] Program development [AB] Functional design [AC] Quality control [AD] Cost control
[B] Engineering firm	[BA] Energy system design [BB] Structural flexibility
[C] General contractor	[CA] Construction quality [CB] Time management [CC] Information management

Step 4: Determination of the relative importance of each attribute for the stakeholders. This may lead to the following values:

A [0.35]	AA [0.20]	→ 0.35 × 0.20 = 0.07
	AB [0.20]	→ 0.35 × 0.20 = 0.07
	AC [0.39]	→ 0.35 × 0.39 = 0.14
	AD [0.21]	→ 0.35 × 0.21 = 0.07
B [0.20]	BA [0.50]	→ 0.20 × 0.50 = 0.10
	BB [0.50]	→ 0.20 × 0.50 = 0.10
C [0.45]	CA [0.52]	→ 0.45 × 0.52 = 0.23
	CB [0.32]	→ 0.45 × 0.32 = 0.14
	CC [0.16]	→ 0.45 × 0.16 = 0.07

Step 5: Determination of how well each object of evaluation serves each value at the lowest level of the value tree. Such numbers, called *single-attribute utilities*, may be the

outcome of measurements, expert judgments, or both. In the present case, assume that they are determined by measuring judgments on a scale as follows:

RE: Quality control (expected)
What is the expected level of quality?
(min.) 1 2 3 4 5 6 7 (max.)

RE: Quality control (delivered)
What is the level of quality achieved?
(min.) 1 2 3 4 5 6 7 (max.)

A score of 6 for the expected quality will provide a utility of 100(6 − 1)/(7 − 1) = 83.33.

A score of 4 for the delivered quality will provide a utility measure of 100(4 − 1)/(7 − 1) = 50.

Step 6: Aggregate utility measures with measures of importance:

		Utility (Measured)		Utility (Aggregate)	
Attribute	Weight	Expected	Delivered	Expected	Delivered
AA	0.07	65.33	70.33	4.573	4.923
AB	0.07	33.33	45.86	2.333	3.210
AC	0.14	83.33	50.00	11.666	7.000
AD	0.07	70.25	50.00	4.900	3.500
BA	0.10	60.66	63.33	6.066	6.333
BB	0.10	72.82	83.33	7.282	8.333
CA	0.52	83.33	50.00	43.332	26.000
CB	0.32	66.67	33.33	21.334	10.666
CC	0.16	66.67	50.00	10.667	8.000
Overall weighted utilities				112.153	77.965

Note that given that the assumptions of preferential independence between the attributes and utility independence are satisfied, the foregoing evaluation not only provides information concerning the extent to which the project objectives have been achieved, but also identifies the attributes where the performance may have suffered. To determine why, for instance, the architectural firm was unable to control the project cost and quality [AD, AC], a second-level evaluation, with task organizations as the stakeholders, can now be performed.

Beside the overall project evaluation, this type of aggregation framework also facilitates the following:

1. Allows the strategic decision maker to stop worrying about tactical issues such as assigning numerical magnitudes to the performance levels of various subproducts

2. Provides him or her with strategic data that comment on the efficiency of various task organizations under specific bargaining powers (w_i's), and for the objectives that are deemed most important for the project (U_j's)

This type of data, in turn, enables the strategic decision maker to control the outcomes of the building process at both the task organization and project level.

1. By allocating the bargaining powers among the individual participants, he or she can control the extent to which they are allowed to maximize their individual returns.
2. By accounting the outcome of each task organization in terms of the final product, on the other hand, he or she can shift the traditional emphasis on suboptimization toward optimization of the building project as a whole.

CONCLUSION

Postconstruction evaluation of buildings currently takes a horizontal view of the building process. It lays emphasis on comparing the performance characteristics of parts of building with the users' functional criteria. It is, in a sense, almost coincidental that the approach enables feedback information to be used to optimize the performance of individual participants in the project team, enabling them to improve their services or products. This approach provides virtually no information enabling the building process as a whole to be bettered. Since the so-called building team is, in fact, a temporary multiorganization, the performance of the building process as a whole is particularly sensitive to the degree of disparity felt by the participants between *the overall project objectives* and *their own objectives.*

The building client's selection of a procurement strategy largely determines the relationship between the participants, including the disparities, which can influence performance so much. To guide these decisions, information must be provided—and it can be provided only by postconstruction evaluation. To achieve this, however, the scope of postconstruction evaluation must be enlarged appropriately. It must include the vertical view of the building process (i.e., the building process must be seen as a unique *series* of product generating subprocesses). Only under such a comprehensive realization of the building process could the appraisal of the building systems and manufactured products used in the project, as well as the assessment of the decision-making processes by which the systems and products were designed or selected, be achieved.

REFERENCES

Building Performance Research Unit (1972). *Building Performance*, Wiley, New York.

Camus, R. (1972). "The Performance Concept and the Evaluation of New Building Products," *Industrialization Forum*, **3**(3): 5–12.

Cherns, A. B., and D. T. Bryant (1984). "Studying the Client's Role in Construction Management," *Construction Management and Economics*, **2**(2): 177–184.

Cohon, J. L. (1978). *Multiobjective Programming and Planning*, Academic Press, New York.

CIB (1972). *CIB Master List for Structuring Documents Relating to Building Elements, Materials and Services*, CIB Report 18, Conseil International du Batiment, Rotterdam.

Davidson, C. H. (1988). "The Building Team," in *Encyclopedia of Architecture: Design, Engineering and Construction*, J. A. Wilkes, and Robert P. Packard (eds.), Wiley, New York, Vol. 1, pp. 509–515.

Davidson, C. H., and R. Mohsini (1987). "Building Procurement: A Strategic Organization and Management Decision," in *Managing Construction Worldwide*, Vol. 1, *Systems for Managing Construction*, P. R. Lansley and P. A. Harlow (eds.), E.&F.N. Spon, London, pp. 28–39.

Edwards, W., and J. R. Newman (1982). *Multiattribute Evaluation*, Sage University Paper Series on Quantitative Applications in the Social Sciences, 07-026, Sage, Beverly Hills, Calif.

Glover, M. (ed.) (1976). *Alternate Processes: Building Procurement, Design and Construction*, Industrialization Forum Occasional Paper 2, Montreal, Quebec/Champaign, Ill.

Haviland, D. S. (1976). *Project Delivery Approaches: An AIA Guide*, American Institute of Architects, Washington, D.C.

Hillebrandt, P. M. (1984). *Analysis of the British Construction Industry*, Macmillan, London.

Ibbs, C. W., and D. B. Ashley (1987). "Impact of Various Construction Contract Clauses," *Journal of Construction Engineering and Management*, **113**(3): 501–521.

Keeney, R., and H. Raiffa (1976). *Decisions with Multiple Objectives: Preferences and Value Tradeoffs*, Wiley, New York.

Mohsini, R. A. (1989). "Performance and Building: Problems of Evaluation," *Journal of Performance of Constructed Facilities*, **3**(4): 235–242.

Mohsini, R., and C. H. Davidson (1986). "Procurement, Organizational Design and Building Team Performance: A Study of Inter-firm Conflict," *CIB '86 Proceedings*, Vol. 8, Washington, D.C., pp. 3548–3555.

Nahapiet, H., and J. Nahapiet (1985). "A Comparison of Contractual Arrangements for Building Projects," *Construction Management and Economics*, **3**(3): 217–231.

Parsons, D. (1972). "Building Performance: Concept and Practice," *Industrialization Forum*, **3**(3): 23–32.

Sidwell, A. C. (1983). "An Evaluation of Management Contracting," *Construction Management and Economics*, **1**(1): 47–55.

15

MULTILEVEL ANALYSIS AND OPTIMIZATION OF DESIGN

Robin S. Liggett

Graduate School of Architecture and Urban Planning
University of California, Los Angeles
Los Angeles, California

William J. Mitchell
Milton Tan

Harvard Graduate School of Design
Cambridge, Massachusetts

In this paper we discuss a knowledge-based computer-aided design system that provides multilevel analysis capabilities and that automatically propagates constraints on design variables from level to level. It also supports formulation and solution of optimization problems at different levels, so that a solution can be approached by solving a sequence of appropriately constrained suboptimization problems. Theory and implementation are discussed, and a detailed case study of application to the design of small house plans is provided.

TOP-DOWN AND BOTTOM-UP DESIGN

> We all know how to cope with something big and complex: divide and rule, i.e., we view the whole as a composition of parts and deal with the parts separately. And if a part is too big, we repeat the procedure. The town is made up from neighborhoods that are structured by streets, that contain buildings, that are made from walls and floors, that are built from bricks, etc., eventually down to the elementary particles. And we have all our specialists along the line, from the town planner via the architect to the solid state physicist, and further [Dijkstra 1989].

Designers often work in abstraction hierarchies: they proceed top-down from an overall conception down to the details, or bottom-up from concrete details up to broad organizational ideas, or in some combination of the two. In top-down design constraints are propagated down the abstraction hierarchy: high-level organizational decisions constrain low-level detailing problems. But in bottom-up design constraints are propagated up the abstraction hierarchy: decisions about details constrain high-level organizational problems. The characteristic difficulty with top-down design, then, is that broad organizational decisions may so constrain the design of the details that by the time you get to them, it is impossible to work them out satisfactorily. Conversely, the characteristic difficulty with bottom-up design is that it may prove impossible to combine the lower-level elements and subsystems that have been established to produce a satisfactory overall organization.

There are related difficulties with application of analysis procedures to designs in progress. In top-down design it may be impossible to use a structural or energy or cost analysis program at an early stage because it requires as input information about details that have not yet been established. Conversely, in bottom-up design, an analysis program may be impossible to use because it requires information about contexts (which establish loads, etc.) that have not yet been established.

Difficulties also show up when optimization procedures are employed. Formulation of an optimization problem in terms of aggregated, high-level variables, and global constraints and objectives may yield lower-level problems that are so constrained as to have only poor solutions, or worse, to have no feasible solutions at all. Conversely, combination of individually optimized low-level subsystems may lead to an overall solution that is far from optimal. (This is related to the well-known problems of reconciling maximization of individual objectives with maximization of the public interest.)

SYNTACTIC LEVELS

The concept of an abstraction hierarchy is closely linked to that of a generative grammar. Such a grammar not only specifies a domain for design exploration, but also assigns a structure to each object in that domain. A phrase-structure grammar to generate simple declarative sentences, for example, assigns a hierarchical structure to sentences as illustrated in Figure 15.1: words are nested within phrases, and simple phrases within more complex phrases. Similarly, the grammatical rules of classical architecture assign a hierarchical structure to a classical order, as shown in Figure 15.2: classical architectural elements are nested within subassemblies, and simple subassemblies are nested within more complex subassemblies.

A grammar is expressed as a set of rules that specify how to put parts together to make a whole. If they tell how to refine a high-level subsystem design into a specific arrangement of lower-level subsystems—a schematic column into a combination of base, shaft and capital, say (Figure 15.3)—applying them will derive designs in top-down fashion. If they tell how to assemble lower-level subsystems into higher-level ones—that the shaft goes on top of the base, and the capital on top of the shaft (Figure 15.4)—applying them will derive designs in bottom-up fashion.

Shape grammars [Stiny 1980] used to specify languages of architectural form do not always assign simple tree structures to the designs that they generate, and they do not

ILLUSTRATIONS (SCALE 1.5:1)

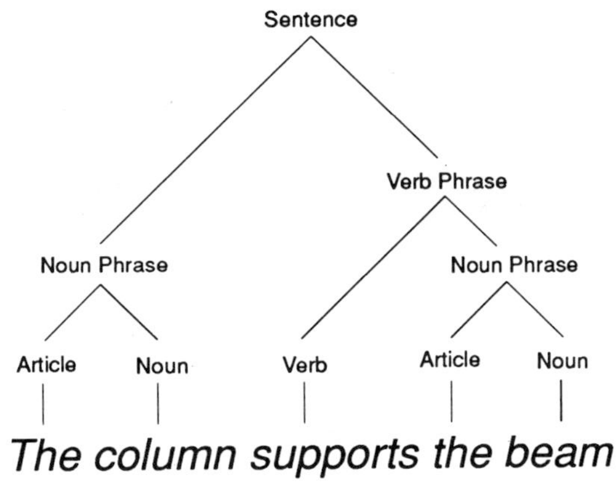

Figure 15.1. Hierarchical structure assigned to a sentence by a phrase-structure grammar.

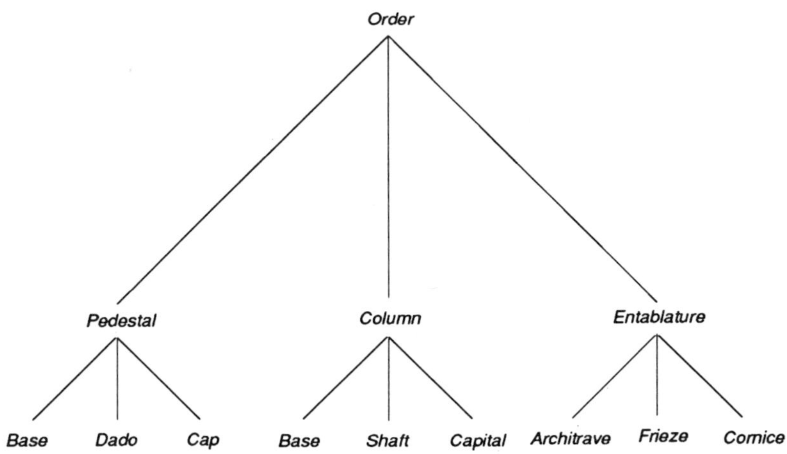

Figure 15.2. Hierarchical structure of a classical order.

necessarily derive designs in strict top-down or strict bottom-up fashion. But the rules of a grammar should reflect the way that a designer wants to look at a problem. If a designer wants to move up and down in a clearly defined abstraction hierarchy, it is appropriate to follow rules that produce such hierarchies. If the parts put together by a shape grammar are parametric objects, application of the rules produces a system consisting of a set of parameters P together with a set of constraints C. Each constraint is an n-ary relation among a subset of P. The designer must assign values to the parameters such that the constraints are satisfied. In general, different designs in the language specified by the shape grammar will have different systems of parameters and constraints. Some of these systems may have feasible solutions, and some may not.

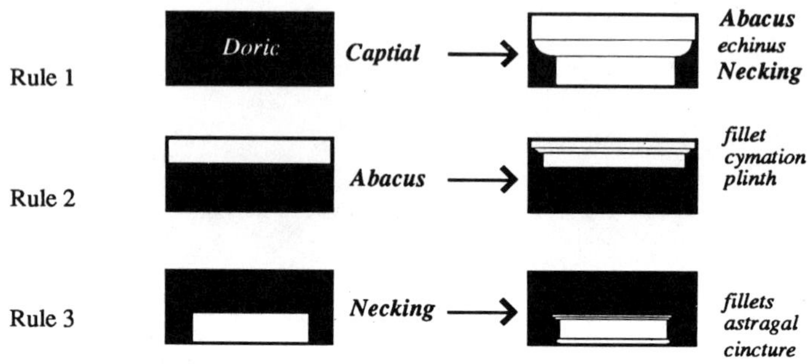

Figure 15.3. Rules for top-down refinement of a schematic design.

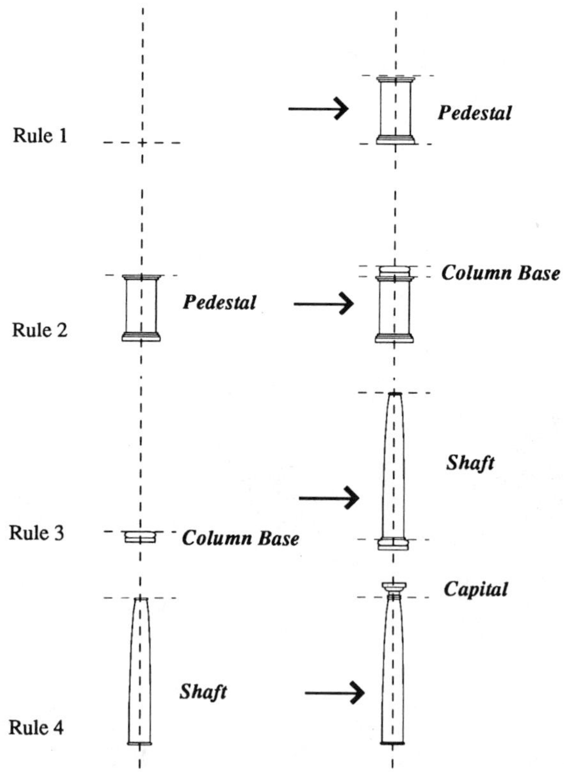

Figure 15.4. Rules for bottom-up assembly of parts.

SEMANTIC LEVELS

The hierarchical structures established by syntactic rules might be arbitrary and meaningless, but they rarely are. We tend to treat them as conceptual hierarchies, as structures with superimposed semantic levels. At the different levels we interpret shapes as different sorts of objects, with different sorts of properties, behaviors, functions, and relationships. Consider, for example, a structural frame. At a high level of abstraction we might depict it with single lines, see it primarily as a collection of linear spatial elements, and be concerned that these elements are in the right places within the building fabric to collect and transfer loads and divide space in an appropriate manner. At a lower level we might see it as a collection of steel webs, flanges, and plates with capacities to resist tension, compression, and shear stresses, and be concerned that these are configured and dimensioned to minimize use of material while keeping stresses and deflections within acceptable bounds. And at a still lower level we might adopt the viewpoint of a materials scientist and focus on the microstructure of steel alloys.

SUBPROBLEM FORMULATION

The frame has to work at all levels: it must provide support at the right places. To do this, the webs and flanges and plates must provide adequate resistance to stresses, and to do this, the microstructure of the material must develop adequate strength. Criteria of feasibility and functional adequacy at each level can be expressed as constraints. The designer at each level must find an appropriate organization of parts and feasible values for the parameters of these parts. But it does not help an architect concerned with placement of structural members to think about molecular structure, and the materials scientist concerned with invention of new alloys does not need to pay much attention to the organization of architectural space. In other words, subproblems at different levels in a well-constructed abstraction hierarchy can, and usually should, be handled relatively independently of each other. The same goes for parallel subproblems at the same level: a cladding system, say, can usually be designed without knowledge of the low-level details of the structural frame that is to support it. As has often been pointed out, partitioning a design problem into a hierarchy of relatively independent subproblems yields important efficiencies in problem solving [Alexander 1967, Mitchell 1990, Simon 1981, Sussman and Steele 1980].

Subproblems in an abstraction hierarchy are usually not completely independent, however. When design solutions are established for some of the subsystems, these solutions propagate constraints that must be satisfied by the designs of remaining subsystems. The point is that in a well-constructed abstraction hierarchy, relatively few constraints are propagated across subsystem boundaries and the paths of propagation are well defined. Thus it is relatively unlikely that the solution of one subproblem will have unexpected and unwelcome side effects complicating the solution of another. Nonetheless, constraints do accumulate. In top-down design the high-level subproblems are often relatively unconstrained, but lower-level subproblems are found to be increasingly tightly constrained as subsystems must be designed to fit precisely determined contexts. In bottom-up design the low-level problems tend to be loosely constrained, but problems of fitting together subsystems to form higher-level subsystems become increasingly difficult.

BIDIRECTIONAL PROPAGATION OF CONSTRAINTS

Thus the strengths and weaknesses of top-down and bottom-up design strategies are complementary, so it is usually prudent to move continually from level to level—sometimes considering the broad picture and sometimes focusing on details—so that elements and subsystems at different levels can gradually be adjusted into good fit with each other. The value of a two-way strategy is nicely illustrated by the example of Palladio's villa designs (Figure 15.5). The evidence of Palladio's drawings show that he habitually worked them out in top-down fashion—beginning with very rough, schematic sketches of plan and elevation arrangements, then refining these down to the details of the moldings and capitals. On the other hand, we know that Palladio followed strict proportioning rules for the architectural orders and that, in a sense, the choice of architectural order (Tuscan, Doric, Ionic, Corinthian, or composite) controlled the entire design: choice of a particular order established a basic dimensional discipline that was propagated throughout. This creates a dilemma. If the designer of a Palladian villa follows a pure top-down strategy, it may turn out to be impossible to fit columns of the desired order into the available space. But if the designer begins by choosing an order, then works bottom-up to the overall organization, the chances of satisfying planning requirements and producing a satisfactory elevation composition are slim. A better strategy is to begin by diving right down to a low level and choosing an order, then to move back up to the top level and work out the rest of the design in top-down fashion—always keeping in mind the dimensional constraints that have been established by the initial choice of order. This provides the advantages of top-down coordination of sub-

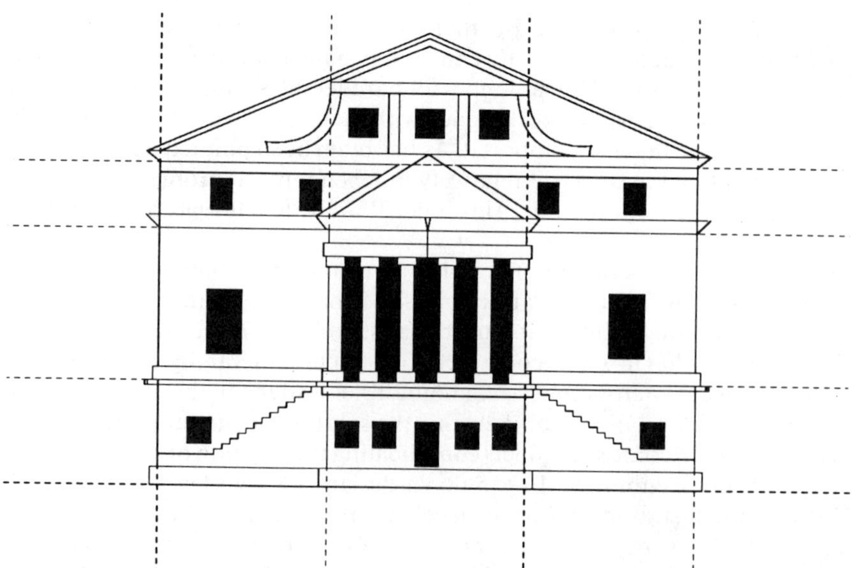

Figure 15.5. Schematic elevation of a Palladian villa. Room width and height decisions determine the space available for the portico, but the rules governing classical orders determine the space required.

systems and details, but assures that no difficulties will arise when porticos and columns eventually have to be fitted into the overall scheme. In general, a designer needs to be able to tie down key design decisions at various levels in an abstraction hierarchy, then go back to fill in other decisions within the framework of constraints established by those initial decisions. This sort of process can be assisted by a computer-aided design system that supports bidirectional automatic propagation of constraints—in other words, that uses information about established lower-level details to constrain higher-level design variables and that uses information about established higher-level organizational decisions to constrain lower-level design variables.

USE OF TOPDOWN TO DESIGN IN AN ABSTRACTION HIERARCHY

Multilevel analysis and optimization and bidirectional constraint propagation from level to level are supported by the Topdown computer-aided design system that we have implemented and used extensively in teaching [Mitchell et al. 1990]. Topdown itself can be described at three levels. From the user's viewpoint it is a highly interactive graphics program, controlled by mouse operations, that enables designs to be assembled in "click-together" fashion. From a programmer's viewpoint it is an environment for specifying shape grammars that generate hierarchically structured designs. At the substructure level it is a piece of software, implemented in Pascal, which maintains a data structure and provides a graphics interface. We consider these in turn.

A designer working at the user level sees the screen shown in Figure 15.6. There are

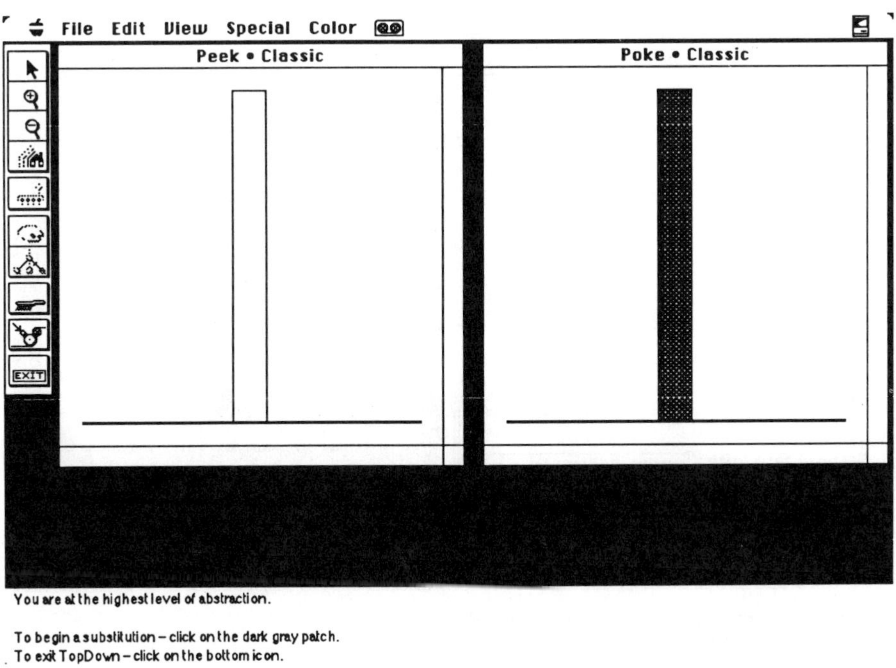

Figure 15.6. Topdown screen, with peek, poke, and message windows.

two graphic windows—the peek window and the poke window. The peek window always shows the current state of the design, depicted as a composition of lines and color-filled polygons. There are the usual pan, zoom, and other display functions. The peek window is for viewing only: the user cannot directly select or operate on graphic objects displayed within it. All design interaction is accomplished via the poke window. The designer only needs to know two basic moves: substitution of a more detailed representation for a less detailed one, and parametric variation of dimensions, angles, colors, and so on. Substitution is accomplished by clicking in the poke window on the part that is to be replaced. This brings up a dialog box showing the substitution options (Figure 15.7). When an option is selected by clicking in the dialog box it is displayed in situ in the poke window. Then, when ok is clicked, the selected option is substituted in the current version of the design that is displayed in the peek window, and the dialog box disappears. Substitutions can be made within substitutions to any depth.

Parametric variation of dimensions is accomplished by clicking on a dimension line in the poke window to bring up a control bar with a slide bar (Figure 15.8). The control bar displays the name and current value of the variable that is being manipulated. The minimum and maximum values between which the slide bar ranges express currently known constraints on that variable. When ok is clicked, the current value of that variable is entered in the data structure and the control bar disappears. At all times a message window is open at the bottom of the screen. This provides a continuous stream of context-sensitive advice and suggestions, critical comments, and results of relevant analyses (such as area and cost calculations). It can, of course, be ignored—but some appropriate advice is always there when the user happens to give it a glance.

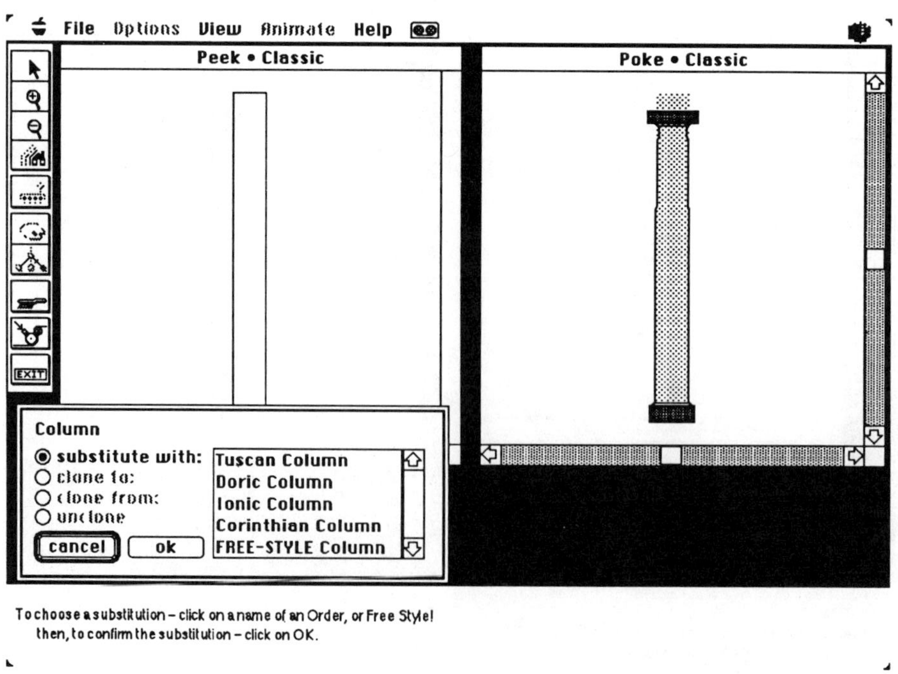

Figure 15.7. Presentation of substitution options.

PROGRAMMING TOPDOWN APPLICATIONS

The programmer of a Topdown application has the task of capturing and encoding knowledge of the shapes and sizes of elements and subsystems of the type of artifact that is of interest, together with knowledge of how to put these parts together. In other words, the programmer must encode the vocabulary and syntactic rules of a parametric shape grammar for this type of artifact [Stiny 1980]. The user is then able to explore designs in the language specified by that grammar by choosing among alternative parts available for substitution and by choosing values for parameters at each level.

The first step in programming an application is to specify the vocabulary of parameterized shapes from which designs are to be assembled. Second, rules specifying which shapes may be substituted for a given shape must be defined. Finally, the positions and sizes of substituted shapes must be related to the positions and sizes of their predecessors by specifying how parameter values will be inherited by substituted shapes. If a detailed description of an elaborate arched window is to be substituted for a simple rectangle in a schematic elevation, for example, the window's height and width should be inherited from that rectangle.

This sort of programming might, in principle, be carried out in many different ways. In current versions of Topdown, which are used for teaching programming, the programmer sees an extension of Pascal. Shapes are represented by parameterized procedures (using the approach described by Mitchell et al. [1987]), and there are facilities for specifying substitution possibilities and paths of inheritance of parameter values.

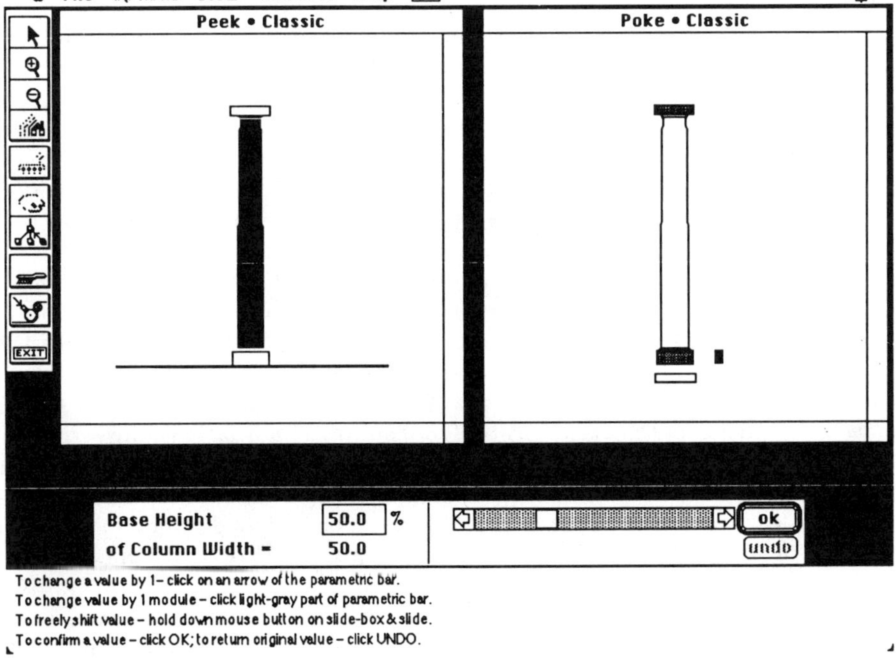

Figure 15.8. Parametric variation by selecting a dimension line and using the slide bar.

The programmer plugs applications into the Topdown shell—a piece of Pascal software that provides the interface between the user and the rules encoded by the programmer. The shell presents shape substitution rules as selectable shapes in the poke window together with substitution options in the substitution dialog box, and it presents dimensioning variables as dimension lines in the poke window which can be selected to invoke the slide bar that is used to assign a value. It also records the user's design decisions in a data structure, and displays the current state of the design in the peek window. Finally, it provides file import, export, and print utilities.

INTEGRATION OF ANALYSIS PROCEDURES

Earlier versions of Topdown (as described, e.g., in Mitchell et al. [1990]) dealt only with abstract shapes: it was up to the user to interpret shapes as architectural elements and subsystems and to keep in mind what they meant. The current version has been extended to deal with functionally interpreted shapes and to integrate knowledge about how to satisfy functional requirements. The function of an element or subsystem of an artifact is whatever useful thing (or things) the element or subsystem provides in that context: a room in a house provides floor area, a column in a structural frame provides support, a window in a wall provides light and air, and so on [Mitchell 1990]. The level of function is the amount of that thing provided—the amount of floor area, the amount of support, the amount of light, the amount of air, and so on. In general, the level of function can be modeled as a mathematical function of the element or subsystem's design parameters. Thus the area provided by a rectangular room, for example, is a function of its length and width.

In a given context, an element or subsystem functions adequately if it functions at the level required by that context—provides enough floor area to accommodate an activity, or enough support to hold up a roof, or enough light and air to make a room habitable. So the designer's task, in dealing with a particular element or subsystem in an abstraction hierarchy, is to adjust the design parameters such that functional adequacy is achieved—that is, the level of function provided matches or exceeds the level of function required. Since the level of function required is determined by the properties of other elements and subsystems, an element or subsystem design problem may be redefined by adjustment of other parts of the system. This means that in top-down design, established decisions about high-level subsystems define the levels of function that must be provided by lower-level subsystems. In bottom-up design, established decisions about low-level elements and subsystems define the levels of function that can be provided to higher-level subsystems, and eventually output by the system as a whole.

Functional adequacy of an element or subsystem must be achieved subject to compliance with internal and external feasibility constraints. Internal feasibility constraints are established by the internal organization—the parts, their properties, and their relationships—characteristic of a particular type of element or subsystem: they limit adaptability to context. A column, for example, must be designed subject to the internal feasibility constraint that some maximum compressive stress for the material is not exceeded. The internal feasibility constraints may also include resource constraints: for example, that a fixed quantity of steel is available for fabrication of the column. External feasibility constraints are established by context: they limit internal adaptability. A column, for example, cannot be made wider than the space in which it must fit, or shorter than the distance over which it must transfer its load.

In top-down design, external feasibility constraints for an element or subsystem are given by previous design decisions, and different structures of internal feasibility constraints will result from different choices of element or subsystem for substitution (e.g., cylindrical, square, or I-shaped column section). In bottom-up design, internal feasibility constraints for an element or subsystem are given by previous design decisions, and different structures of external feasibility constraints will result from different decisions about how to combine lower-level subsystems into higher-level ones. The programmer of a Topdown application can associate analysis procedures with a subsystem at any level. These are invoked whenever the subsystem is altered, and report (in the message window) on the feasibility and functional adequacy of the subsystem currently under consideration. Diagnostic rules can also be included, so that if there is a difficulty with feasibility or functional adequacy, there can be additional messages saying why and providing suggestions on how to fix it.

DOWNWARD PROPAGATION OF CONSTRAINTS

A Topdown programmer may choose not only to generate messages when constraints are violated, but also to enforce compliance with constraints. This can be accomplished in several ways—the simplest of which is to provide for downward propagation of constraints. As noted earlier, to assure that elements and subsystems fit together properly, the values of certain parameters of low-level elements and subsystems in a Topdown application are usually made dependent on the values of parameters of higher-level subsystems. The length of a beam, for example, might be made dependent on the distance between two supporting columns. Thus, when the distance between columns is widened, the beam is automatically lengthened to maintain fit. However, there may be internal constraints on the dependent values: if a beam's section and material are fixed, the beam will collapse if it is made too long.

In principle, parameters of subsystems at different levels might be symmetrically related or asymmetrically related. In other words, the length of the beam might be made dependent on the spacing of the columns, the spacing of the columns might be made dependent on the length of the beam, or the user might be allowed to set either length or spacing first. The downward propagation approach depends on consistently asymmetrical relationship of parameters: lower-level parameters may be dependent on higher-level parameters, but higher-level parameters are never dependent on lower-level parameters. So Topdown in its simplest form implements a "military" structure of command and control in which the direction of causality is rigidly fixed. Commands always come down from above. At each level they are obeyed, and further commands are propagated to lower levels. At the lowest level, commands are simply obeyed—no matter what the cost. This imposes a discipline on the design process—but one that can have high payoffs in terms of clarity and efficiency.

When a user makes a substitution in order to refine a design to the next level of detail, the choice of subsystem type to substitute automatically establishes the internal constraints on its parameters, and design decisions made at higher levels establish the external constraints. When a parameter is selected for adjustment, the range of movement of the slide bar is determined automatically by these constraints. The designer's task is to adjust the parameters within these limits to achieve functional adequacy.

UPWARD PROPAGATION OF EVALUATIONS

The simplest form of Topdown, with its "military" structure of command and control, takes no account of constraints at lower levels than the one currently under consideration by the user: it is simply assumed that the lower levels will somehow adapt and perform. A slightly more enlightened view of command and control is that commands are propagated down the hierarchy, but complaints may be propagated back up the hierarchy if these commands turn out to be difficult or impossible to execute satisfactorily. This structure can also be implemented in a Topdown application.

With this version of constraint propagation the user attempts, by trial-and-error adjustment of parameters, to find a good subsystem design at each level—one in which there is an appropriate fit between internal structure and the context in which the subsystem must function. When a parameter is selected for adjustment, the slide bar is automatically constrained so as not to allow violation of any external constraints. As parameters are adjusted Topdown automatically maintains a running report (in the message window) of internal constraints that are currently violated (Figure 15.9). Constraint reports refer only to the parts of the subsystem currently under consideration, not

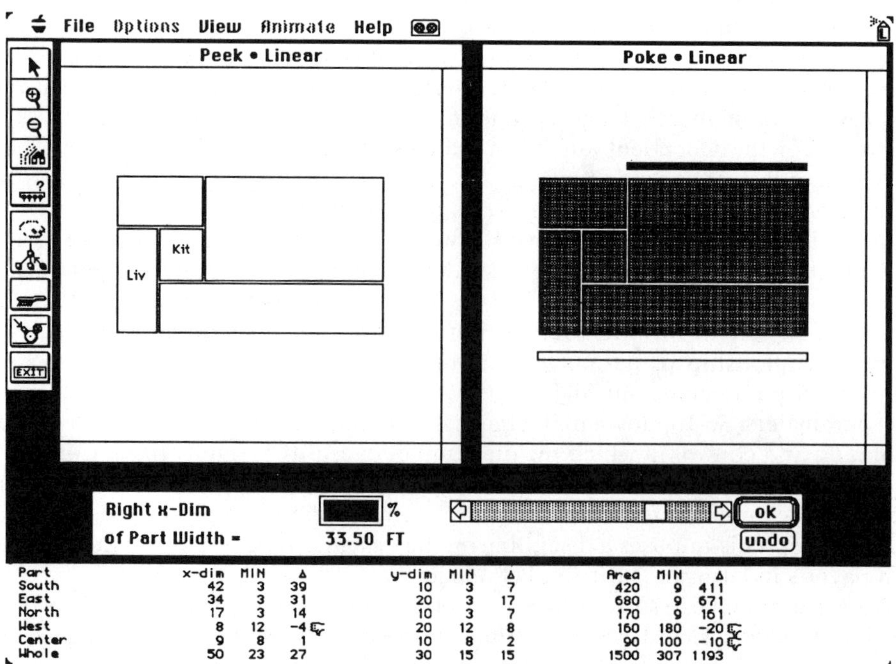

Part	x-dim	MIN	Δ	y-dim	MIN	Δ	Area	MIN	Δ
South	42	3	39	10	3	7	420	9	411
East	34	3	31	20	3	17	680	9	671
North	17	3	14	10	3	7	170	9	161
West	8	12	−4 👎	20	12	8	160	180	−20 👎
Center	9	8	1	10	8	2	90	100	−10 👎
Whole	50	23	27	30	15	15	1500	307	1193

Figure 15.9. Constraint violation report in the message window. The thumbs-down symbol indicates an unsatisfied constraint.

to the parts of these parts. In other words, the constraint violations at lower levels that cause the constraint violations at the current level are hidden behind an abstraction barrier. Of course, the user can always cross this abstraction barrier by moving down to a lower level.

If a feasible design cannot be found at a given level, the user can either try substituting another subsystem type to establish a different set of internal constraints or else can back up a level and try manipulating the parameters of the higher-level subsystem to loosen the external constraints. If this proves fruitless, the user can back up another level and repeat this procedure, and so on to the top of the abstraction hierarchy. The total design problem is solved when two conditions are met: the complete system (viewed from the top of the abstraction hierarchy) fits the given context and meets the overall functional requirements, and subsystem design has been carried through successfully down to the lowest level—feasibility has been shown at all levels, and there are no unresolved details.

USE OF CONSTRAINT SOLVERS TO MAINTAIN FEASIBILITY

A still more enlightened concept of command and control recognizes not only that the lower levels can voice complaints, but also that they have "rights" which cannot be violated. Thus when a command comes down from above, obedience to it might cause violation of these rights. So a conflict arises, and a conflict mediation mechanism is needed. Topdown also provides for this. It allows for lower-level constraints to be defined as "rigid" and for constraint-solving algorithms to be used to resolve conflicts automatically. Then whenever a parameter at any level is adjusted, Topdown readjusts parameters at other levels to maintain consistency with all constraints. The slide bar for the current parameter is constrained to the range that will not produce unresolvable violations. If any change would produce violations, the slide bar is locked in place.

Numerous constraint-solving algorithms are now available. Some are specialized and efficient, while others are more general but less efficient, so different algorithms are appropriate for use in different contexts. Most appropriate for use with Topdown applications are incremental constraint solvers that maintain an evolving feasible solution and modify it appropriately as constraints are added to or deleted from the problem [Freeman-Benson et al. 1990].

MULTILEVEL OPTIMIZATION

Finally, a command and control philosophy might recognize that the lower levels not only have rights but also objectives of their own. The task is then to negotiate a design in which objectives at every level are taken into account, and some appropriate balance is achieved. In other words, "private" satisfaction must be balanced against the "public" good. The most sophisticated version of Topdown provides for this by allowing an objective function to be associated with each subsystem, and then employing formal optimization procedures to produce optimal subsystem designs.

Usually, in subsystem design, the objective is either to maximize some measure of performance or to minimize use of some resource, subject to internal and external feasibility constraints. Let us assume, for example, that a steel-framed pavilion is to be designed in a context where steel is expensive, energy for air conditioning is expensive, as much floor area as possible is to be provided, natural light in the interior is considered desirable, and there is a limit on total cost. These considerations will be expressed and considered at different levels in the abstraction hierarchy. At the level of detailed design of steel members the objective will be to minimize the weights of members, and at the level of frame design the objective will be to minimize the total weight of steel used. Minimization of steel usage will not be a direct objective at the level of layout, but layout decisions will constrain lower-level problems of support placement and member design. Alternatively, if the frame design is fixed at an early stage to assure economy, this will constrain layout possibilities. At the level of overall layout, maximization of total floor area will be a direct objective. At the level of fenestration design, the designer must try to make the windows as large as possible to provide enough light to the depths of the interior: this will be difficult if a deep layout and low ceiling have been adopted to maximize floor area without increasing air conditioning costs too much, but it will be easy if the section is tall and narrow. At the highest level the design must be adjusted to satisfy the constraint on overall cost: this might be done by first establishing feasible designs for all the lower-level subsystems, then trying to adjust top-level variables to bring cost down to an acceptable level without violating lower-level feasibility constraints. The problem is restructured in complex ways if, say, natural light comes to be considered undesirable, or there is significant change in the costs of steel or air conditioning.

Thus Topdown with optimization provides an effective way to approach complicated, multicriteria problems. Objectives are separated and articulated at different levels. A designer can explore their interrelationships, and the possible trade-offs between them, not by conflating semantic levels and running some mysterious global optimization procedure but by quickly trying out designs for particular subsystems with particular objectives, then investigating the implications of these for other subsystems with other objectives. Furthermore, the effects of setting varying objectives at some level can be explored. What happens to the higher-level organization and the details if, for example, the objective at the overall layout level is changed from that of maximizing the depth of the plan to that of minimizing it? And what happens if the objective at the frame design level changes from that of minimizing the amount of steelwork to that of maximizing the clear span?

AN EXAMPLE: SMALL HOUSE PLANS

To illustrate the uses of analysis, constraint propagation, and optimization in Topdown, the task of designing small house plans was chosen. Some time ago it was shown that this type of problem could be solved by first enumerating structures (patterns of rooms) that satisfied adjacency requirements, then later applying standard linear, nonlinear, or dynamic programming methods to dimension rooms such that length, width, area, perimeter, and proportion constraints were satisfied and overall floor area was minimized [Mitchell et al. 1976, Mitchell 1977]. In a Topdown application this type of problem is solved interactively by working down from the overall shape to the fine details of internal layout. At each step, alternative internal structures are considered for substitution, and values are assigned to dimensioning variables such that constraints are satisfied and an economical layout is achieved. Each subproblem in the hierarchy of abstraction turns out to be so simple that solutions are usually fairly obvious and the user can quickly explore ranges of feasible alternatives.

The grammar employed to generate plans is illustrated in Figure 15.10. The starting shape is a rectangle, which can be of any size or proportion. There is a rule for dividing rectangles vertically, a rule for dividing rectangles horizontally, and a rule for replacing a rectangle by a pinwheel of five rectangles. These rules suffice for producing dissections of the original rectangle into smaller rectangles. Finally, a schematic house plan can be produced by substituting any one of the following rooms for an abstract rectangle: circulation, utility, living, dining, kitchen, bathroom, bedroom.

Let us consider a simple case of the problem first, using a version of Topdown that supports upward propagation of evaluations. Assume that there are no pinwheels in the plan (i.e., horizontal and vertical divisions of rectangles only) and that each room in the plan has a minimum allowable length, a minimum allowable width, a maximum allowable length, and a maximum allowable width. The hierarchical structure will be a binary tree, as shown in Figure 15.11. When the space has been subdivided and an attempt is made to substitute a room into a rectangle that is too small, a constraint violation

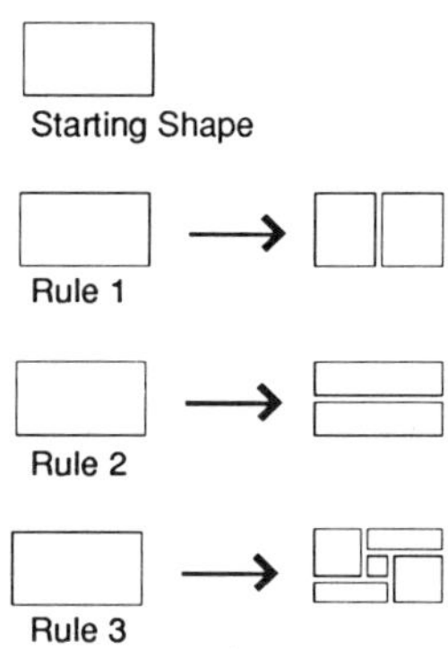

Figure 15.10. Rules for subdividing a rectangular plan.

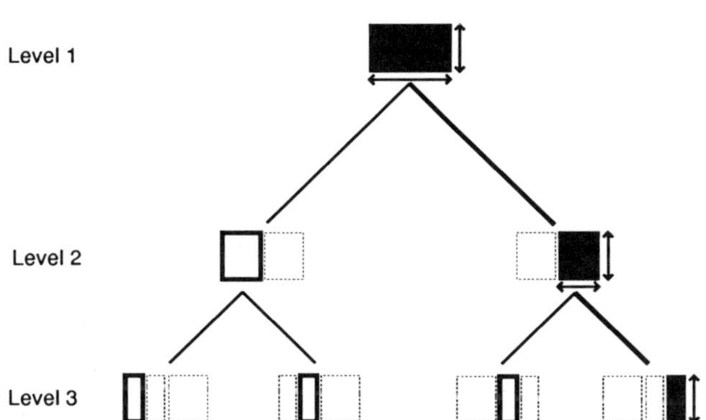

Figure 15.11. Binary tree structure of a simple plan, showing how dimensioning variables at each level can be adjusted as the user backs up the abstraction hierarchy.

message appears. The user can attempt to satisfy the violated length or width constraint by shifting a horizontal or vertical wall. If this can be done without violating constraints on other rooms, there is no further difficulty. If it cannot be done, the user can go up to the next level of abstraction and redimension the rectangle containing the affected rooms: Topdown calculates the minimum and maximum dimensions required for this larger rectangle. If sufficient space cannot be created at this level without causing constraint violations in some adjacent rectangle, the user can go to a still higher level, and so on to the top of the hierarchy. In this process Topdown can always calculate the minimum and maximum dimensions required for the rectangle currently under consideration, it is always obvious how to satisfy these requirements, and the user can quickly and easily produce a complete plan that satisfies all the room length and width requirements and has minimal overall length and width.

Selection of a pinwheel plan (Figure 15.12) creates a slightly more complex problem. There is no simple hierarchy of rectangular spaces within rectangular spaces, and constraints on rectangle sizes are cyclically related. For example, changing dimension a, b,

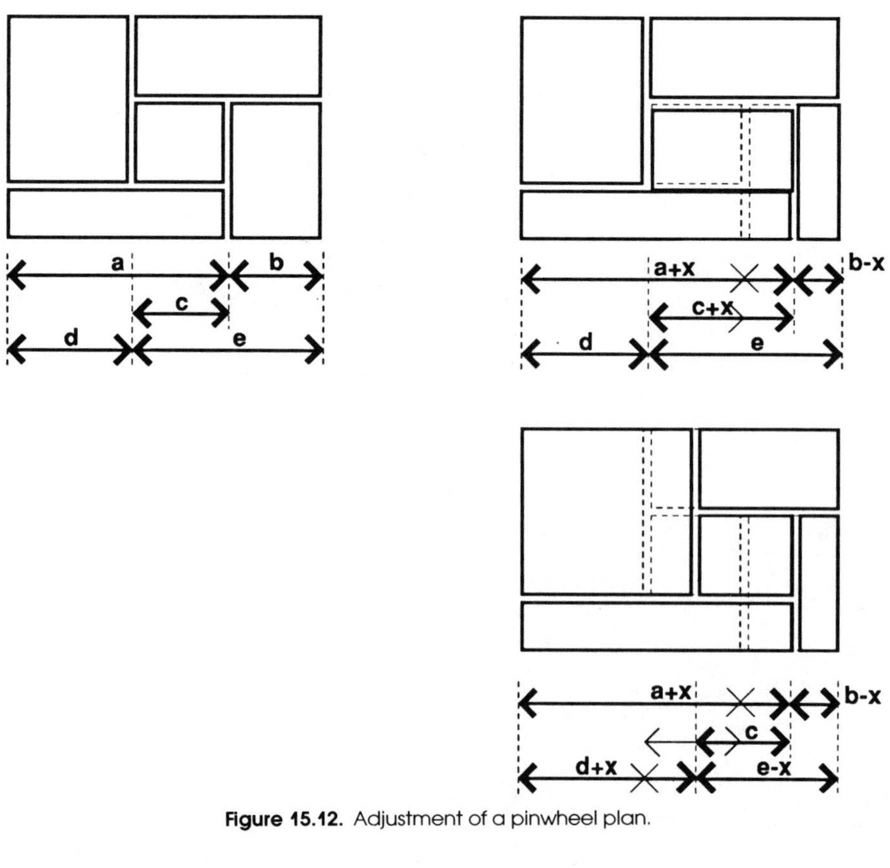

Figure 15.12. Adjustment of a pinwheel plan.

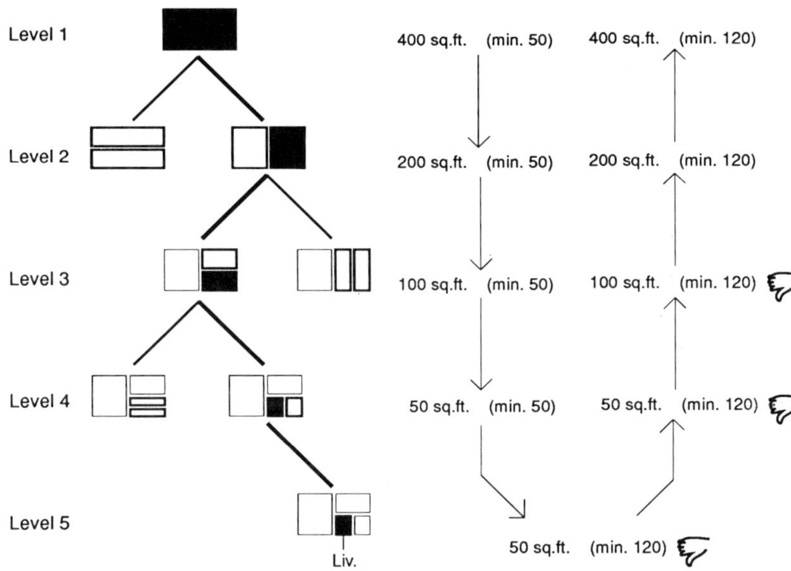

Figure 15.13. Subdividing and adjusting a plan with area constraints.

d, or e will change dimension c, but c can also be kept constant by making equal changes to b and d. So when the user backs up the abstraction hierarchy to create more space, it is not so obvious which variable should be selected and changed to accomplish the desired result, and a constraint-solving algorithm could usefully be employed to maintain feasibility by adjusting other variables whenever a variable is manipulated. However, this is by no means essential: good solutions can quickly be found by trial and error.

Inclusion of minimum and maximum floor area constraints on rooms, in addition to length and width constraints, also makes a problem more complex. Figure 15.13 illustrates the process of first working top-down to subdivide the space, then assigning rooms to rectangles to establish low-level constraints, then backing up the abstraction hierarchy to find a way to satisfy these low-level area constraints. In this process, Top-down can calculate upper and lower bounds on the total area that should be provided at each level, but these may or may not turn out to be actual minimal and maximal areas: it will depend on the actual configuration and dimensioning at the lower levels. Trial-and-error adjustment still yields good solutions to small problems fairly quickly, but for large problems it would be useful to apply constraint-solving or optimization algorithms capable of handling nonlinear constraints.

LIMITATIONS OF STRICT HIERARCHICAL STRUCTURES

The advantages of hierarchical subdivision of design problems for interactive solution should, by now, be obvious. There are, however, classes of problems that do not yield to this approach. In a properly constructed hierarchy, for example, there is no "horizontal" propagation of constraints between subsystems at the same level of abstraction. If the design of a subsystem affects the design of another subsystem at the same level, their relationship can be resolved (if it is resolvable) by backing up to the next higher level and working it out there. But the situation shown in Figure 15.14 can arise: the design of a low-level subsystem interacts with the design of another low-level subsystem such that the relationship spans a large portion of the abstraction hierarchy. In this case it becomes difficult to resolve the relationship at a higher level. Worse still, there may be cyclic relationships (as with the pinwheel plan) over large portions of the hierarchy. The Topdown approach works effectively only when horizontal and cyclic relationships do not span large portions of the hierarchy. However, a remarkably wide range of practical design problems can be structured such that this condition holds.

CONCLUSION

The Topdown system facilitates quick, confident design of standard types of artifacts by applying rules about what these artifacts are made of and how they are put together. These rules specify both overall organizational alternatives and methods of low-level detailing, and they are arranged to support the design strategy of working top-down through a hierarchy of subproblems. The subproblems are separated from each other and reduced to manageable size by careful deployment of abstraction barriers—a technique that has proved highly successful in software development. In earlier versions of Topdown the rules were purely syntactic: it was entirely up to the user to interpret and evaluate the arrangements of shapes that were produced. The version described here goes a step further and provides for automatic functional interpretation and evaluation of designs. This allows a designer to gain a detailed, structured understanding of the interrelationships between form and function through rapid, highly interactive exploration within a richly informative environment.

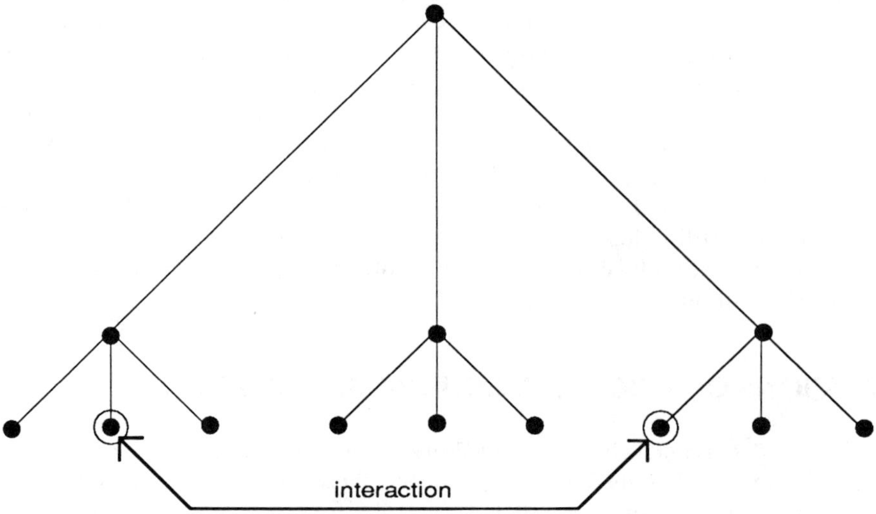

Figure 15.14. Constraint that spans a large portion of the abstraction hierarchy.

REFERENCES

Alexander, C. (1967). *Notes on the Synthesis of Form*, Harvard University Press, Cambridge, Mass.

Dijkstra, E. W. (1989). "On the Cruelty of Really Teaching Computing Science," *Communications of the ACM*, **32**(12): 1398–1404.

Freeman-Benson, B. N., J. Maloney, and A. Borning (1990). "An Incremental Constraint Solver," *Communications of the ACM*, **33**(1): 54–63.

Mitchell, W. J. (1977). *Computer-Aided Architectural Design*, Van Nostrand Reinhold, New York.

Mitchell, W. J. (1990). *The Logic of Architecture*, MIT Press, Cambridge, Mass.

Mitchell, W. J., P. Steadman, and R. S. Liggett (1976). "Synthesis and Optimization of Small Rectangular Floor Plans," *Environment and Planning B*, **3**(1): 37–70.

Mitchell, W. J., R. S. Liggett, and T. Kvan (1987). *The Art of Computer Graphics Programming*, Van Nostrand Reinhold, New York.

Mitchell, W. J., R. S. Liggett, and M. Tan (1990). "Top-Down Knowledge-Based Design," in *The Electronic Design Studio*, M. McCullough, W. J. Mitchell, and P. Purcell (eds.), MIT Press, Cambridge, Mass.

Simon, H. A. (1981). *The Sciences of the Artificial*, 2nd ed. MIT Press, Cambridge, Mass.

Stiny, G. (1980). "Introduction to Shape and Shape Grammars," *Environment and Planning B*, **7**: 343–351.

Sussman, G. J., and G. L. Steele, Jr. (1980). "Constraints: A Language for Expressing Almost-Hierarchical Descriptions," *Artificial Intelligence*, **14**: 1–39.

PART FIVE

ENERGY EVALUATION

16

HUMAN THERMAL RESPONSE DESIGN MODEL FOR NONORTHOGONAL SPACES

Loukas N. Kalisperis
Mitch Steinman

Department of Architecture
The Pennsylvania State University
University Park, Pennsylvania

Luis H. Summers

Department of Civil, Environmental and Architectural Engineering
University of Colorado at Boulder
Boulder, Colorado

Mechanization and industrialization of society have resulted in most people's spending the greater part of their lives in enclosed environments. Optimal design of indoor artificial climates is therefore of increasing importance. Wherever artificial climates are created for human occupation, the aim is that the environment be designed so that individuals experience thermal comfort. Current design methods for radiant panel heating systems do not adequately account for the complexities of human thermal comfort, because the systems monitor air temperature alone and do not account for thermal neutrality in complex enclosures. Thermal comfort for a person is defined as that condition of mind which expresses satisfaction with the thermal environment. Thermal comfort is dependent on mean radiant temperature and operative temperature, among other factors. In creating artificial climates for human occupancy, the designer must take into account the interaction of the person with the heated surfaces, as well as the surface-to-surface heat exchange. Early work in the area provided an elaborate and difficult method for calculating radiant heat exchange for simplistic and orthogonal enclosures. In this paper we present a new, improved method developed for designing radiant panel heating systems based on human thermal comfort and mean radiant temperature.

Through automation and elaboration, this method overcomes the limitations of the early work. The design procedure accounts for human thermal comfort in nonorthogonal as well as orthogonal spaces, based on mean radiant temperature prediction. The limitation of simplistic orthogonal geometries has been overcome with the introduction of the MRT-correction method and inclined surface-to-person shape factor methodology. The new design method increases the accuracy of calculation and prediction of human thermal comfort and will allow designers to simulate complex enclosures, utilizing the latest design knowledge of radiant heat exchange to increase human thermal comfort.

INTRODUCTION

The growing mechanization and industrialization of society have resulted in most people's spending by far the greater part of their lives, often more than 95 percent, in an artificial climate. This fact has generated an increased interest in designing environmental conditions within which people habitate. Indoor climate is defined as the collective whole of all the physical properties in an enclosure that influence a person via heat loss and respiration. Wherever artificial climates are created for human occupation, the aim is that the environment be designed so that each person is in thermal comfort: that condition of mind which expresses satisfaction with the thermal environment. Due to biological variance, it will not be possible to satisfy everyone within the same environment. Thus, creating optimal thermal comfort for the group, the condition in which the highest possible percentage of the group is in thermal comfort, is the designer's goal. Thermal comfort is influenced greatly by the air temperature of the enclosure, mean radiant temperature, activity level, thermal resistance of clothing, relative air velocity, and water vapor content. The mean radiant temperature in relation to a person in a given body posture and clothing placed in a given room is defined as that uniform temperature of black surroundings which will give the same radiant heat exchange from the person as the actual case under study. Mean radiant temperature is directly correlated to the geometrical configuration of the enclosure and the location of the person within it.

Perhaps the most important reasons that the environment in many buildings is not satisfactory are, first, the various problems that arise are not well understood, and second, there is a serious lack of knowledge concerning the design methods that can be employed to utilize the buildings themselves to improve internal conditions. Poor methods often result from inadequate knowledge of factors that affect human thermal comfort, other than air temperature and humidity. The traditional approach used by mechanical systems that attempted to control the environment was to utilize convective air systems which monitored air temperature alone. Recently, there has been renewed interest in radiant technologies and energy conservation through the use of radiant panels and passive solar designs. With the development of new technologies, new techniques are needed to ensure individual thermal comfort. Methods to assess mean radiant temperature and operative temperature must be designed for the new technology. Operative temperature is defined as the uniform temperature of an imaginary enclosure in which the radiant and convective heat transfer from the human body equals the radiant and convective heat transfer in the actual nonuniform enclosure.

Currently, designers are using design methods for determining heating loads based on inappropriate convective air specifications. Designing radiant panel heating systems based on the design parameters of conventional convective systems results in overestimated loads and equipment needs. Problems with design methods currently being used for radiant systems include using inappropriate convective specifications for determining heating loads and failing to account for thermal neutrality. Radiant panel heating systems traditionally have been sized based on the heat loss of the enclosure at a prescribed air temperature. Usually, the prescribed air temperature is the same as that which would be used in the design of a conventional convective system [ASHRAE 1987]. This procedure virtually ignores the energy-conserving benefits of radiant systems. Radiant heating allows the space-air temperature, the average temperature of the room air, to be set a few degrees lower than for convective heating while still maintaining acceptable levels of thermal comfort. Designing a radiant panel system in a conventional manner can lead to the overestimation of loads and equipment needs, as well as possible overheating of spaces. On the other hand, choosing a space-air temperature that is too low can lead to undersized equipment, resulting in the occupant's discomfort. This problem was analyzed by Zmeureanu et al. [1988], using a detailed transient model. They compared a conventionally designed radiant system with one designed using thermal-comfort criteria. The results show that the conventional system overestimated peak loads and daily loads by approximately 72 percent.

Traditional methods do not provide spatial uniformity of human thermal comfort because space-air temperature is maintained at the thermostat location alone, which is a single point within the environment. This method does not account for radiant temperature asymmetry due to diverse surface temperatures as a result of the environmental complexity of nonorthogonal space. Recently, how to predict human thermal comfort has been an area of extensive research activity [Kalisperis et al. 1988, Olesen 1985, Fanger 1982, Olesen and Mortensen 1980]. Essential in predicting human thermal comfort is the concept of thermal neutrality. In addition, for comfort to be achieved, local body thermal discomfort must not be caused by local convective cooling (draft), by vertical air temperature gradients, or by asymmetric thermal radiation. Design standards using thermal comfort were set forth by ASHRAE 55-81, NKB [1981] guidelines (Nordic Committee on Building Regulations), and ISO 7730 (International Organization for Standardization). Research [Fanger 1982] has shown that mean radiant temperature is an important factor in determining thermal comfort.

Despite recent advances, there is still great resistance in the industry to the use of thermal comfort and mean radiant temperature prediction as a basis for designing radiant panel heating systems. Current methods of determining thermal comfort in the design stage, such as those suggested by Maloney [1985] and used in the computer program BLAST and the thermal comfort equation method [Fanger 1982], are complex and time consuming, while traditional convection heating design methods are simple and easy to use. Thus inaccurate traditional methods continue to be the standard for indoor climate design in complex environments. The need is imperative for a design method, based on new knowledge, that can be utilized in the design process.

P. O. Fanger [1982] was the first to establish the thermal comfort equation based on the body's energy balance. Evaluating and correlating experimental data gathered at the Technical University of Denmark, Fanger also developed a manual method for generating the shape factor between a person and a surface necessary for calculating mean radiant temperature.

One method for evaluating the mean radiant temperature for a person placed in an enclosure consisting of N surfaces, each of which are isothermal and have temperatures

$T_1, T_2, \ldots, T_n$, and which are considered blackbodies (reflections negligible), and also of shape factors $F_{p-1}, F_{p-2}, \ldots, F_{p-n}$. This method is given in the form

$$T^4_{mrt} = T^4_1 F_{p-1} + T^4_2 F_{p-2} + \cdots + T^4_n F_{p-n} \tag{16.1}$$

Calculations based on Fanger's work are complex and time intensive, thus preventing its wide utilization as a design methodology. The method is only applicable to rectilinear enclosures. But complex, nonorthogonal spaces constitute part of the designed and built physical environment. Additionally, the method for calculating the necessary room geometry is so cumbersome and time consuming that its use is prohibitive for the designer constrained by considerations of invested time versus improvement of design. Current computer methodologies are also inadequate for optimum design of radiant panel heating systems.

A new design procedure based on recent findings is proposed in this paper. At the present time no computer programs are available that use thermal comfort and mean radiant temperature as design variables. Such programs would eliminate the complexity of calculation for the designer and increase the speed of design. The current programs (BLAST, NBSLD, DOE-II) primarily determine energy consumption of various radiant panel heating systems, using space-air temperature as a means of control. These programs have limited design application since they make no provision for the initial estimation of panel design, and the design process is iterative and requires generation of alternatives to optimize panel design. Each alternative panel design requires an energy simulation, a problem that designers have found prohibitive. To solve these problems and allow the implementation of this important knowledge of factors affecting human thermal comfort, designers need new methodologies that are applicable to more complex geometries, incorporating parameters such as shape factors, radiation exchange, and new design methods.

SHAPE FACTOR CALCULATION

Calculation of the mean radiant temperature requires determining the shape factors between the person and the different surrounding surfaces. In general, a shape factor for radiant heat exchange between a person and a room surface is defined as the fraction of diffusely radiated energy leaving the human surface that is incident on the surface in question. Raber and Hutchinson [1947] developed shape factors through extensive experimentation. Fanger [1982] developed geometrical radiation data for the human body through extensive experimentation by photographing human subjects from many directions in an effort to develop more accurate shape factors. These shape factors have been correlated in nondimensional graphs for standing and sitting persons in relation to vertical and horizontal surfaces [Fanger 1982]. We have extended the analytical procedure to determine shape factors from a person to an inclined surface. Such a method will allow for the resolution of thermal comfort in nonorthogonal spaces. The general derivation is based on positioning a person in the orthogonal coordinate system (X, Y, Z), as shown in Figure 16.1.

The standing person (P) faces an inclined rectangle above and to the left of his or her center, at an angle Φ. When considering a finite surface area of the person to a differential element (dA) on the tilted rectangle, the reciprocity equation is [Howell 1989]

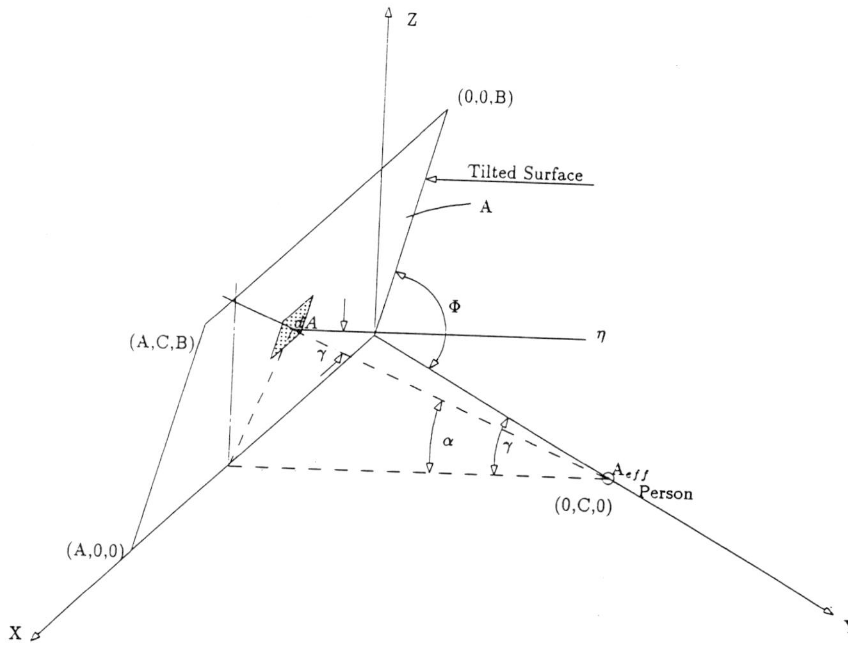

Figure 16.1. Development of the shape factor between a person and a rectangle at a tilted angle φ.

$$A_{\text{eff}} dF_{p-dA} = dA\, F_{dA-p}$$

Integrating the reciprocity equation, substituting, converting to polar coordinates, and nondimensionalizing yield

$$F_{P-A} = \frac{1}{\pi} \int_0^{a/c} \int_0^{b/c} \frac{F_p(\alpha,\beta)}{[(\frac{x}{c})^2 + (\frac{z}{c})^2 + (1 - \frac{z}{c\tan\Phi})^2]^{3/2}}\, d\left(\frac{x}{c}\right)\, d\left(\frac{z}{c}\right) \tag{16.2}$$

where $F_P(\alpha,\beta) = \dfrac{A_p}{A_{\text{eff}}}$ = mean projected area factor at azimuth (α) and altitude (β)

F_{p-A} = shape factor from person to area A

dF_{p-dA} = differential shape factor from a person to a differential element dA

F_{dA-P} = shape factor from a differential element dA to a person

A_p = projected area of person, determined experimentally

dA = differential element on the surface in question

The effective radiation area of a person's entire body (A_{eff}) is the actual surface area of the person that is available to radiate to other surfaces. This effective radiation area is typically less than the Dubois area [Fanger 1982]. Equation (16.2) represents the final form of the equation that was used in this study to generate shape factors from a standing person to an inclined surface in front of and above his or her center, and to the rear and below his or her center (case 1).

Mean projected area factors (F_p) have been determined experimentally by Fanger [1982] for standing persons (Figure 16.2). As explained by Fanger [1982], the projected area (A_p) was determined by photographing a person with a spherical path about him or her and then scaling the photographs to determine the projected area at each particular point. For each individual subject, the projected area factors were calculated as a function of (α,β) for four posture/clothing combinations. Mean values for males and females, nude and clothed, were calculated. When the location of a standing person is known, but his or her orientation to a surface is not, the mean shape factor ($\bar{F}_{P-A}$) may be used by taking the mean of Equation (16.2), which yields

$$\bar{F}_{P-A} = \frac{1}{2\pi^2} \int_0^{a/c} \int_0^{b/c} \int_0^{2\pi} \frac{F_p(\alpha,\beta)}{\left[\left(\frac{x}{c}\right)^2 + \left(\frac{z}{c}\right)^2 + \left(1 - \frac{z}{c\tan\Phi}\right)^2\right]^{3/2}} \, d\left(\frac{x}{c}\right) \, d\left(\frac{z}{c}\right) \, d\alpha \qquad (16.3)$$

Q_{ij} = the interpolated point at α and β

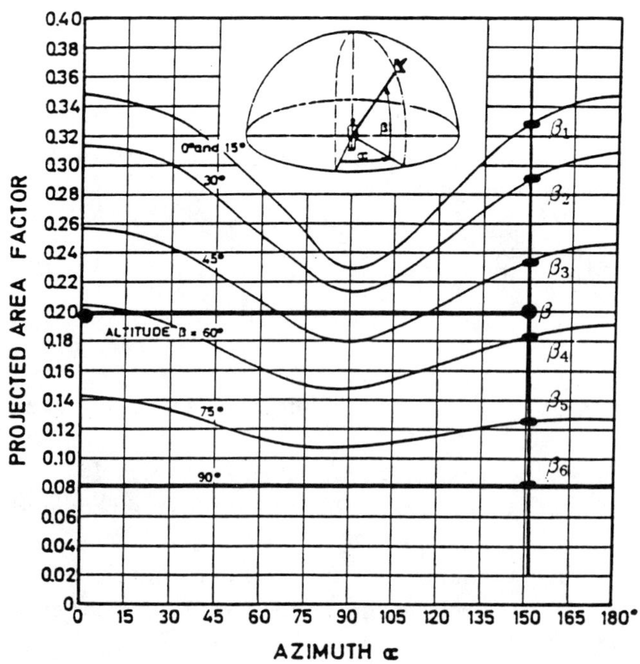

Figure 16.2. Projected area factor curve for a standing person, nude and clothed, using the natural spline method. (Adapted from F. O. Fanger, *Thermal Comfort*, Robert E. Krieger, Malabar, Fla., 1982.)

Figure 16.3 shows an example where a person is surrounded by tilted walls at 45° that are in front of and behind him or her, and the shape factor curves generated for this situation. For this study, Equations (16.2) and (16.3) were solved numerically with the aid of a digital computer. Shape factors were generated for a standing person in relation to an inclined surface for each of the cases above and the mean shape factor cases. Using the natural cubic spline method, equations were developed for the projected mean shape factor curves for each altitude angle (β) for standing persons (see Figure 16.3). Each altitude angle (β) curve was divided into equal segments along the azimuth angle (α) from 0° to 180° (see Figure 16.2). Each interval of the altitude angle (β) curve was represented by a third-degree polynomial of the form

$$f_{pj}(\alpha)=a_j+b_j(\alpha-\alpha_j)+c_j(\alpha-\alpha_j)^2+d_j(\alpha-\alpha_j)^3 \qquad j=1, m \quad \textit{intervals}$$

where a_j = azimuth angle at jth interval
a_j,b_j,c_j,d_j = constants for the polynomial equations at j interval
$f_p(\alpha)$ = projected area factor at azimuth (α) in the jth interval

The equation above can be solved as a set of linear equations and the matrix equation $AX = B$ as defined by Burden and Faires [1981]. The accuracy of this method can be improved by increasing the number of j intervals. Interpolation between the jth points at the altitude (β) curves was employed using Neville's iterated interpolation technique [Burden and Faires 1981]. An array (Q_{ij}) is constructed such that

$$Q_{ij} = \frac{(\beta-\beta_{i-j})Q_{i,j-i}-(\beta-\beta_i)Q_{i-j,j-i}}{\beta-\beta_{i-j}} \tag{16.4}$$

where β = altitude angle to be interpolated at a (α)
β_{i-j},β_i = altitude angle that is known to correspond to an azimuth angle (α) and projected area factor ($f_{pj(\alpha)}$)
Q_{ij} = interpolated projected area factor at α and β

The value of the interpolated point Q_{ij} depends on the degree of the polynomial selected in the array. The accuracy of this interpolative technique increases directly with the number of known points (β_{i-j}) and polynomial degree. Equations (16.2) and (16.3) were solved as a double and triple integral, respectively, using the composite Simpson rule [Burden and Faires 1981]. The area of the tilted surface is broken into $n \times m$ Simpson intervals, whereby the function was evaluated at each of these intervals. The projected area factor ($f_{pj(\alpha)}$) was interpolated at each interval, using the methods described previously. The integral was evaluated for various tilt angles (Φ) and at specified limits a/c and b/c.

The results generated are similar to the graphs developed by Fanger and can be used in the same way. Although this technique eliminates the limitations associated with simplistic room geometry, the problem of elaborate hand calculation remains. For the method to be widely utilized by designers, it must be easy to use as well as applicable to all situations. This method solves only part of the problem of environmental complexity in nonorthogonal space. To predict thermal comfort in an enclosure accurately, the radiant heat exchange for surface to surface must also be resolved within the nonorthogonal space. The MRT correction method significantly improves existing methods for radiant heat exchange in nonrectilinear enclosures.

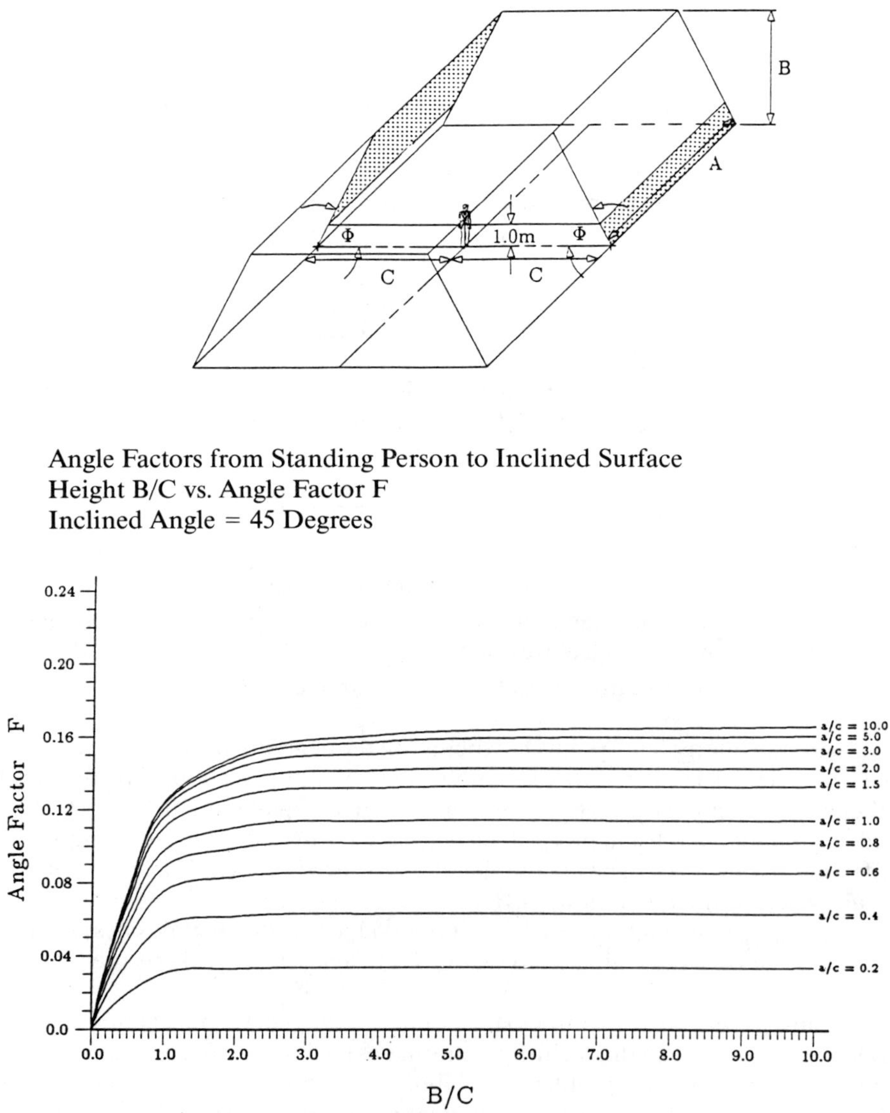

Figure 16.3. Shape factors of person to inclined surfaces in front of and above a person's center and in back of and below his or her center for 45° angle.

RADIANT EXCHANGE METHODS

Current Methods

Building load models can be evaluated using a technique known as the *heat-balance method.* Numerous energy-load calculation computer programs available today make use of this method. The general approach is to write an energy boundary equation for each enclosing surface, plus one for the room air mass. These sets of equations can then be solved for the unknown surface and air temperatures. The general surface boundary equations yield

$$q_{\text{cond}_i} + q_{\text{conv}_i} + q_{\text{rad}_i} = 0 \qquad i = 1,\ N \tag{16.5}$$

where
q_{cond_i} = total energy flux conducted in surface by *i*, *Btu-h/ft*^{2}
q_{conv_i} = total energy flux convected to and from surface *i* by room air, *Btu-h/ft*^{2}
q_{rad_i} = total net radiation flux to and from surface *i*, *Btu-h/ft*^{2}
N = total number of surfaces in enclosure

In this paper we deal mainly with the last term of Equation (16.5), namely, the net radiation flux term Q_{rad_i}. The gray, diffuse radiosity model [Sparrow 1978] represents a balanced exchange of energy within a multisurfaced enclosure in which there is not net radiation increase, such that

$$q_{\text{rad}_{i,\text{exact}}} = B_i - \sum_{i=1}^{N} B_j F_{ij} \qquad i = 1,\ N \tag{16.6}$$

where:
$q_{\text{rad}_{i,\text{exact}}}$ = net radiation leaving surface *i* in a gray diffuse enclosure
B_i, B_j = radiosity of surface *i* or *j*, which is the amount of reflected and emitted energy leaving surface *i* or *j*
F_{ij} = shape factor between surface *i* and *j*

We refer to this term as the "exact" radiosity model, similar to that proposed by Walton [1980]; this usage is in contrast to the MRT method, which is an approximation of the "exact" radiosity model. The MRT method for calculating radiant temperature of surfaces in a surface-to-surface radiant heat exchange should not be confused with the mean radiant temperature of a person, as explained above. Solving Equation (16.6) for each surface in terms of the exact radiation is very time consuming. Various loads programs (BLAST, NBSLD) have made approximations so that the sets of equations may be solved as a matrix solution. The NBSLD algorithm creates a radiation imbalance and a violation of the first law of thermodynamics. The MRT method attempts to correct the radiation imbalance by distributing radiation proportionally within the room, however, it does so incorrectly. The degree of error for both methods depends on room geometry and surface temperature variations [Kalisperis et al. 1989a].

George Walton [1980] proposed an algorithm for improving the accuracy of the Gauss–Seidel iteration by using the mean radiant temperature (MRT) method developed by Davies [1978]. He accomplished this by proposing that "the radiant interchange in a room can be adequately modeled by assuming that each surface radiates to a fictitious surface which has an area, emissivity, and temperature giving about the same heat transfer from the surface as in the real multi-surface case. The radiant heat exchange between two parallel gray surfaces, one real surface i, and the other a fictitious surface f yields the general form of the 'MRT equation' " [Walton 1980, Hollman 1976, Maloney 1985, Maloney et al. 1989]:

$$q_{\text{rad}_i,\text{mrt}} = \sigma \mathbf{F}_{if}(T_i^4 - T_{fi}^4) + q_{\text{int}_i} + q_{\text{sol}_i} \qquad i = 1, N \tag{16.7}$$

An approximation is made that

$$(T_i^4 - T_{fi}^4) = 4T^3_{\text{avg}_i}(T_i - T_f) \tag{16.8}$$

where substituting Equation (16.7) and adding a radiation balance term q_{bal_i}, which accounts for the net imbalance of the method, the "MRT/balance" equation becomes

$$q_{\text{rad}_i,\text{mrt/bal}} = 4\sigma \mathbf{F}_{if} T^3_{\text{avg}_i}(T_i - T_{fi}) + q_{\text{int}_i} + q_{\text{sol}_i} + q_{\text{bal}_i} \tag{16.9}$$

This is the final form of the MRT method's radiation flux term and can be substituted directly into the general surface boundary equation (16.5). This is an order 1 equation, since the T_{avg_i} term is considered a constant at each iteration, vastly improving computational speed over the exact method's order 4 equation. At the same time, and most important, there can be significant errors in the MRT equation (16.7) as an approximation to the "exact" radiosity model or (16.9), since the $\mathbf{F}_{if}$ term does not involve exact shape factors.

Although the accuracy and speed of the solution was improved significantly by the MRT method, it has limitations: it is accurate only for simple geometric enclosures, and may not be so for varied and diverse surface temperatures or complex geometric enclosures. There are assumptions embedded in this method—that each surface views all other surfaces in the enclosure in the radiant heat exchange model, causing the methodology to generally predict lower temperatures for hot surfaces, and higher temperatures for cold surfaces in the enclosure. These assumptions may limit the method's applicability to rectilinear enclosures with nondiverse length/width ratios and small surface temperatures differences [Walton 1980]. The MRT method allows surfaces to be viewed that, in the actual physical environment, do not view one another, such as in complex enclosures (L-shaped rooms, alcoves, and central core, for example). Further analysis of the error in surface-to-surface radiant heat exchange calculations can be found in Kalisperis et al. [1989b].

The MRT-Correction Method

The authors have developed an improved method of radiant surface-to-surface heat exchange that produces significantly improved accuracies over existing approximate methods. The MRT-correction method incorporates the "mean radiant temperature" method [Walton 1980]. It is additionally proposed that with error analysis and numerical methods correction, components can be applied to the MRT algorithm. The correction components account for an enclosed environment's surface-to-surface exact shape factors. The MRT-correction method approaches the accuracy of the "exact" radiosity model by incorporating surface-to-surface shape factors while maintaining computational speed. The correction components for the two most common situations are determined. The first involves an enclosed environment in which all surface emittances are equal and range from 0.0 to 1.0 for all surfaces, while the second involves enclosure cases where all surface emittances are equal, except one surface for which the emittances range from 0.0 to 1.0.

Surface Emittances Equal from 0.0 to 1.0. The correction component for enclosed environments in which inside surface emittances (ε_i) vary from 0.0 to 1.0 was determined. Typical case studies were performed in which inside surface temperatures were specified, and all effects of internal and solar loads were neglected to facilitate the comparison between methods. Correction curves were generated, indicating the behavior of various case studies, so that a correction component could be developed and applied to the actual loads program analysis.

Figure 16.4 is a representative surface showing a comparison between the exact, MRT, and MRT-correction methods, where the inside surface net radiation (Q_{rad_i}) varies with the inside surface emittance (ε_i)). Also, correction curves (Q_{c_i}) are shown that are the arithmetic difference between the exact and MRT radiation curves. Notice that all surface emittances (ε_i) vary equally in this analysis and do not vary independently of one another. The correction curves all show a maximum error when $\varepsilon_i = 1.0$, which then follows a smooth curve that converges to an error of zero at $\varepsilon_i = 0.0$. Also, Q_{c_i} will always be below or above the abscissa line.

To determine MRT-corrected at any emittance, we add the correction component to the MRT equation at ε_i:

$$Q_{\text{rad}_{i,\text{mrtcor}}}\Big|_{\varepsilon_i} = Q_{\text{rad}_{i,\text{mrt}}}\Big|_{\varepsilon_i} + Q_{c_i}\Big|_{\varepsilon_i} \qquad i=1,\ N \tag{16.10}$$

To determine the correction curve (Q_{c_i}), correction endpoints will be determined at $\varepsilon_i =$ 1.0, while the correction endpoints at $\varepsilon_i = 0.0$ will always be zero. The characteristic of the correction curves between end points have to be determined statistically since it cannot easily be determined during the actual loads program analysis. The correction component Q_{c_i} at the endpoint $\varepsilon_i = 1.0$ is

$$Q_{c_i}\Big|_{\varepsilon_i=1.0} = A_i(q_{\text{rad}_i,\text{exact}} - q_{\text{rad}_i,\text{mrt}}) \qquad i = 1,\ N \tag{16.11}$$

At $\varepsilon_i = 1.0$ (blackbody case) Equation (16.6) (not including internal or solar loads) becomes

$$q_{\text{rad}_{i,\text{exact}}} = \sigma \sum_{j\neq i}^{N} F_{ij}(T_i^4 - T_j^4) \qquad i = 1,\ N \tag{16.12}$$

$$q_{\text{rad}_{i,\text{mrt}}} = \sigma(T_i^4 - T_{fi}^4) \qquad i = 1,\ N \tag{16.13}$$

Since $T_{fi} = \sum_{j\neq i}^{N} A_j T_j / \sum_{j\neq i}^{N} A_j$ *when* $\varepsilon_i = 1.0$. Substituting Equations (16.12) and (16.13) into (16.11) and simplifying, the final form of the endpoint, $\varepsilon_i = 1.0$, is

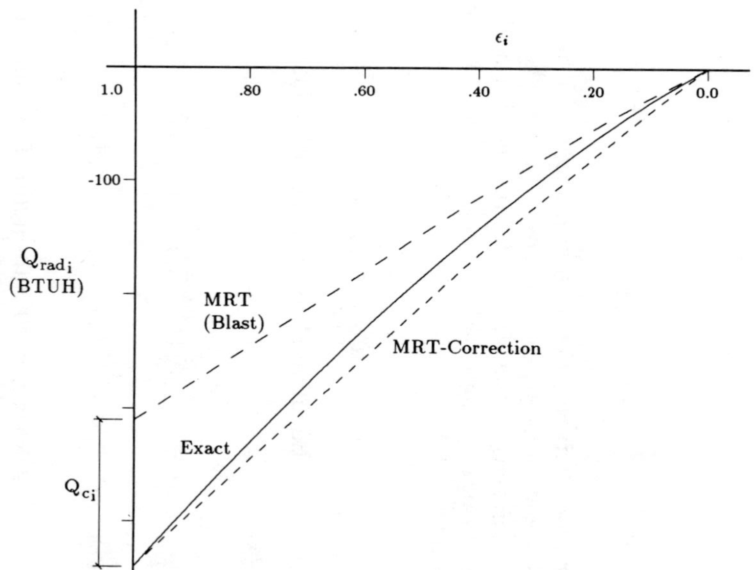

Q_{rad_i} *vs.* ϵ_i *North Wall,* $i = 1$

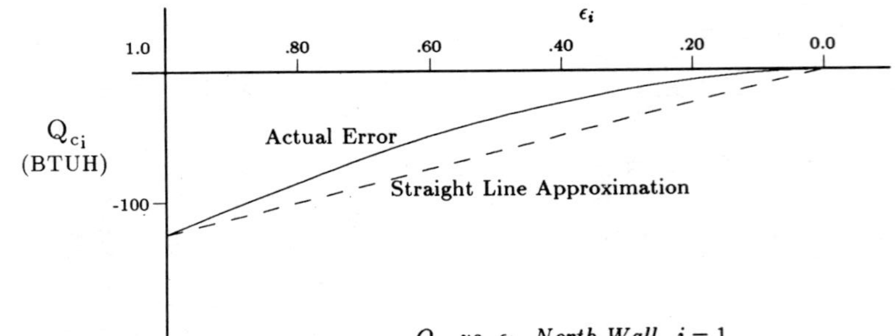

Q_{c_i} *vs.* ϵ_i *North Wall,* $i = 1$

Figure 16.4. Correction analysis: emittances (ε_i) equal 1.0 to 0.0.

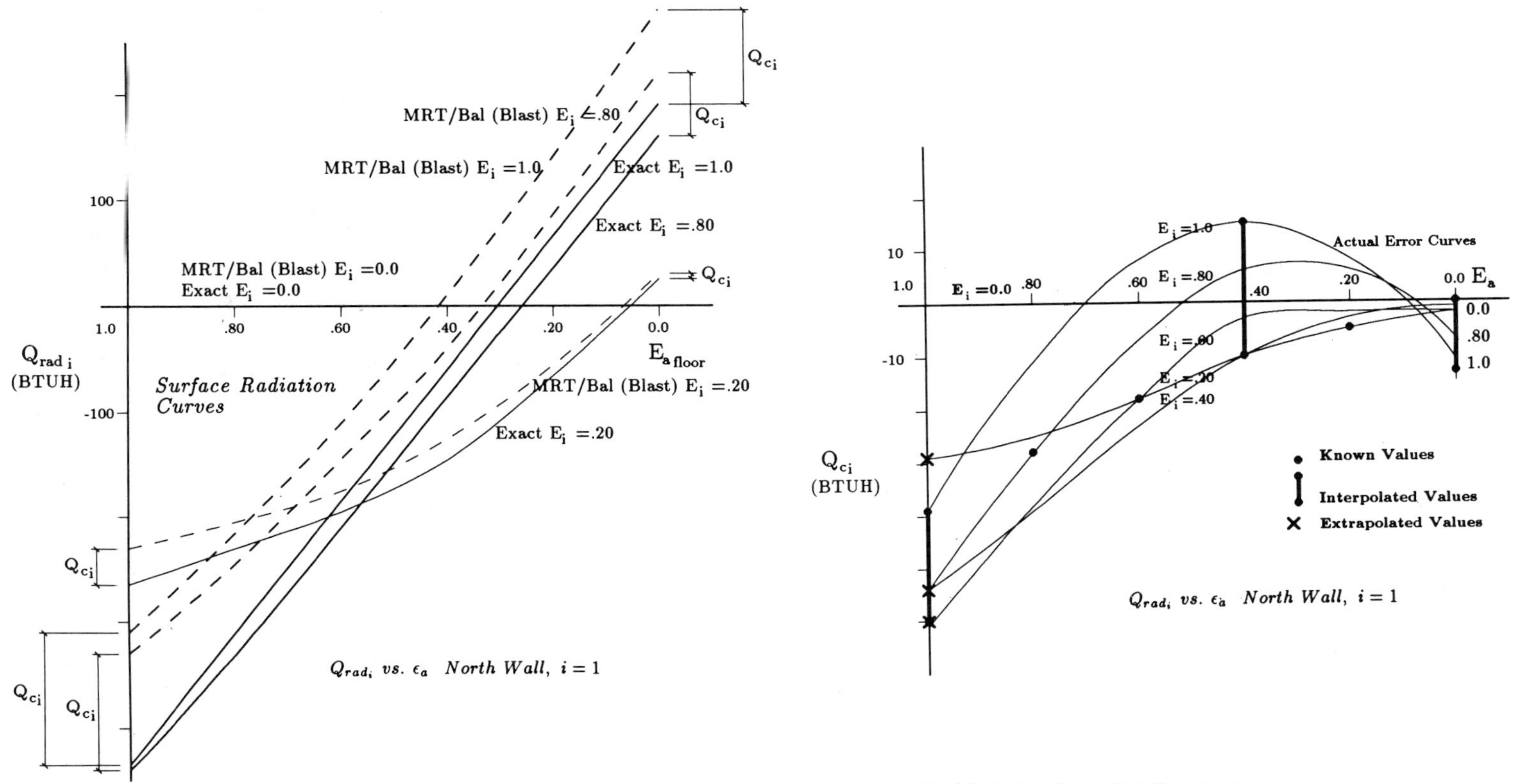

Figure 16.5. Correction analysis and numerical solution to determine characteristic curves.

$$Q_{ci}\Big|_{\varepsilon_i=1.0}=\sigma A_i\Big[\Big(\sum_{j\neq i}^{N} A_j T_j \Big/ \sum_{j\neq i}^{N} A_j\Big)^4-\sum_{j\neq i}^{N} F_{ij}T_j^4\Big] \qquad i=1,\ N \tag{16.14}$$

The characteristic of the correction curves can be determined statistically by finding the most representative characteristic of a group of curves with equal endpoints. For this analysis, which in itself has yielded highly accurate results, we have used a straight-line approximation to describe the curves' characteristic (see Figures 16.4 and 16.5). The Q_{rad} term is the net radiation to or from the surface. Figures 16.4 and 16.5 indicate the difference between Q_{rad_i} for the "exact" radiosity model and the MRT method. The difference between the two methods, Q_{c_i}, is shown plotted against emittance values. Therefore, the correction component at any emittance ε_i becomes

$$Q_{ci}\Big|_{\varepsilon_i}=\varepsilon_i Q_{ci}\Big|_{\varepsilon_i=1.0} \qquad i=1,\ N \qquad \text{straight-line approximation} \tag{16.15}$$

The correction component can now be added to the MRT method to yield the final form of the MRT-correction equation (MRT-correction/balance equation if the balance term is included):

$$q_{\text{rad}_i,\text{mrtcor/bal}}=4\sigma \mathbf{F}_{if}T^3_{\text{avg}_i}(T_i-T_{fi})+q_{\text{int}_i}+q_{\text{sol}_i}+q_{\text{bal}_i}+q_{ci}\Big|_{\varepsilon_i} \qquad i=1,\ N \tag{16.16}$$

Equation (16.16) can now be substituted into the surface-boundary equation. The order 1 solution has been maintained with the addition of exact shape factors F_{ij}, found in the correction component q_c, which are considered constants in the iteration. The solution of that heat-balance equation will then be the same as that used by the MRT method. The boundary equation is then solved for T_i. Surface temperatures T_i are evaluated one at a time, with values A_{fi}, ε_{fi}, $\mathbf{F}_{if}$, and F_{if} constants and values T_j, T_{fi}, T_{avg_i}, q_{bal_i}, *and* q_{ci} updated as each surface temperature is evaluated. A prespecified number of iterations is performed to converge.

Surface Emittances Are Equal, Except One Surface Emittance Varies from 0.0 to 1.0. Figure 16.5 shows correction analysis results attained by comparing exact and MRT methods for the case in which all surface emittances remain equal (ε_i = constant), while one surface emittance varies from 1.0 to 0.0 (ε_a). Also, correction curves (Q_{c_i}) show the arithmetic difference between the exact and MRT methods. Figures 16.4 and 16.5 show that at $\varepsilon_a = 1.0$, $Q_{ci\neq a}$ decreases directly as ε_i varies from 1.0 to 0.0. Determining each endpoint $Q_{ci\neq a}$ at $\varepsilon_a = 1.0$ directly would approach an exact solution method, making it unsuitable for this analysis. However, from the previously shown condition (i.e., all surface emittances are equal from 1.0 to 0.0), one point on each curve can be determined, namely, when $\varepsilon_a = \varepsilon_i$ (bold points). Endpoints at $\varepsilon_a = 1.0$ can then be extrapolated from each of the known points. Also referring to the endpoints at $\varepsilon_a = 0.0$, it is shown that in each case Q_{c_i} varies from a maximum value to zero. The endpoint Q_{c_i} at $\varepsilon_a = 0.0$ and ε_i 1.0 can be determined, while all other values along $\varepsilon_a = 0.0$ can be interpolated. Since the correction curves may lie above or below the abscissa line (+ or −), a midpoint value is required where $\varepsilon_i = 1.0$ and $\varepsilon_a \approx 0.50$. All other correction values where $\varepsilon_i < 1.0$ and $\varepsilon_a \approx 0.50$ can also be interpolated. The correction endpoints $Q_{ci=a}$ can be solved in a similar fashion as shown in type 1, where all surface emittances are equal using the endpoint analysis of Equation (16.16), and using interpolation techniques. Midpoint analysis will not be necessary.

As described above, the correction value q_{c_i}, once found, can be used in the MRT-correction/balance equation (16.14) and substituted into the surface boundary equation (16.5). The order 1 solution has been maintained with the addition of *exact* shape factors. Surface temperatures T_i are evaluated one at a time, with values A_{fi}, ε_{fi}, and F_{if} being constant and values T_j, T_{fi}, T_{avg_i}, q_{bal_i}, and q_{c_i} updated as each surface temperature is evaluated. It has been found that the type 1 situation can be applied to most cases where the surface emittances vary between 0.80 and 1.0.

DESIGN METHOD

Incorporating the surface-to-person and surface-to-surface methods discussed above, we present a new method for designing radiant panel heating systems, using accepted thermal comfort criteria and mean radiant temperature as a basis for decision making. The required design space-air temperature and proper sizing of panels are determined at the coldest point in the room in order to maintain optimum comfort conditions. Uniformity of thermal comfort within an enclosure is displayed graphically using this method. The contour map generated indicates enclosure operative temperatures, mean radiant temperatures, predicted percentage dissatisfied, and radiant temperature asymmetry. The design method is an analytical tool that can be used to optimize radiant panel configuration within a defined complex enclosure. In addition, the procedures used in this method are applicable to conventional convection systems for rooms with high window-to-wall ratios. Hot-water-panel performance algorithms can be integrated into the method and are confirmed by ASHRAE charts. Available product test data for radiant panels can be used in place of panel algorithms when required. The automated procedures are applicable to the complex enclosures of modern habitats.

Inadequacies due to the time constraints of the current methods and the need to incorporate current research into the design process have led to the development of the new design method. As outlined in this paper, this method not only utilizes current knowledge of human thermal comfort and mean radiant temperature, but can also be easily incorporated into the design process. The new method includes automated design procedures which evaluate thermal comfort and mean radiant temperature at an instantaneous period of time at peak design winter conditions. Special cases, such as industrial and process buildings that have large radiative equipment within their spaces, should be evaluated by other methods.

The automated design procedure assures acceptable thermal comfort at all points in the room for *uniformity* of thermal comfort. Uniformity of thermal comfort can be achieved by proper placement of radiant panels, proper placement of glass windows, adjustments of insulations in walls, roof, and floors, and adjustments of glass types.

Graphic representation is incorporated into the program to help the designer achieve uniformity of thermal comfort. The program generates a scaled floor plan illustrating positions of windows, doors, heaters, and face temperatures of primary surfaces. Contour lines are projected over the floor plan showing operative temperatures, PPD (predicted percentage dissatisfied) and MRT (mean radiant temperature).) PPD is the experimental correlation developed by Fanger [1982] to determine percentage of people uncomfortable in any given thermal environment. The designer then has a visual representation of thermal comfort deficiencies, thus making possible generation of additional design alternatives with the given time constraints (Figure 16.6).

This computer program has been developed for automatic generation of mean radiant temperature at any given point in an enclosure. While the program incorporates Fanger's work, it accounts for surface-to-surface radiant heat exchange in addition to person-to-surface exchange. The algorithm of radiant interchange, incorporated in the program, is dependent on shape factor analysis. The analysis assumes diffusely emitting and reflecting surfaces [Moszczienski 1982]. The program uses Fanger's method to establish radiation factors between a person and vertical or horizontal surfaces, as well as the new method for calculating person-to-inclined-surface shape factors, and incorporates the MRT-correction analysis for radiant exchange.

A coordinate system is established with an (X, Y, Z) coordinate frame about the enclosure. Room surface enclosures are termed *primary*, while *secondary* surfaces include any subsurfaces that serve a different function from the primary surfaces, and consequently, exhibit separate radiative behavior. The program input describes the room's geometry, thermal properties of surfaces, primary-to-secondary surface relations, and air temperatures (indoor and outdoor) [Kalisperis 1985].

A two-stage computer program (MRT33GRAPH) based on the automatic generation of shape factors was developed to depict graphically mean radiant temperature distribution conditions within an enclosure. To assist the designer in establishing thermal-comfort conditions in an enclosure, a graphics postprocessor was developed based on scaling and contouring techniques. This program yields a scaled floor plan giving the positions of windows, doors, and radiant heaters, if any, and face temperatures of the primary surfaces after radiant heat-exchange balance. A typical two-dimensional contour map indicating mean radiant temperature distribution is shown in Figure 16.6. Through visual evaluation, mean radiant distribution contours can assist the designer in achieving optimum and uniform thermal comfort conditions, thus increasing comfort and decreasing the operating cost of heating equipment. Additionally, the program calculates the average percent of people dissatisfied (PPD) within an enclosure, as well as the predicted mean vote (PMV) based on Fanger's method for thermal comfort. Predicted mean vote is the index that makes it possible to predict the sensation of thermal comfort for any given combination of the environmental parameters [Fanger 1982]. The results are displayed graphically in Figure 16.7.

CONCLUSION

Original work by P. O. Fanger pointed out essential factors missing previously in designing for human thermal comfort; these factors account for radiant heat exchange and the person in the enclosure. Although it identified the problem, this early work did not provide a convenient method for calculating or utilizing information about radiant heat exchange. Thus, although the problem was known, designers could not readily adopt the new methodology. The work outlined above solves several of the shortcomings of the early work. The limitation of simplistic room geometry has been overcome by the generation of shape factor from person to inclined surfaces and the MRT-correction method.

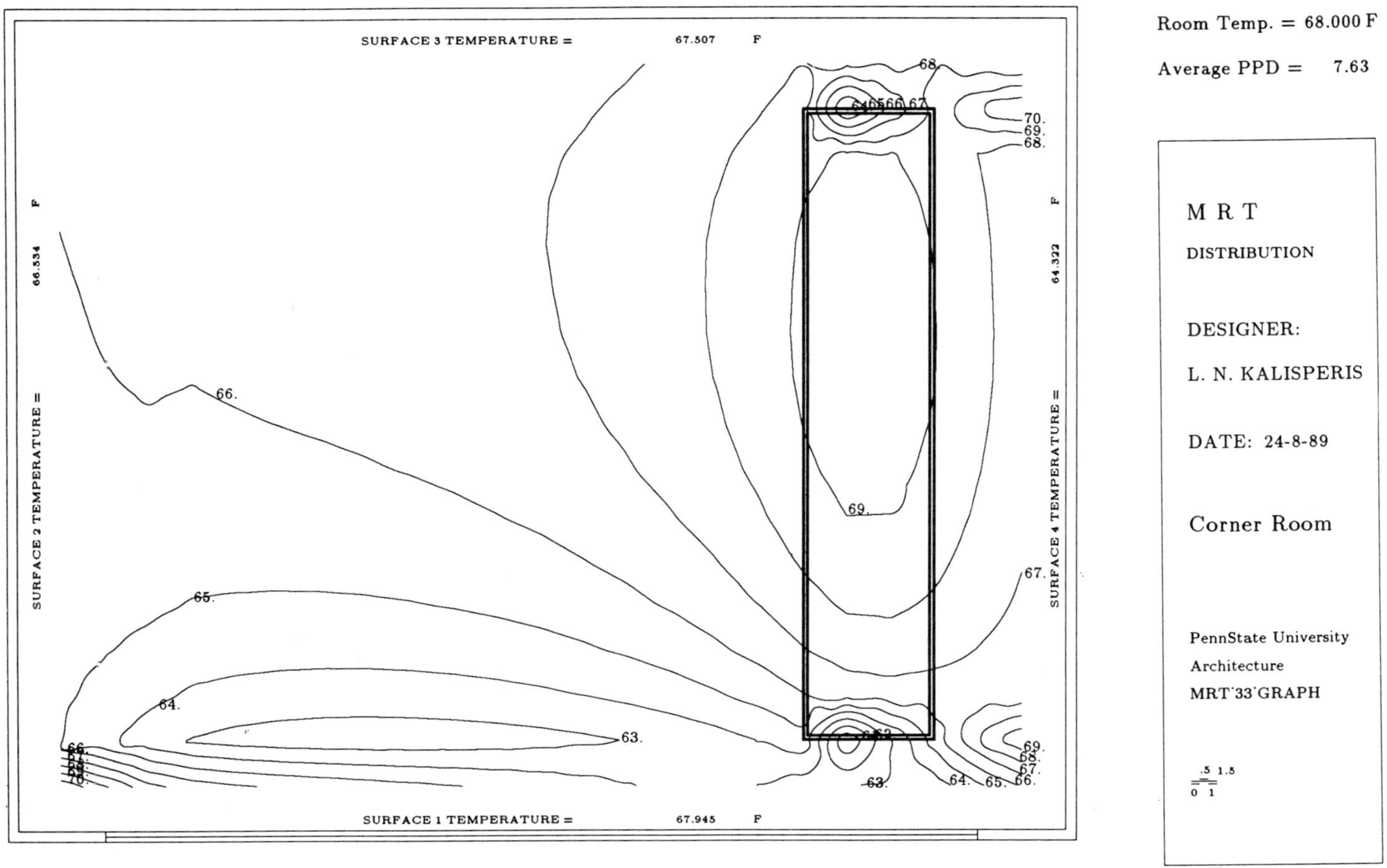

Figure 16.6. Output from MRT33Graph representing mean radiant temperature distribution (contour data).

Room Temp. = 68.000 F

Average PPD = 5.19

SURFACE 3 TEMPERATURE = 64.303 F

SURFACE 2 TEMPERATURE = 64.508 F

SURFACE 4 TEMPERATURE = 69.118 F

0.11	0.10	0.10	0.09	0.09	0.09	0.08	0.08	0.08	0.08	0.07	0.07	0.07	0.07	0.07	0.07	0.06	0.06	0.06	0.05	0.05	0.04	0.02
0.11	0.09	0.09	0.08	0.08	0.07	0.07	0.07	0.06	0.06	0.06	0.06	0.06	0.05	0.05	0.05	0.05	0.04	0.04	0.04	0.03	0.02	0.00
0.10	0.09	0.08	0.07	0.07	0.06	0.06	0.05	0.05	0.05	0.05	0.05	0.04	0.04	0.04	0.04	0.03	0.03	0.03	0.02	0.01	0.00	-0.01
0.10	0.08	0.07	0.07	0.06	0.05	0.05	0.05	0.04	0.04	0.04	0.04	0.03	0.03	0.03	0.03	0.02	0.02	0.02	0.01	0.01	0.00	-0.02
0.09	0.08	0.07	0.06	0.05	0.05	0.04	0.04	0.03	0.03	0.03	0.03	0.02	0.02	0.02	0.02	0.01	0.01	0.01	0.00	0.00	-0.01	-0.03
0.09	0.07	0.06	0.05	0.04	0.04	0.03	0.03	0.02	0.02	0.02	0.02	0.01	0.01	0.01	0.01	0.01	0.00	0.00	0.00	-0.01	-0.02	-0.03
0.08	0.07	0.05	0.04	0.04	0.03	0.02	0.02	0.02	0.01	0.01	0.01	0.00	0.00	0.00	0.00	0.00	-0.01	-0.01	-0.01	-0.02	-0.03	-0.04
0.08	0.06	0.05	0.04	0.03	0.02	0.01	0.01	0.01	0.00	0.00	0.00	-0.01	-0.01	-0.01	-0.01	-0.01	-0.02	-0.02	-0.02	-0.02	-0.03	-0.04
0.08	0.06	0.04	0.03	0.02	0.01	0.00	0.00	-0.01	-0.01	-0.01	-0.01	-0.02	-0.02	-0.02	-0.02	-0.02	-0.02	-0.03	-0.03	-0.03	-0.04	-0.05
0.07	0.05	0.03	0.02	0.01	0.00	-0.01	-0.01	-0.02	-0.02	-0.02	-0.03	-0.03	-0.03	-0.03	-0.03	-0.04	-0.04	-0.04	-0.04	-0.04	-0.04	-0.05
0.06	0.04	0.02	0.01	0.00	-0.01	-0.02	-0.03	-0.03	-0.04	-0.04	-0.04	-0.04	-0.05	-0.05	-0.05	-0.05	-0.05	-0.05	-0.05	-0.05	-0.05	-0.06
0.06	0.03	0.01	-0.01	-0.02	-0.03	-0.04	-0.05	-0.05	-0.06	-0.06	-0.06	-0.06	-0.06	-0.07	-0.07	-0.07	-0.07	-0.07	-0.06	-0.06	-0.06	-0.06
0.05	0.02	-0.01	-0.03	-0.04	-0.05	-0.06	-0.07	-0.07	-0.08	-0.08	-0.08	-0.09	-0.09	-0.09	-0.09	-0.09	-0.09	-0.08	-0.08	-0.08	-0.07	-0.07
0.05	0.01	-0.03	-0.05	-0.07	-0.08	-0.09	-0.10	-0.11	-0.11	-0.11	-0.12	-0.12	-0.12	-0.12	-0.12	-0.12	-0.11	-0.11	-0.10	-0.10	-0.08	-0.07
0.04	-0.01	-0.06	-0.09	-0.11	-0.13	-0.14	-0.14	-0.15	-0.15	-0.16	-0.16	-0.16	-0.16	-0.16	-0.16	-0.16	-0.16	-0.15	-0.14	-0.13	-0.10	-0.08
0.03	-0.04	-0.10	-0.14	-0.17	-0.18	-0.19	-0.20	-0.21	-0.21	-0.21	-0.21	-0.22	-0.22	-0.22	-0.22	-0.21	-0.21	-0.20	-0.19	-0.16	-0.12	-0.08
0.05	-0.07	-0.20	-0.24	-0.27	-0.28	-0.29	-0.29	-0.30	-0.30	-0.30	-0.30	-0.31	-0.31	-0.31	-0.31	-0.31	-0.30	-0.30	-0.28	-0.25	-0.14	-0.04

SURFACE 1 TEMPERATURE = 64.881 F

M R T

DISTRIBUTION

DESIGNER:

L. N. KALISPERIS

DATE: 24-8-89

Corner Room

PennState University

Architecture

MRT'33'GRAPH

.5 1.5

0 1

Figure 16.7. Output from MRT33Graph representing predictive mean distribution.

The method proposed here can be used as a design tool for dealing with environmental complexity in nonorthogonal spaces. The designer can define the complex environment, simulate for thermal comfort, and graphically represent occupant response to the environment, with respect to dissatisfaction level and thermal neutrality, as well as climatic conditions. As it accounts for a wider range of environmental variables, this method should be used instead of traditional methods in the design of complex environmental systems.

REFERENCES

ASHRAE (1987). *ASHRAE Handbook 1987: HVAC Systems and Applications*, American Society of Heating, Refrigerating and Air-Conditioning Engineers, Inc., Atlanta, Ga.

Burden, R. L., and J. D. Faires (1981). *Numerical Analysis*, 3rd ed., Prindle, Weber & Schmidt, Boston, pp. 97, 122, 191.

Davies, M. G. (1978). "On the Basis of Environmental Procedures," *Building and Environment*, **13**(1).

Fanger, P. O. (1982). *Thermal Comfort: Analysis and Applications in Environmental Engineering*, R. E. Krieger, Melbourne, Fla.

Hollman, J. P. (1976). *Heat Transfer*, McGraw-Hill, New York.

Howell, J. R. (1982). *Radiation Configuration Factors*, McGraw-Hill, New York, pp. 2–4.

Kalisperis, L. N. (1985). "Design Patterns for Mean Radiant Temperature Prediction," Department of Architectural Engineering, The Pennsylvania State University, University Park, Pa.

Kalisperis, L. N., M. Steinman, and L. H. Summers (1988). "The MRT Correction Method," unpublished paper, Department of Architecture, The Pennsylvania State University, University Park, Pa.

Kalisperis, L. N., M. Steinman, and L. H. Summers, (1989a). "Angle Factor Determination from a Person to Inclined Surfaces," *ASHRAE Transactions*, **94**, Pt. 1.

Kalisperis, L. N., M. Steinman, and L. H. Summers (1989b)."The MRT-Correction Method: An Improved Method for Radiant Heat Exchange," *ASHRAE Transactions*, **96**, Pt. 1.

Maloney, D. M. (1985). "Development of a Radiant Heater Model and the Incorporation of Thermal Comfort Considerations into the BLAST Energy Analysis Program," M.S. thesis, University of Illinois at Urbana–Champaign.

Maloney, D. M., C. Pedersen, and T. Witte (1989). "Development of a Radiant Heating System Model for BLAST," *ASHRAE Transactions*, **96**, Pt. 1.

Moszczienski, S. (1982). "Computer Simulation of Radiant Heating in Cubical Enclosures," Department of Architectural Engineering, The Pennsylvania State University, University Park, Pa.

NKB (1981). *Indoor Climate*, Technical Report 41, Nordic Committee on Building Regulations, Stockholm.

Olesen, B. W. (1985). "Local Thermal Discomfort," *Bruel & Kjaer Technical Review*, Bruel & Kjaer, Lyndy, Denmark.

Olesen, B. W., and E. Mortensen (1980). "Thermal Comfort in a Room Heated by Different Methods," *ASHRAE Transactions*, **86**(1): 34–48.

Raber, B. F., and F. W. Hutchinson (1947). *Panel Heating and Cooling Analysis*, Wiley, New York.

Sparrow, E. M. (1978). *Radiation Heat Transfer*, McGraw-Hill, New York.

Walton, G. N. (1980). "A New Algorithm for Radiant Interchange in Room Loads Calculations," *ASHRAE Transactions*, **86**, Pt. 2, No. 2599.

Zmeureanu, R., P. Fazio, and F. Haghighat (1988). "Thermal Performance of Radiant Heating Panels," *ASHRAE Transactions,* **94**, Pt. 2.

17

EVALUATION OF DESIGNS FOR ENERGY PERFORMANCE USING A KNOWLEDGE-BASED SYSTEM

Bharati Jog

School of Architecture and Interior Design
University of Cincinnati
Cinicinatti, Ohio

Principles of knowledge-based (or expert) systems have been applied in different knowledge-rich domains such as geology, medicine, and very large scale integrated circuits (VLSI). There have been some efforts to develop expert systems for evaluation and prediction of architectural designs in this decade. In this paper we present a prototype system, Energy Expert, which quickly computes the approximate yearly energy performance of a building design, analyzes the energy performance, and gives advice on possible ways of improving the design. These modifications are intended to make the building more energy efficient and help cut down on heating and cooling costs. The system is designed for the schematic design phase of an architectural project. We also discuss briefly the reasoning behind developing such a system for the schematic design rather than the final design phase.

INTRODUCTION

During the process of architectural design the designer needs to analyze and evaluate designs according to different criteria. These evaluations help a designer in improving the solutions and predicting the final product. Some of these criteria, such as structural feasibility and energy and fire-safety performance, are measurable and thus computable. These quantitative evaluations are by and large based on sets of rules and building regulations. Thus principles of rule-based expert systems can be applied to develop these evaluators.

In this paper a prototype evaluation system designed to analyze the energy performance of a building design at the schematic design stage is presented. The system gives advice in the form of directives to the designer on improving the energy performance of a building. The knowledge base in this system includes facts (data) about building design and the environment of the site, and rules that use these facts as the basis for decision making. Throughout this paper we refer to this prototype system as *Energy Expert.*

We present the purpose of the project in the next section. The methodology of the system is given in the third section. In the fourth section the implementation details are briefly described. Then the structure of the advisory system is presented. The graphical output is given in the sixth section. The last section is a summary.

PURPOSE OF THE PROJECT

An architectural design process mainly involves three phases: predesign, design, and construction supervision. The design phase is broadly divided into two parts: schematic design and final design. The schematic design is iterative. At this stage a sketch design is created and evaluated for its performance against many criteria. Based on these evaluations, the sketch design is modified or an altogether different sketch design is created. During this process, several alternatives may be created. After considering different trade-offs, the most suitable alternative is developed into a detailed design at the final design stage.

At the final design stage, final decisions about exact structural details and building materials are made. Thus at the end of this stage, a detailed and accurate evaluation of the design for energy performance, fire-safety requirements, structural soundness, and cost analysis can be carried out. At the schematic design stage these evaluations can only be approximate. One may therefore argue that since the evaluations at the schematic design stage are only approximate, they need not be done at this stage, saving cost and time involved in the process. However, we feel that these approximate analyses and predictions at the schematic design stage are essential. In the next two paragraphs, we discuss briefly the reasoning behind this premise.

The decisions (such as design concepts, configuration of different functions) made at the schematic design stage are carried over to the later phases. Thus decisions made at this stage by and large affect the final product. Although it is possible, it is not typical to go back and change the major design concepts at a later stage. The use of approximate evaluation methods at an early design stage such as schematic design prunes totally inefficient designs early and prevents them from being developed into final designs.

There is a growing demand to design buildings against some overall energy use constraints such as Btu per square foot per year [Hart et al. 1980]. As such standards become widely accepted, it will be helpful to analyze designs against these constraints at the schematic design stage. This will reduce the risk of redesigning the project if it fails to meet such tests after construction documents have been made. Such redesigning could be extremely costly.

The evaluations of a design at the final design stage is accurate and thus lengthy. However, those at the schematic design stage are approximate and can be very quick. There are already many programs, such as CALPAS3 [Berkeley Solar Group 1984], that produce a detailed energy analysis based on the accurate building and site data. The goal of this project was to develop a prototype that would quickly calculate the approximate energy performance of a building and give directives on improving the designs at the schematic design stage.

During the last decade researchers in universities and industry have tried to develop evaluation programs for architects. Invariably these programs are developed either on mainframes such as VAX 8600, Amdahl 5860, or workstations such as Apollo. Examples of such programs include the ARCH:FIRESAFETY program [Ozel 1985] developed to check the fire-safety requirements of a building at the initial stage of the project and a prototype system for an integrated design evaluation [Kim 1986] implemented on an Apollo workstation. Both programs were developed at the University of Michigan. Skidmore, Owings & Merrill's Architectural and Engineering Series [IBM 1989], which includes structural and energy analysis applications, runs on IBM RT workstations. Most small architectural firms and schools cannot afford these systems. As a result, these programs are used primarily at those particular sites where they are developed and where such machines are available. While developing our system this aspect was considered. The tools chosen to implement this prototype system are affordable to small organizations. Architects are known to appreciate graphical data more than the quantitative output. Therefore, the system has been designed to generate graphical results.

METHODOLOGY

In this section we present briefly the methodology of *Energy Expert. Energy Expert* is based on the method described in *Energy Graphics* [Hart et al. 1980]. This manual method of energy analysis was developed by Booze Allen & Hamilton Inc. to assist federal personnel in designing energy-efficient federal buildings. *Energy Expert* collects data in three ways: from AutoCAD [Autodesk Inc. 1986a] drawing files, interactively from the user, and from a site database created for *Energy Expert.* In the absence of AutoCAD files, the relevant data are collected interactively from the user. Using these data, *Energy Expert* quickly calculates the approximate energy performance of a building for each of the four seasons in a year and displays the energy performance graphically. Based on the user-given fuel rate, an approximate cost of heating and cooling is calculated and displayed graphically. *Energy Expert* also gives directives to the user on how to make the building more energy efficient.

In *Energy Expert* the thermal performance of a building in terms of heat gain and loss is considered as the measure of its energy performance. The four basic components of thermal performance that are considered in calculating the energy analysis of a building in *Energy Expert* are (1) thermal heat gain, (2) solar heat gain, (3) envelope heat gain and loss, and (4) ventilation heat gain and loss.

IMPLEMENTATION DETAILS

Before implementing *Energy Expert*, certain decisions about the system requirements had to be made. These decisions included the type of the computer, the drafting tool, programming language, and the evaluation tool to be used for this implementation. To develop the prototype, an IBM PC AT with an enhanced color display adapter and 640K memory, running the DOS operating system was chosen as this machine has been popular in architectural firms and schools. Turbo Prolog [Borland International Inc. 1986] was chosen from the available backtracking languages such as Arity Prolog because it supports turtle graphics, windows, and color display. The drafting tool selected is AutoCAD, as it is popular among architects. As mentioned before, the evaluation method is based on *Energy Graphics* [Hart et al. 1980]. To perform energy analysis of a building, *Energy Expert* requires certain knowledge about the building design and the environment of the site. In the following subsections we describe these two types of necessary information.

Site Data

A database of 70 U.S. cities which contains the necessary environmental information required for *Energy Expert* was created. There are five important characteristics of a site: the temperature of the city (dry-bulb temperature), insolation (direct and diffused sunlight), earth temperature, wind patterns, and humidity (wet-bulb temperature). In this prototype we have restricted ourselves to the two most significant factors: the temperature of the site and its insolation. In a full implementation, all five determinants would need to be considered.

Theoretically, the climatic information should be detailed and available for all 365 days of the year. But carrying out lengthy computations on so much data would be too time consuming. Since the system needs to perform an approximate and quick energy analysis, the year is divided into four distinct seasons: winter, spring, summer, and fall. Information about an average day in each season for each of the 70 cities is stored in the database.

The temperature data (the highest and lowest temperatures of an average day in each of the four seasons) were obtained from *Climatological Data* [National Oceanic and Atmospheric Administration 1982]. The insolation information was obtained directly from *Hourly Solar Radiation Data for Vertical and Horizontal Surfaces on Average Days in the United States and Canada* [Kusuda and Ishii 1977]. The data for each city are stored in the form of predicates. The predicates for temperature data look as follows:

```
place (city, state, latitude, longitude, timezone,
       winter_highest, winter_lowest, spring_highest,
       spring_lowest, summer_highest,
       summer_lowest, fall_highest, fall_lowest).
```

The predicates giving the radiation data are as follows:

```
radiation (city, state, season, L).
```

where **L** is a list of solar radiations. In the prototype the user is prompted for the city and state name of the building site. If the city specified is not found in the program, the user needs to insert his or her own data about the site.

Building Data

In *Energy Expert*, the information about a building design is collected from the user and/or AutoCAD files. This information is (1) floor and roof areas, (2) floor-to-roof height,

(3) the allowable interior temperature range, (4) exterior wall areas in each direction and their *U* values, (5) window areas in each direction and their *U* values, (6) average number of occupants, and (7) fuel costs. Using these data the energy performance of the building design in each season is calculated.

To collect information interactively from the user, a user interface has been developed. There are several types of dialogues that can be embedded in the user interface. Examples of these types include questions and answers, form filling, menu selection, function keys, and so on. In the prototype, a *question-and-answer* technique is used. For consistency, throughout the program this technique is used in only two forms. In the first type the answer of a question is either "yes" or "no." If the answer is "yes," the user needs to enter "y." Any other key is taken as "no." For example:

```
Do you have an AutoCAD data file? (y/any key)>
```

In the second form, the user is prompted to enter a value. For example:

```
Enter the U value of the wall>
```

Currently, extraction of information from AutoCAD has been considered only to check for its feasibility. Instead of extracting all the necessary information from AutoCAD, only the windows, floor and roof information has been considered. A set of different types of window and door entities has been developed using AutoLISP [Autodesk Inc. 1986b]. While drafting building plans, the user is expected to use these entities instead of drafting windows and doors using lines and arcs. When the user selects a type of window to draft, he or she needs to give the position of the window, width and height of the window, and the *U* value of the glass. This information is stored as attributes of that entity in the database. In the same way information about a door is stored if it is a glass door on an external wall.

When a floor plan is complete, the user gives a command called *shellinf.* This routine has been written in AutoLISP. It collects the necessary information about windows and external glass doors from the attributes and writes to a file *data.dat.* This command also prompts the user to locate external corners of the floor and roof plan, calculates areas, and writes to the file *data.dat.* This is a text file with one value on each line in a definite order. This file is read by *Energy Expert.* If the user has created this data file, the information above is fetched from it while the remaining information is collected interactively from him or her. In the absence of this file all the information is collected from the user.

ADVICE IN *ENERGY EXPERT*

Energy Expert is an advisory system. Many expert systems diagnose a problem in a given domain, but they may not necessarily give solutions to solve the problem. *Energy Expert* analyzes the approximate energy performance of a building and based on this performance, gives directives to improve the building design together with possible modifications in the design.

Energy Expert is a rule-based evaluation system in which the rules are checked against knowledge (a collection of facts) about the current situation. When a condition given by the IF portion (antecedent) of a rule is matched with the facts, the action (consequent) specified by the THEN part is performed. This is referred to as *firing a rule.*

In *Energy Expert*, the energy performance of the current alternative building design is calculated. This provides the knowledge about the design. Since the knowledge about the performance of the building alternative (i.e., design solution) is calculated in this system, forward chaining was the most appropriate system for *Energy Expert.* In a forward chaining system, when a rule is fired, new facts may be added to the knowledge base. Matching the IF portion of the rules with the new facts can produce a forward inference chain. The forward inference chain can be used to give the reasoning behind the decision.

In the next subsections the structure of the advisory system of *Energy Expert* is presented followed by an example.

The Advisory System

As mentioned previously, using the site and building information heat gains and losses are calculated in each season. *Energy Expert* prompts the user to enter a range of comfortable (i.e., allowable) temperatures in the building. When heat gains in a building are such that interior temperature remains within this allowable temperature range, no artificial cooling is required. Such gains are referred to as *allowable heat gains.* When the interior temperature rises beyond the upper limit of the allowable temperature range, head gain is in excess of what is allowed and referred to as *excess heat gain.* The objective of the advisory system is to maintain the interior building temperature within the comfort range while keeping the energy costs as low as possible. Based on this range, the seasonal energy performance is tested for six conditions, numbered 1 to 6. The conditions and the respective numbers (identifiers) are as follows:

```
condition(1, heat_loss_all_day).
condition(2, some_heat_loss).
condition(3, some_allowable_heat_gain).
condition(4, some_excess_heat_gain).
condition(5, all_allowable_heat_gain).
condition(6, all_excess_heat_gain).
```

The matched conditions and the corresponding condition numbers are kept in the database as the current energy performance of the building design. There are six rules based on the six conditions above. A rule gets fired matching the conditions of the energy performance. The logical structure of the rules is as follows:

```
rule(rule_no)    :-    conditions(L),
                       action(rule_no).
```

where **rule_no** is a unique identifier for the rule and the corresponding **action**, and **L** is a list of condition numbers. The rules are as follows:

```
rule(1)    :-    conditions([1]),
                 action(1).
```

```
rule(2)   :-   conditions([2,3]),
               action(2).

rule(3)   :-   conditions([2,3,4]),
               action(3).

rule(4)   :-   conditions([5]),
               action(4).

rule(5)   :-   conditions([3,4]),
               action(5).

rule(6)   :-   conditions([6]),
               action(6).
```

Once a rule is fired, the system tries to satisfy the goal **action**. The logical structure of the predicate **action(rule_no)** can be expressed as follows:

```
action(rule_no)     :-
      get_subconditions(rule_no, subcondition_no),
      possible_solutions(rule_no, subcondition_no, [soln_nos]),
      get_directives([soln_nos]).
```

where **rule_no** stands for the rule number of the fired rule and **subcondition_no** stands for the number (identifier) of a matched subcondition.

The logical structure of the predicate **get_subconditions** is as follows:

```
get_subconditions(rule_no, subcondition_no) :-
      check_subcondition(subcondition_no).
```

where **rule_no** stands for the rule number of the fired rule, and **subcondition_no** stands for the number (identifier) of a matched subcondition.

It is assumed that besides the main conditions there can be some subconditions that may affect the advice. For example, when the outside temperature is higher than the inside temperature, the heat inside the building can be removed by using artificial means. However, this solution increases the cost of cooling. Since the objective of the advice is to suggest improvements in the design keeping the energy costs down, this alternative cannot be used. So the system has to check whether the outside temperature is higher than the inside temperature. In the prototype some sample subconditions have been developed. For example:

```
subcondition(1) :- exterior_temperature > interior_temperature.
subcondition(2) :- exterior_temperature <= interior_temperature.
```

If a subcondition is satisfied, **subcondition_no** is instantiated to the corresponding subcondition number (identifier); otherwise, it is instantiated to zero.

After trying to satisfy the goal **get_subconditions**, the system tries to satisfy the next goal, **possible_solutions**. The clauses **possible_solutions(rule_no, subcondition_no, [soln_nos])** which are in the knowledge base are designed so that given the combination of a rule (**rule_no**) and subcondition (**subcondition_no**), a list of possible solution numbers (**[soln_no]**) is provided. In *Energy Expert*, the predicate storing the solution numbers and the corresponding solutions looks as follows:

```
soln(soln_no, solution).
```

The solutions and their corresponding numbers (identifiers) are as follows:

```
soln(1,  increase_internal_heat_generation).
soln(2,  decrease_internal_heat_generation).
soln(3,  increase_solar_heat_gain).
soln(4,  decrease_solar_heat_gain).
soln(5,  increase_envelope_heat_gain_and_loss).
soln(6,  decrease_envelope_heat_gain_and_loss).
soln(7,  increase_ventilation_heat_gain_and_loss).
soln(8,  decrease_ventilation_heat_gain_and_loss).
soln(9,  store_heat).
soln(10, store_cold).
```

After trying to satisfy the goal **possible_solutions**, the system tries to satisfy the third goal, **get_directives**. The logical structure of this routine can be expressed as follows:

```
get_directives([soln_nos])
   {For each soln_no
    directive_list(soln_no, [directives]),
    {For each directive
     modification_list(directive, [modifications]).
    }
   }
```

Thus, for each of the solutions in the list **[soln_nos]**, a list of directives (**[directives]**) is obtained and for each of these directives a list of possible building modifications is represented by the list **[modifications]**.

The knowledge base of the system contains the clauses **directive_list**. The structure of these clauses is as follows:

```
directive_list(soln_no, [directives]).
```

where the list of directives corresponding to a **soln_no** is given by **[directives]**. The following is an example of these clauses:

```
directive_list(3, [increase_exposure_to_direct_solar_radiation,
                   consider_active_solar_heating,
                   increase_absorptivity]).
```

The knowledge base contains the clauses for possible modifications to achieve the given directives. The structure of the predicate **modification_list** is as follows:

```
modification_list(directive, [modifications]).
```

where the list of modifications corresponding to the directives is given by **[modifications]**. An example for a directive and the corresponding modifications is as follows:

```
modification_list(increase_exposure_to_direct_solar_radiation,
                  [slope_roof_towards_south,
                   use_clear_glazing, avoid_shading]).
```

The fired rules and their corresponding solution numbers are asserted (stored) in the working memory. The rules are independent and do not communicate with each other. When multiple rules are fired, the system applies a simple conflict resolution strategy to select one rule. The rule with the most conditions is selected and the corresponding solutions, directives, and modifications are displayed on the screen. This is the advice about the possible changes in the building design for better energy performance. In a fuller implementation of *Energy Expert* more directives and modifications can be included in its knowledge base. Also, there can be some directives and modifications that may conflict with each other when combined. Hence a more general conflict resolution strategy and some heuristics can be included to pick up directives and modifications that provide a maximum benefit in the given situation.

An Example

Suppose that the energy performance of a building is such that during fall there is some allowable heat gain and some excess heat gain. In this case, conditions 3 and 4 are satisfied. So rule 5 gets fired and the system takes the corresponding action. The calling routine is as follows:

```
rule(5)    :-   conditions([3,4]),
                action(5).
```

While processing **action(5)**, the goal **get__subconditions(5, subcondition__no)** is checked in order to find the corresponding subcondition number (i.e., **subcondition__no**). Suppose that in this example the exterior temperature is less than the maximum allowable temperature and thus subcondition 2 is satisfied. Hence **subcondition__no** is instantiated to 2.

The system, then, tries to satisfy the goal

```
possible_solutions(5, 2, [soln_nos]).
```

In this example, the solution numbers (**[soln__nos]**) for rule 5 and subcondition 2 are given by the list **[4,5,7,10]**. The solution numbers (**soln__no**) and the corresponding solutions are given by the predicates **soln**:

```
soln(4,  decrease_solar_heat_gain).
soln(5,  increase_envelope_heat_gain_and_loss).
soln(7,  increase_ventilation_heat_gain_and_loss).
soln(10, store_cold).
```

When the predicate **possible_solutions** succeeds, the next goal **get_directives** is tried. For each **soln_no** in the solution list **[4,5,7,10]** the corresponding directives are retrieved from the knowledge base by matching **soln_no** in the clause **directive_list**. These clauses are given as follows:

```
directive_list(4,  [decrease_surfaces_exposed_to_radiation,
                    increase_thermal_resistance,
                    inrease_reflectance,
                    consider_solar_cooling]).

directive_list(5,  [increase_exposure, increase_U,
                    increase_infilteration]).

directive_list(7,  [increase_ventilation_rate,
                    increase_temperature_differential]).

directive_list(10, [actively_store_cold,
            passively_store_cold_by_increasing_thermal_mass]).
```

For each directive in the lists above, the corresponding modifications are retrieved from the knowledge base by matching the directives in the clause **modification_list**. The matching modifications of the directives **decrease_surfaces_exposed_to_radiation**, and **increase_thermal_resistance** for example are as follows:

```
modification_list(decrease_surfaces_exposed_to_radiation,
                  [reduce_surface_area_to_volume_ratio,
                   orient_building_to_minimize_insolation]).

modification_list(increase_thermal_resistance
           [consider_double_roof,use_multiple_layer_glazing]).
```

These solutions, directives, and the modifications are displayed as advice on the screen.

GRAPHICAL OUTPUT

In *Energy Expert* results are displayed graphically on the screen. The graphical output was limited by the language tool (Turbo Prolog) and the time available to implement it. However, the graphs are sufficient to provide information about energy consumption at the schematic design phase.

Energy Expert displays the seasonal energy performance on the screen for each of the four seasons, one at a time. The screen is divided into three windows. The upper left window displays the line graph of the total heat gain and loss on an average day in a season. The upper right window contains two pie charts of cooling and heating costs in that season. The lower window displays the advice. Only the lower window has a border. The space bar is used to check the performance of the following seasons. Figure 17.1 illustrates the graphical output.

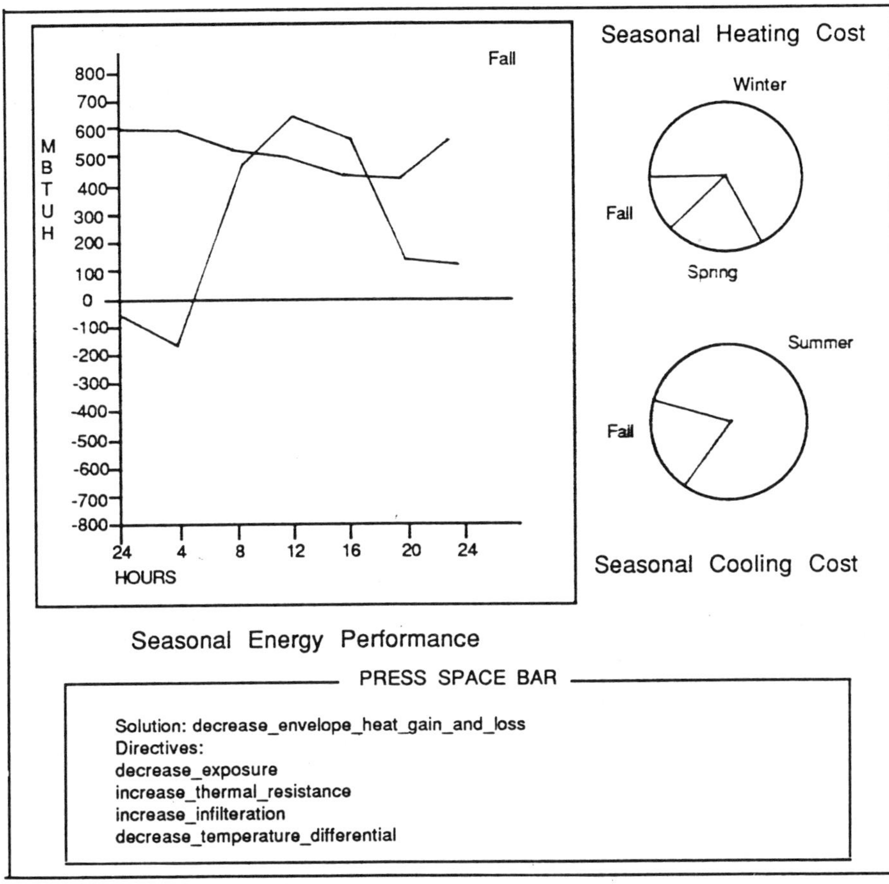

Figure 17.1. Graphical output. Reprinted with permission from *Comput., Environ. and Urban Systems,* **13**, B. Jog and M. Zemankova, Energy Expert: An Expert System for Architects, copyright © 1989 Pergamon Press PLC.

CONCLUSION

In this paper we presented a prototype evaluation system, *Energy Expert*, which is designed to be used at the schematic design stage to obtain a quick and approximate energy analysis and for advice on possible modifications in the building design to improve the energy performance. The results are displayed by simple graphical output.

Ideally, the program should store the user-given information of a design in its database. This can be used for the next iteration of design, so that the user has to insert only the changed information. Because of the time constraints, currently this feature is not implemented. This could be done by storing the user-given information in a file and reading this file in the next iteration. The system needs to be field tested thoroughly with a variety of data to determine its usefulness before expanding its scope. Currently, more rules cannot be added to the database without affecting the way the system infers its conclusions. The advice in the system is based mainly on directives discussed in *Energy*

Graphics [Hart et al. 1980]. More directives need to be added in the database after discussions with experts in the field to provide nontrivial advice. That may necessitate using some heuristics to choose appropriate directives. Also, the current site database needs to be expanded. On the whole, *Energy Expert* provides an interesting exploration into the possibilities of developing an evaluation system for architects.

REFERENCES

Autodesk Inc. (1986a). *AutoCAD Reference Manual*, Sausalito, Calif.

Autodesk Inc. (1986b). *AutoLISP Programmer's Reference*, Sausalito, Calif.

Berkeley Solar Group (1984). *CALPAS3 User Manual*, Berkeley, Calif.

Borland International Inc. (1986). *Turbo Prolog, Owner's Handbook*, Scotts Valley, Calif.

Hart, K. G., J. M. Kurtz, and W. I. Whiddon (1980). *Energy Graphics: A New Approach to Energy-Conscious Design*, Booze Allen & Hamilton Inc., U.S. Department of Energy, Washington, D.C.

IBM (1989). *Reference Guides: Energy Application and Structural Application*, Architecture & Engineering Series.

Kim, U. (1986). "Model for an Integrated Design Evaluation System Using Knowledge Bases," in *ACADIA '86 Workshop Proceedings*, J. A. Turner (ed.), University of Houston, Houston, Tex.

Kusuda, T., and K. Ishii (1977). *Hourly Solar Radiation Data for Vertical and Horizontal Surfaces on Average Days in the United States and Canada*, Building Science Series 96, National Bureau of Standards, Washington, D.C.

National Oceanic and Atmospheric Administration (1982). *Climatological Data*, National Climatic Center, Ashville, N.C.

Ozel, F. (1985). "Using CAD in Fire Safety Research," in *ACADIA '85 Workshop Proceedings*, P. G. McIntosh (ed.), Arizona State University, Tempe, Ariz.

18

COMBINED PROCEDURAL AND HEURISTIC METHOD TO ENERGY-CONSCIOUS BUILDING DESIGN AND EVALUATION

Edna Shaviv

Faculty of Architecture and Town Planning
Technion–Israel Institute of Technology
Haifa, Israel

Yehuda E. Kalay

Department of Architecture
School of Architecture and Planning
State University of New York at Buffalo
Buffalo, New York

In this paper we describe a methodology for supporting computationally all phases of energy-conscious design and evaluation, by combining procedural simulation and knowledge-based heuristic methods in one integrated system. The methodology is based on partitioning the design process into discrete phases and identifying the energy-conscious informational characteristics of each phase. These informational characteristics are expressed in the form of design variables (parameters) and the relationships between them. The methodology allows for testing the applicability of heuristic rules in nonstandard cases by context-sensitive application of default values and judicious selection of rules of thumb, thereby improving the overall predictive powers of the evaluation.

INTRODUCTION

According to a study published by the Israel Ministry of Energy and Infrastructure [1984], unless energy conservation practices are applied the energy needs of buildings in Israel are expected to rise by 300 percent by the year 2000 from its 1980 level (from 0.8 million metric tons of oil equivalents to 2.5 million metric tons). A similar study commissioned by the U.S. Congress in 1981 estimates that $200 billion could be saved in the same period by applying energy-conscious design practices alone [Solar Conservation Study 1981]. This source also claims that energy-conscious design is the most cost-effective alternative to managing U.S. energy needs in the future.

It is curious, therefore, that whereas energy conservation developments in building technology and materials, and the knowledge accumulated by researchers, designers, and by builders, have been growing steadily, their actual application in practice has not. Furthermore, although a host of energy-related computer programs that purport to put the accumulated energy-related experience and research results at the designers' disposal in a practical and readily applicable manner were developed, their impact on making actual design decisions have, so far, been marginal. Their failure to improve significantly energy-related design decisions may be attributed to their inappropriateness for use in various design phases.

Energy evaluation programs have been based on three different physical and computational models. The first model uses procedural tools for *precise* simulation of heat transfer, ventilation, lighting, and other attributes of the built environment, based on first principles (deep knowledge). It provides designers with precise information concerning the expected thermal performance of design alternatives, in terms of the inhabitants' comfort and the building's energy cycle. It requires detailed information about the building's location, orientation, geometry, and material composition. Therefore, this model is applicable to evaluating finished buildings, or for use in later design phases where most of the values associated with relevant design parameters have already been decided. At that stage of the process, the designer can experiment with different value assignments to selected parameters (e.g., material, thermal mass, etc.), leaving the other parameters unaffected. Examples of systems based on this model are ENERGY [Shaviv and Shaviv 1977, 1978a,b, Shaviv 1977, 1979], DOE-2 [LBL and LASL 1982], TARP [Walton 1983], and SERIRES [Palmiter et al. 1982].

The second model is based on procedural tools that *approximate* heat transfer, solar energy collection, energy storage, and so on. This model is not based on first principles, but rather on some approximation calculations (e.g., regression formulas) and simplified assumptions (e.g., steady-state analysis). It provides designers with a rough evaluation of the expected thermal performance of different design alternatives. Like the precise simulation model, the approximate model requires detailed data of the building parameters. It is, nevertheless, faster than the precise model and is intended to be applied manually during intermediate design stages. Examples of systems that are based on this model are all the steady-state calculations of heat lost [ASHRAE 1972], like the LCR or the SLR models that were developed in Los Alamos for passive solar heating [Balcomb et al. 1984].

The third model, first suggested by Balcomb et al. [1984], is founded on knowledge-based, heuristic tools for gross evaluation of thermal properties of design alternatives, rules of thumb, and accumulated experience (wide knowledge). It provides designers with generalized appraisal of their choice of building orientation, geometry (massing), and choice of site treatments. The input required by this model is approximate and gross compared to the input required by the procedural model. The knowledge-based evaluation model can therefore be applied at the early formative stages of the design process, where the impact of design choices on the energy performance of the building is more significant than choices made in later design phases [Shaviv 1977, Kalay et al. 1985, 1987, Schmitt 1987, Maher and Fenves 1985]. Examples of systems based on this model include ARCADE [Antony 1987], a program developed by Schmitt at CMU [Schmitt 1987], and an expert system developed by Gero and Coyne [1984].

The approximate procedural model was developed for manual application and recommended by many researchers for use in intermediate design phases. Shaviv (1989) claimed that its utility, however, is questionable. This model is similar to the exact simulation model in that it requires a full description of the building, with all the attendant difficulty of preparing such data. It differs from exact simulation only in that it calculates energy consumption on an annual, monthly, or daily basis rather than on an hourly basis. It thus simplifies the required climatological data and the energy calculations themselves. These advantages diminish, however, when computers are used to perform the simulation that allow storage of hourly climatological data for each region as easily as daily data, and can handle accurate simulation calculations efficiently, quickly, and at low cost.*

In addition to requiring the same amount of work for preparing building data as accurate simulations do, approximation models are really too complicated to be solved manually (particularly by architects) and may lead to wrong conclusions. Their deficiency stems from using simplified models, which ignore the time lag of heat transfer in buildings constructed from heavy materials that store energy. To demonstrate this point, we have performed many simulation runs, showing that the relationship between various design parameters are not linear (see Figures 18.1 and 18.2). Furthermore, while in one case changing a certain parameter may improve the design, in another case the same change may degrade the solution.

The analysis of different design alternatives was carried out by means of the dynamic simulation program ENERGY, which was developed by Shaviv and Shaviv [1977, 1978a,b]. This program solves simultaneously the heat transfer equations through all exterior walls, taking into account the thermal mass of each external wall as well as of the internal partitions. The thermal behavior of the building is checked during winter and summer, and the energy consumption of the building is examined in both cases. In addition, the program calculates the maximum in-house summer temperature without air conditioning (Tmax-NV or Tmax-V) or with a fan only (Tmax-F). Most of these results, while of considerable importance to the user of the building, are not available in the approximate models. For example, Figure 18.1 represents the thermal performance of a middle-floor apartment with a large southern window in Jerusalem (a region of cold winters and mild summers). It shows that there is no need to cool this apartment mechanically in the summer. Moreover, improving the *R* value of the walls reduces significantly the energy consumption for heating (as an example, compare JE-SL with

*Special care was taken in writing the simulation program ENERGY to develop an efficient algorithm that solves the implicit scheme of the time-dependent heat flow equation. Consequently, it takes less than 10 seconds to run a full-scale building simulation using this program on a Silicon Graphics Personal IRIS workstation, supporting our claim that workstations that are available today are sufficiently powerful to eliminate the need for approximate energy calculations.

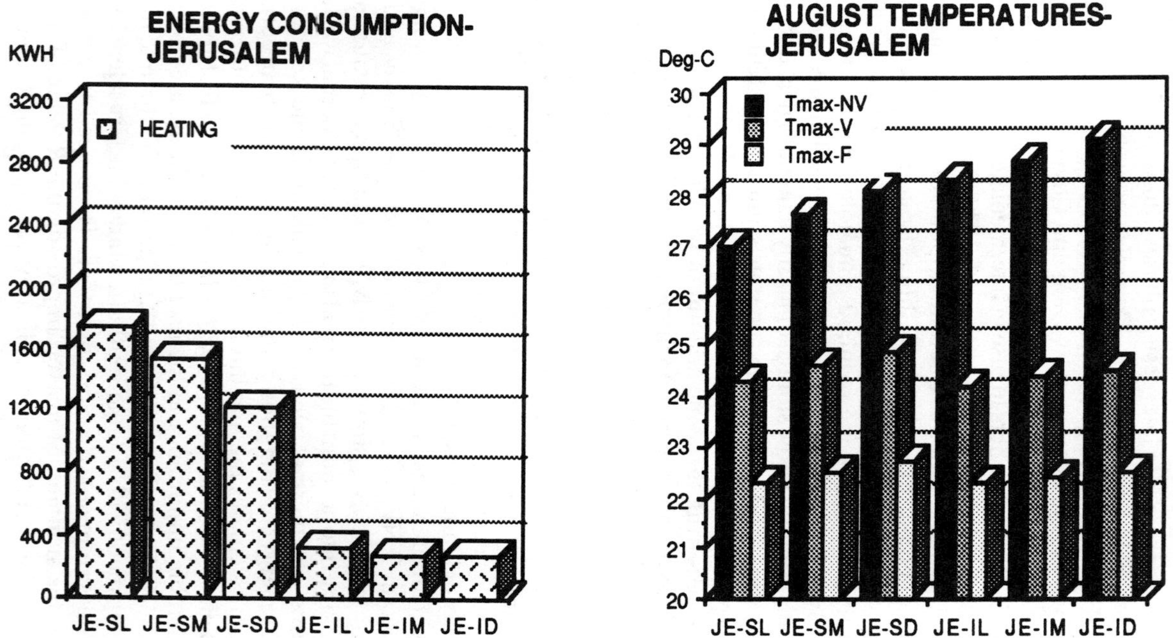

Figure 18.1. Energy consumption and thermal performance of a middle-floor apartment with a large southern window in Jerusalem. S, standard insulation; I, improved insulation; L, light color; M, medium color; D, dark color; JE, Jerusalem.

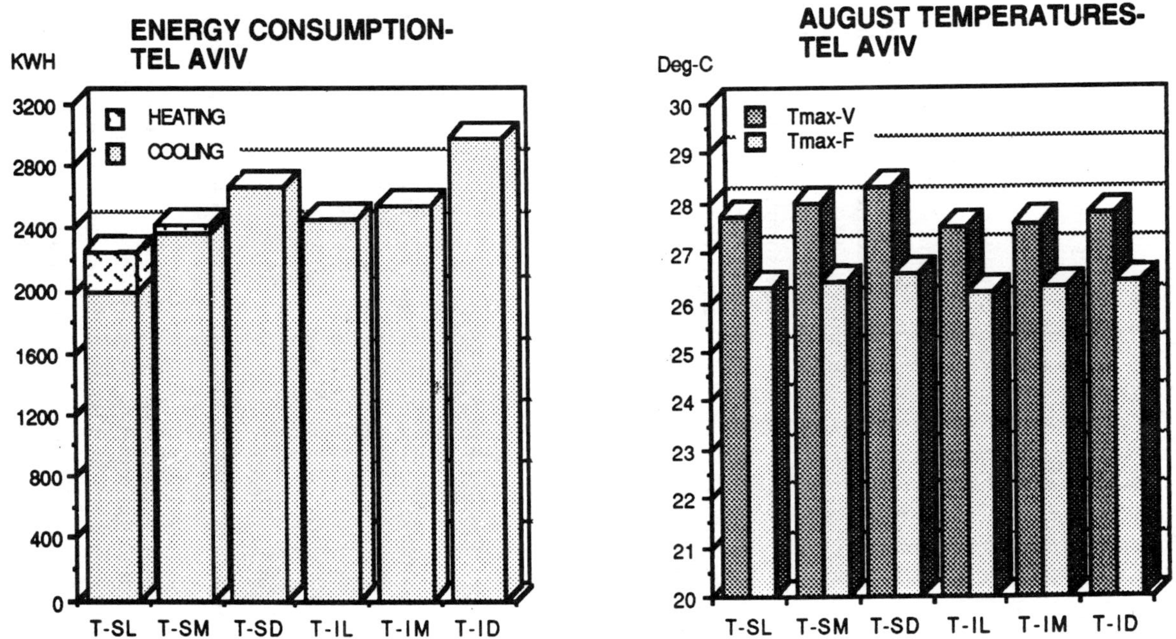

Figure 18.2. Energy consumption and thermal performance of a middle-floor apartment with a large southern window in Tel Aviv (T).

JE-IL). In the summer, improving the *R* value of walls in a poorly ventilated building (Tmax-NV) will actually deteriorate the thermal performance of the building, because the heat that was stored in the thermal mass is emitted to the unvented rooms, and because the improved insulation prevents its loss through the building's envelope. Such conclusions could not be reached using approximate, steady-state calculations.

Figure 18.2 depicts the thermal performance of same building in the climate of Tel Aviv (a region of mild winters and hot-humid summers). It shows that there is almost no need to heat this apartment in the winter (as there is a large window facing south). Improving the *R* value of the walls will increase the energy consumption for cooling if the building is not well ventilated (T-SL compare with T-IL).

Figures 18.1 and 18.2 also show that in summer conditions, light-colored buildings behave thermally better than dark-colored ones. However, this factor is not significant in the well-ventilated building (Tmax-V and Tmax-F) or in the well-insulated building (JE-IL, JE-IM, and JE-ID). In the hot climate of Tel Aviv, light color is always preferable. On the other hand, in Jerusalem the energy consumption for heating the building can be reduced by painting the walls dark (JE-SL compared to JE-SD).

We presented the examples above to show that the complexity of evaluating the thermal performance of buildings warrants the use of accurate simulation models rather than approximated ones. Furthermore, due to the strong interdependencies between different design variables, a great deal of expertise is required to reach good conclusions without actually running many simulations. We believe, therefore, that integration of heuristics, in which such experience can be stored, and procedural simulations methods from which this expertise can be derived and through which it can be tested, will support all phases of energy-conscious design and evaluation processes, including the intermediate design stages.

ENERGY-CONSCIOUS DESIGN PROCESS

The process of designing energy-conscious buildings can be viewed as a sequence of decisions made at different levels of abstraction, each successive level more detailed and specific than the former one. These levels of abstraction correspond to discrete design phases, which include [Cross 77, Jones 70, Mitchell 77]:

1. **Briefing:** statement of user needs.
2. **Pre-Conceptual design:** feasibility study and determination of detailed program requirements.
3. **Conceptual design:** exploring different schematic design alternatives that agree with the programmatic requirements. This phase is concerned primarily with geometry and orientations, without considering material compositions.
4. **Preliminary design:** determining material compositions and building details.
5. **Detailed design:** exploring different detailed design alternatives. This phase deals with structure and material composition considerations.
6. **Design documentation:** preparing building documents.

An energy-conscious design process comprises the choice, coordination, and assignment of appropriate values to many design parameters. The process is complicated by the high degree of interdependencies between various design parameters, where the choice of one value limits the range of assignable values to other parameters. Furthermore, maximization or minimization of some values will in many cases lead to suboptimal overall solutions. Rather, trade-offs must be made to achieve a good overall balance. For instance, in Tel Aviv a light-colored, standard-insulated building behaves thermally as well as a dark-colored, better insulated building (Figure 18.2 T-SL compare with T-ID). It would obviously be cheaper to paint the building white rather than improve its insulation.

Early in the design process, most parameters have not yet been assigned specific values. The designer therefore is free to choose almost any values for the parameters of current concern (e.g., form, orientation, site location, etc.). It is difficult to guide the choice by conventional, procedural methods, for lack of information. The only plausible alternative to arbitrary assignment of default values (so a simulation can be run) is the application of context-sensitive past experiences in the form of rules of thumb (heuristics). For example, if the designer wishes to run an energy simulation on the building form he or she chose at the schematic design phase, the designer must determine the materials from which the building will be constructed, even though he or she may not have considered them yet. To assist the designer, the system could provide him or her with reasonable defaults, based on the recommended *R* value for the climatic zone in which the building is located [Shaviv and Peleg 1990].

As the design progresses, the values assigned earlier limit the designer's freedom in choosing the values that are assignable to parameters considered later. For example, in Jerusalem, if the walls were assigned a material composition of low *R* value, they must be painted in a dark color to reduce the energy consumption for heating the building in winter conditions (see Figure 18.1, JE-SD compare with JE-SL). Having assigned a dark color to the walls, the designer must be careful to design the building with good ventilation for summer conditions, to avoid overheating. On the other hand, if the building is well insulated, colors have a negligible influence on its energy consumption (Figure 18.1 JE-SD compare with JE-SL), hence the designer is free to choose any color for the walls (Figure 18.1 JEIL, JE-IM, and JE-ID).

The difficulty arises in intermediate design phases when the quantity and nature of formerly assigned values is not sufficient to allow the application of procedural simulation, yet it is too constraining to allow free, unchecked application of heuristics. In such instances (which comprise the bulk of the design process), the applicability of heuristics must be tested against the framework of parameters that have already been determined, by means of additional, specific rules, or by means of partial simulations. Alternatively, default values could be added to allow the application of procedural methods. The choice of these defaults must nevertheless be guided by heuristics, since the framework may still be too loose to allow for their unarguable determination. This paper focuses on these intermediate design phases and outlines a methodology for aiding the choice of values assignable to design parameters using both heuristic and procedural methods.

Furthermore, the information provided by the procedural method (consisting primarily of quantitative, rather than qualitative data) is not of the kind needed at early design phases, where the massing of the building, its orientation, proportions, and many of its other major characteristics are yet to be determined. Consequently, the information provided by procedural simulation in early design phases is often inappropriate and devoid of hints on how to improve the general energy performance of the building.

Heuristic models attempt to capture, in the form of rules of thumb, the accumulated experience of designers. Such so-called "expert systems" evaluate conceptual design alternatives in a manner similar to the designer himself or herself, who (in the case of energy-conscious design) must determine important orientation, massing, openings, and other gross building characteristics on the basis of sketches and other incomplete information. For example, a knowledge-based system (KBS) could suggest orienting the building 15 degrees eastward of the south to achieve better energy performance all year round in very hot climates [Shaviv 1981]. Such information is very different from the quantitative results provided by exact procedural methods and is more appropriate at the early formative phases of the design process. Nevertheless, the heuristics used by expert systems of this kind are formed by generalizations, based on average cases. They may be totally inappropriate when applied in particular, nonstandard cases, where the designer is most in need of assistance. In fact, the recommendations produced by such systems may actually be contrary to the warranted ones. For example, if the orientation heuristic cited earlier were derived from a hot (and not very hot) climate sample alone, it would have recommended orienting the building due south instead of southeast [Shaviv 1981]. Even for the same hot climatic zone, this recommendation may be modified in particular cases, such as by adding windows to the eastern or western elevations [Shaviv and Peleg 1990]. We can summarize this argument by pointing out that heuristic methods are suitable for evaluating the expected energy performance of typical design alternatives at the very early design phases but are not suitable for evaluating the actual expected performance of detailed, particular design solutions.

Consequently, although procedural and heuristic methods are each applicable and useful at particular phases of the design process, neither one can support all design phases all by itself. Furthermore, neither procedural nor heuristic models alone are applicable in intermediate design phases, where the bulk of the design decisions occurs. In such phases the design alternative is too well developed to benefit from the application of generalized expert systems, but not developed well enough to admit procedural systems (either precise or approximate methods). In such cases the designer is often left unaided, yet must make decisions that may significantly affect the thermal performance of the solution.

To support energy evaluation at design phases that are neither informationally rich nor poor, it follows that a combined heuristic–procedural method is needed (see Figure 18.3). In this paper we describe such a method, which combines both simulation and heuristics into one system where the advantages of each method strengthen each other and where their drawbacks cancel out.

INFORMATIONAL CHARACTERISTICS OF DESIGN PHASES

Each design phase can be informationally characterized in terms of the design objective to be achieved at that phase (i.e., design goals) and the constraints any solution that meets these goals must abide by [Kalay et al. 1987]. The energy-related design parameters that characterize design phases include site location, building's geometry, openings, building materials, mechanical systems, and energy conservation policies. In each phase of the design process, the designer assigns certain values to some of these design parameters.

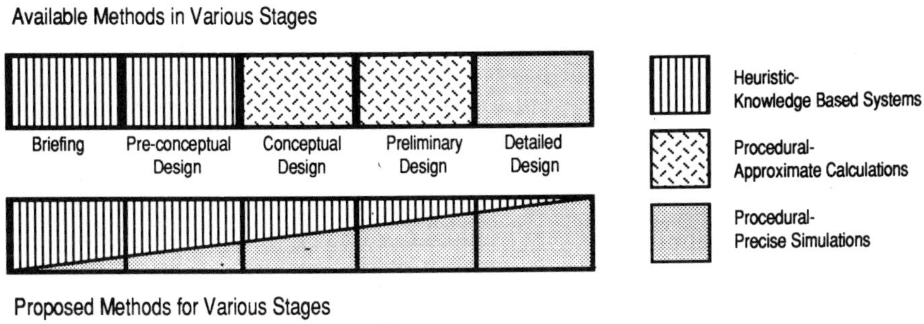

Figure 18.3. Schematic comparison of available versus proposed energy evaluation tools in different design phases.

Together, they determine the overall energy performance of the finished building. Therefore, the selection of individual values at particular design phases is complicated by the high degree of interdependencies between the parameters: determining the value of one design parameter may limit the choice of available values that can be assigned to other design parameters.

Design parameters are related both physically and temporally. For instance, the choice of a certain orientation may require the design of a large window to compensate for lack of adequate ventilation and lighting. The choice of a large window size may, in turn, force a change in the building proportions, to fit the window in a particular wall. These geometrically (or physically) linked parameters are typically assigned values at the same phase of the design process. Temporal relationships, on the other hand, link design parameters that are determined in different design phases. For instance, the choice of a particular orientation and window size at an early design phase may determine the choice of thermal mass, which is typically determined in much later phases.

Selection of the values that ought to be assigned to design parameters at each design phase depends on three types of information:

1. **Preconditional design information:** What is known about the evaluated design alternative at this design phase?
2. **Postconditional design information:** What does the designer want to learn about the design alternative from an energy evaluation point of view, and in what form should this information be presented?
3. **Missing design information:** Information that is neither available nor concerns the designer at that design phase, but influences other design parameters and should therefore be considered at the same design phase.

The first source of information comprises design parameters whose values have been determined in earlier design phases. The second source comprises design parameters whose values are being considered by the designer at the particular design phase. The third source of information is the most intricate, as it is neither known to the designer, nor does it concern him or her, yet it should. For example, at an early design phase the designer may already have determined the site location of the building and may currently be concerned with such parameters as massing and orientation. He or she may not be currently concerned with color schemes, building materials, and shading devices. Nevertheless, all these considerations together determine the nature, scope, and degree of accuracy of design information needed to make current design decisions on the issues that do concern the designer.

The discretization of the design process into phases provides informational preconditions for energy evaluation of each design phase (in terms of what is known about the building at that phase) and informational postconditions that determine the type, extent, and form of the information to be provided by the system to the designer. Together, the pre- and postconditions form a unique context for filling in the missing information. The difference between the available information, as determined by the preconditions, and the information that should be provided by the postconditions, can be bridged by defaults or heuristics that apply in the particular design phase. Such missing information can be filled in by the user in response to specific questions, or by defaults that are stored in the system's database, or by heuristics that are stored in the system's knowledge base. The choice of the proper mode for filling in missing information is based on the preconditions of the design phase, the desired postconditions, and the availability of data or heuristics.

ENERGY-RELATED DESIGN PARAMETERS

The energy-related design parameters needed for simulating the thermal performance of buildings include its floor area and its volume, the area and heat capacity of the isothermal mass, the initial wall temperature conditions, and the internal heat gain according to a given schedule. The building data include information about the mode of operation and size of mechanical equipment (thermostat settings and hours during which the heater or cooler operate) and the number of air changes during day and night (due to infiltration and natural or forced ventilation). The building data also contain information about the number of walls, including all opaque elements in the external envelope (such as roofs and floors). For each external element, the heat capacity, heat conductivity, and thickness of each layer must be provided, as well as the albedo and emissivity of the element and its shading factors. The external elements data also comprise geometrical design parameters such as the azimuth, inclination, and area of the element. The data incorporate the type, azimuth, inclination, shading factor, and area of each window located on this element as well. Table 18.1 shows part of the many energy-related design parameters that should be assigned values in order to run a simulation with the ENERGY simulation model.

Energy-related design parameters can be considered from several points of view, including their physical manifestation in the building, their effect on the building's energy load, and the design phase in which they are most likely to be determined. Physical manifestations include the building's geometry (area, volume, proportions, orientation, inclination, etc.), envelope (materials, dimensions, composition, color, texture, etc.), openings (area, location, orientation, details, materials, shading, etc.), internal partitions, mechanical systems, and the building-use policies (hours of occupancy, thermostat settings, etc.), as can be seen in Table 18.1.

The physical building components can affect the building's energy load in different ways (see Table 18.2). They affect ventilation and air infiltration (through openings),

TABLE 18.1 Part of the Energy-Related Design Parameters That Form the Input for the Simulation Model ENERGY

```
BUILDING'S GEOMETRY

180.0                                                TOTAL FLOOR AREA
504.0                                                VOLUME OF BUILDING
6                                                    NO. OF EXTERNAL WALLS
0.0                                                  AZIMUTH OF REFERENCE WALL

INTERNAL PARTITIONS

300.0                                                AREA OF INTERNAL MASS
8400.0                                               HEAT CAPACITY OF INTERNAL MASS
18.3                                                 INITIAL TEMP. DISTRIBUTION

MECHANICAL SYSTEMS AND BUILDING USE POLICIES

 250.                                                HEAT GAINS: CONSTANT
1000.       18.0     22.0                            HEAT GAINS: SCHEDULE
10000       07.0     22.0     18.0     19.0          HEATER: SCHEDULE; TEMP.SET
 6000       07.0     22.0     24.0     28.0          COOLER: SCHEDULE; TEMP.SET
 2 18.0     07.0     0.75      4.0                   VENT: TYPE, SCHEDULE, ACH: DAY; NIGHT

ENVELOPE

   1                          WALL NO 1
   3                                                 NUMBER OF LAYERS

500.0            1.40          0.02                  (c,λ,d) 'OUTSIDE PLASTER'
 75.0            0.23          0.20                  (c,λ,d) 'YTUNG BLOCK 800'
450.0            0.87          0.01                  (c,λ,d) 'INSIDE PLASTER'

180. 90.                                             WALL: AZIMUTH, INCLINATION
.65 .95                                              WALL: ALBEDO, EMISSIVITY
37.56                                                AREA OF WALL
 90. 90.   90. 90.                                   SC-SUMMER: DIR,DIF; SC-WINTER: DIR DIF
 1                                                   NO OF WINDOWS AND SOLAR ELEMENTS

WINDOWS AND SOLAR ELEMENTS    (ON WALL NO 1)

1          18.00                                     WINDOW:TYPE;AREA
10.  50.  90. 90.                                    SC-SUMMER:DIR,DIF;SC-WINTER:DIR DIF
18.  07.  5.  3.5                                    U WINDOW:SCHEDULE;DAY;NIGHT
```

heat gains (by solar radiation and internal heat gains), heat lost (through the building envelope), and heat storage (by thermal mass). It can be readily ascertained that ventilation and infiltration (V+I) are affected by several design parameters. The same is true for heat gains (H.G.), energy lost (E.L.), and thermal storage (T.S.). It is clear that when different design parameters influence heat gains, there will be a trade-off between these design parameters. However, a trade-off might also exist between heat gains and heat losses, as was presented in the example of Figure 18.1, where the energy consumption of standard-insulated dark wall was compared to energy consumption of a well-insulated white wall. The dark wall gained more solar energy, but the better insulated wall lost less. As a result, both design alternatives consumed the same amount of energy for heating.

The problem is even more complicated, as design parameters may influence the energy balance in different ways. For instance, the building envelope materials act on the building insulation and its thermal storage. Their different ways of influence can

TABLE 18.2 Physical Effects of Energy-Related Design Parameters on the Energy Balance of the Building[a]

PHYSICAL EFFECT	V+I.	H.G.	E.L.	T.S
TOTAL FLOOR AREA			+	+
VOLUME OF BUILDING	+		+	
NO.OF EXTERNAL WALLS			+	+
AZIMUTH OF REFERENCE WALL	+	+		
AREA OF INTERNAL MASS				+
HEAT CAPACITY OF INTERNAL MASS				+
INITIAL TEMP. DISTRIBUTION				+
HEAT GAINS: CONSTANT		+		
HEAT GAINS: SCHEDULE		+		
HEATER: SCHEDULE; TEMP.SET		+		
COOLER: SCHEDULE; TEMP.SET		+		
VENT: TYPE, SCHEDULE, ACH: DAY; NIGHT	+			
WALL NO 1				
NUMBER OF LAYERS			+	+
MATERIALS (c,λ,d)			+	+
" " "			+	+
" " "			+	+
WALL: AZIMUTH, INCLINATION		+		
WALL: ALBEDO, EMISSIVITY		+		
AREA OF WALL			+	+
WALL:SC-SUMMER,WINTER		+		
NO OF WINDOWS AND SOLAR ELEMENTS	+	+		
WINDOW: TYPE; AREA		+	+	
WINDOW: SC-SUMMER,WINTER		+		
U WINDOW: DAY;NIGHT		+	+	
U WINDOW: SCHEDULE		+	+	

[a]V.+I., ventilation and infiltration; H.G., heat gains; E.L., energy loss; T.S., thermal storage.

hardly be isolated. Even when the physical effect on the thermal performance of the building is known, we are not always able to point out the exact reason why that happens, unless we can isolate an energy-related design parameter that affects physically only one performance criterion. For example, it is possible to change the building envelope materials in such a way that they influence only the insulation characteristics of the envelope, without alternating its thermal mass properties. In the case shown in Figures 18.1 and 18.2, the better insulated building also provides a different thermal storing capacity.

While it is likely that many of the building components are considered at more than a single design phase, in different degrees of detail, their first consideration is often of primary importance. It is possible, therefore, to identify the design phase in which each design parameter is considered first. For example, the gross proportions and orientation of the building are most likely to be determined at an early design phase, with the type of glazing of the windows (reflective, triple pane, etc.) at a much later phase. Table 18.3 shows the design phase at which various energy-related design parameters would normally be considered first and the design phases in which they are reconsidered.

TABLE 18.3 Design Phases[a] at Which Various Energy-Related Design Parameters Are First Considered ✚, and Reconsidered +

DESIGN PHASE	a.	b.	c.	d.	e.	f.
TOTAL FLOOR AREA	✚		+	+		
VOLUME OF BUILDING		✚	+	+		
NO.OF EXTERNAL WALLS			✚	+		
AZIMUTH OF REFERENCE WALL			✚	+		
AREA OF INTERNAL MASS				✚	+	
HEAT CAPACITY OF INTERNAL MASS					✚	
INITIAL TEMP. DISTRIBUTION		✚				
HEAT GAINS: CONSTANT	✚					
HEAT GAINS: SCHEDULE	✚					
HEATER: SCHEDULE; TEMP.SET		✚			+	
COOLER: SCHEDULE; TEMP.SET		✚			+	
VENT: TYPE, SCHEDULE, ACH: DAY; NIGHT		✚			+	
WALL NO 1						
NUMBER OF LAYERS				✚	+	
MATERIALS (c, λ, d)				✚	+	
" " "						
" " "						
WALL: AZIMUTH, INCLINATION			✚	+		
AREA OF WALL			✚	+		
WALL:ALBEDO,EMISSIVITY					✚	
WALL:SC-SUMMER,WINTER				✚	+	
NO OF WINDOWS AND SOLAR ELEMENTS			✚	+		
WINDOW: AREA			✚	+		
WINDOW: TYPE				✚	+	
WINDOW: SC-SUMMER, WINTER		✚			+	
U WINDOW: DAY; NIGHT				✚	+	
U WINDOW: SCHEDULE		✚			+	

[a] The design phases are: a, briefing; b, preconceptual; c, conceptual; d, preliminary; e, detailed design.

INTEGRATING HEURISTICS AND SIMULATION FOR THE CONCEPTUAL DESIGN PHASE

Table 18.3 also shows why it is so difficult to use a simulation model in early design phases, where values of many building components have not yet been assigned, because typically they would be considered only at later design phases. The guestion arises: How should we, or the system, assign values to these design variables such that simulation could be performed?

For demonstration purposes, we shall discuss in detail the use of the combined procedural and heuristic method for energy-conscious building design during the schematic–conceptual phase of design. First, let us investigate the information presented in Table 18.3 to identify the informational characteristics of the conceptual design phase (phase c in Table 18.3). This will allow us to determine the energy-related pre- and postconditions of this design phase.

The informational characteristics of the conceptual design phase are presented in Table 18.4. It shows that most of the geometrical parameters in this design phase have

TABLE 18.4 Informational Characteristics of the Conceptual Design Phase[a]

INFORMATIONAL CHARACTERISTICS OF DESIGN PARAMETER	a+b.	c.	d+f.
TOTAL FLOOR AREA	**+**	+	+
VOLUME OF BUILDING	**+**	+	+
NO.OF EXTERNAL WALLS		**+**	+
AZIMUTH OF REFERENCE WALL		**+**	+
AREA OF INTERNAL MASS			**+**
HEAT CAPACITY OF INTERNAL MASS			**+**
INITIAL TEMP. DISTRIBUTION	**+**		
HEAT GAINS: CONSTANT	**+**		
HEAT GAINS: SCHEDULE	**+**		
HEATER: SCHEDULE; TEMP.SET	**+**		+
COOLER: SCHEDULE; TEMP.SET	**+**		+
VENT: TYPE, SCHEDULE, ACH: DAY; NIGHT	**+**		+
WALL NO 1			
NUMBER OF LAYERS			**+**
MATERIALS (c, λ, d)			**+**
" " "			
" " "			
WALL: AZIMUTH, INCLINATION		**+**	+
AREA OF WALL		**+**	+
WALL:ALBEDO,EMISSIVITY			**+**
WALL:SC-SUMMER,WINTER			**+**
NO OF WINDOWS AND SOLAR ELEMENTS		**+**	+
WINDOW:AREA		**+**	+
WINDOW:TYPE			**+**
WINDOW:SC-SUMMER,WINTER	**+**		+
U WINDOW:DAY;NIGHT			**+**
U WINDOW:SCHEDULE	**+**		+

[a] + under a+b, information that is available at the conceptual phase; + under c, information that concerns the designer at the conceptual phase; + under c, information that is reconsidered at the conceptual phase; + under d+e, information that is not available at the conceptual phase; + under d+e, information that will be reconsidered at later design phases.

already been determined, or at least they have been considered. These geometrical variables include the building orientation, which also defines the azimuth of each wall, the number of external walls, their area and inclination, and the number and area of all windows located in each wall. Also, the total floor area and the volume of the building may have been considered. At this phase the designer has already assigned certain values to parameters that were set during the early design phases a+b. To apply a simulation model at this phase, it is necessary to assign default values to design parameters that would normally have been determined only in phases d+f. These variables include the recommended *U* value for the building's envelope (including different wall compositions), types of windows and their glazing, the color scheme of the building, and the area and heat capacity of the thermal mass.

Two questions arise: First, what values should be assigned at the present conceptual design phase to parameters that would not be considered until later design phases? Second, how can values be assigned simultaneously to multiple parameters at the conceptual design phase? To answer these guestions, a preliminary study of the mutual influences of different design parameters, which should be considered during the conceptual design phase c and during the advanced phases d+e, was carried out. In this study, the knowledge base for a standard single-family house in two climatic zones in Israel was developed by running many simulations [Shaviv and Peleg 1989]. This knowledge base contains case-specific defaults of the design parameters that are to be determined in phases c and d+f. It defines the possible range of values for each design parameter that is considered at all design phases. It also includes the rules that guide the designer as far as when and how to run a simulation, in order to provide rigor to the heuristics used by early design phase expert systems.

The preliminary study that was carried shows that all design parameters can be characterized according to two main categories:

1. A design parameter that has a strong influence on the energy performance of the building
2. A design parameter that has a weak influence on the energy performance of the building

In each category we choose to separate our treatment of parameters that influence the shape of the building from those that do not, because architects (and their clients) would prefer to have the freedom to select the form according to their own taste, whenever possible. The design parameters can also be characterized according to the strength of their interdependencies. A change made to a parameter that is strongly interdependent must be accompanied by a thorough analysis to ascertain its effects on other parameters.

In Figure 18.4 we present some results that were obtained from the study that was carried out for a house in Tel Aviv climate. The design parameters in both cases belong to the subcategory of geometrical ones (the building's proportions and its orientation). In the first case (Figure 18.4*a*) one can see that changing the building's proportions (the ratio of the length of southern and northern walls to the length of eastern and western walls, in a rectangle floor plan, is changed from 0.8 to 1.6) has a small effect on the energy consumption of the building. On the other hand, a change in the orientation of the building, so that the large window is not directed due south, affects the energy consumption of the building significantly (Figure 18.4*b*). Other design parameters were examined in the same way. The characteristics of the energy-related design parameters, as found by the preliminary study, are summarized in Table 18.5. According to the characteristics of the energy-related design parameters presented in Table 18.5, the following policies are suggested:

1. **Phase c:** In this phase there are only geometrical design parameters.
   a. Strong influence
      (1) **Geometrical parameters:** The policy suggests that combined case-specific defaults be assigned by the KBS and tested by simulation–evaluation. In other words, the context sensitivity of the particular plan to the defaults recommended by the KBS should be ascertained by precise simulation. This will ensure that heuristic rules of thumb which are applicable only to average cases will not adversely affect nonstandard cases. A context-sensitivity analysis should also be carried out whenever the designer changes the recommended default values that were assigned to one design parameter which strongly influences energy performance. The following parameters belong to this subcategory: the orientation of the building (which defines the orientation of the windows and the solar elements) and the area of the windows and solar elements. It was found that the interdependencies be-

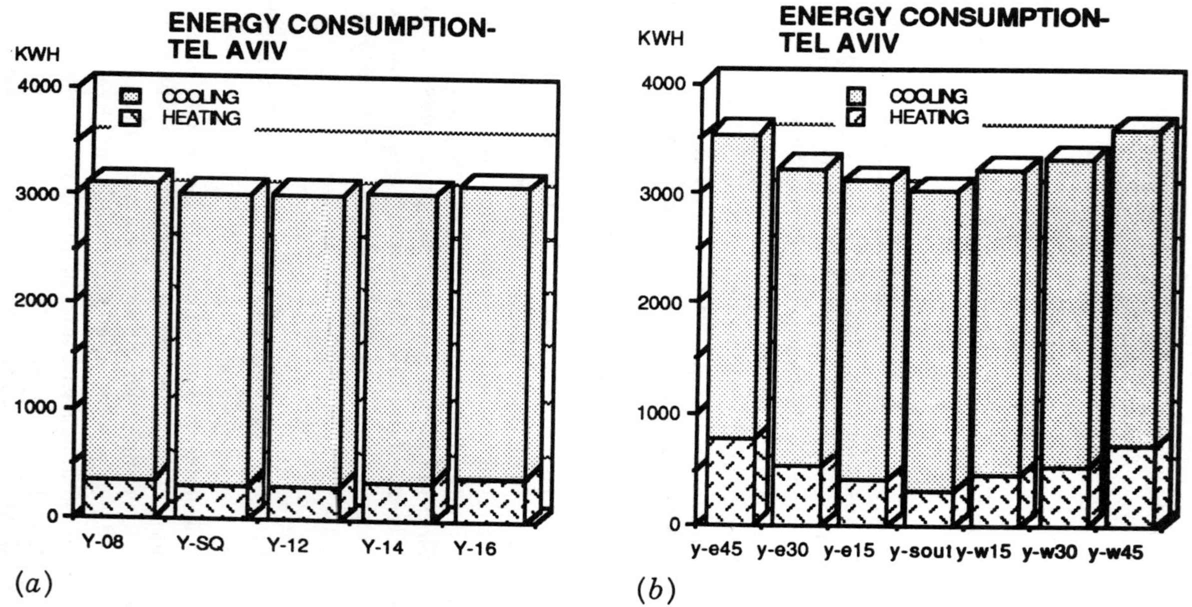

Figure 18.4. (*a*) The influence of the building's proportion on its energy consumption. (*b*) The influence of the building's orientation on its energy consumption. Both cases are for a standard-insulated house with a large southern window (Tel-Aviv climate).

TABLE 18.5 Characteristics of Energy-Related Design Parameters Considered in Phases c and d+e

DESIGN PHASE: c CHARACTERISTICS	INFLUENCE GEOMETRY	YES YES	YES NO	NO YES	NO NO
NO.OF EXTERNAL WALLS				+	
AZIMUTH OF REFERENCE WALL		+			
WALL: AZIMUTH, INCLINATION				+	
WALL'S AREA IN EACH ORIENTATION (BUILDING'S PROPORTIONS)				+	
WINDOWS AND SOLAR ELEMENTS: NO				+	
WINDOWS AND SOLAR ELEMENTS: AREA		+			
DESIGN PHASES: d+e CHARACTERISTICS	**INFLUENCE GEOMETRY**	**YES YES**	**YES NO**	**NO YES**	**NO NO**
INTERNAL MASS: AREA		+			
INTERNAL MASS: HEAT CAPACITY			+		
WALL COMPOSITION:LAYERS MATERIALS (c,λ,d)			+		
WALL: ALBEDO, EMISSIVITY					+
WALL: SC-SUMMER, WINTER					+
WINDOW: TYPE			+		
U WINDOW: DAY; NIGHT			+		

tween these two geometrical design parameters are very strong. Therefore, whenever the first one is changed, the second design parameter should be reevaluated [Shaviv and Peleg 1990].

b. Weak influence

(1) **Geometrical parameters:** The policy suggests that design parameters can be assigned values freely (e.g., graphically), by the designer, according to taste. These data will be interpreted by the system in the format required by the simulation model to allow a more rigorous evaluation of a specific design case, whenever needed. The following parameters belong to this subcategory: the building's proportions (which are defined by the area, azimuth, and inclination of each different external wall) and the number of windows and solar elements (but not their area).

2. Phases d + e

a. Strong influence

(1) **Nongeometrical parameters:** The policy suggests that case-specific defaults be assigned by the KBS. The following parameters belong to this subcategory: the heat capacity of the internal mass and all the building's external elements (walls, roofs, first floors, windows), including type, depth, and materials.

(2) **Geometrical parameters:** The policy suggests that case-specific heuristic rules be used to assign values. Data can be introduced graphically by the designer and should be continuously checked against the recommendations that are stored as heuristic rules in the KBS. This subcategory includes the area of internal mass.

b. Weak influence

(1) **Nongeometrical parameters:** The policy suggests that reasonable defaults be assigned by the KBS. This subcategory includes the color scheme and the shading coefficient of external elements.

According to the suggested methodology, a KBS for the schematic-conceptual design of passive solar buildings was developed by Shaviv and Peleg [1990]. This KBS uses simulations as a context-binding tool, and provides specificity to energy-related heuristic rules. By virtue of combining procedural and heuristic methods, the presented approach is able to test the applicability of heuristic rules, which are based on generalized experience, in nonstandard cases. The combined KBS–simulation tool comprises a more realistic design and evaluation tool for early design phases than former, pioneering systems do, such as ARCADE [Antony 1987] and the Gero-Coyne systems [Gero and Coyne 1984]. The design tool that was developed proves that the methodology suggested in this work can be realized.

CONCLUSION

Computer-aided design tools that have been developed for evaluating the energy-related performances of design alternatives are based either on procedural or on knowledge-based, heuristic methods. Each of these methods is applicable and useful at particular phases of the design process, but neither alone can assist all design phases. Consequently, many design phases where important energy-related decisions are made are not well supported by any computational methods.

Simulation techniques are based on procedural mathematical tools and provide accurate expected energy performance evaluation of fully articulated design alternatives. They are therefore very useful at later design phases. However, their accuracy and need for much detailed information limits their use at early design phases, when the information they need have not yet been determined, and the information they provide is too specific (e.g., quantitative rather than qualitative results). If simulation methods are to be used at early design phases, default values must be assigned to many design parameters.

Knowledge-based methods, on the other hand, derive their power from heuristics. Therefore, they need only gross design information and provide generalized appraisal of the design alternative. As such, they are suitable for evaluating the expected energy performance of typical design alternatives at the very early design phases, but are not suitable for evaluating the actual expected performance of detailed particular design solutions.

Approximate methods, which are generally suggested for the intermediate design stages, are based on procedural mathematical tools, such as simulation techniques, but provide only approximate energy performance evaluation of fully articulated design alternatives. Hence they suffer from the same drawbacks of both simulations and knowledge-based methods.

The methodology presented in this paper for intermediate design stages as well as for all other phases is based on combining the advantages of procedural approaches with those of knowledge-based approaches, to achieve the following goals:

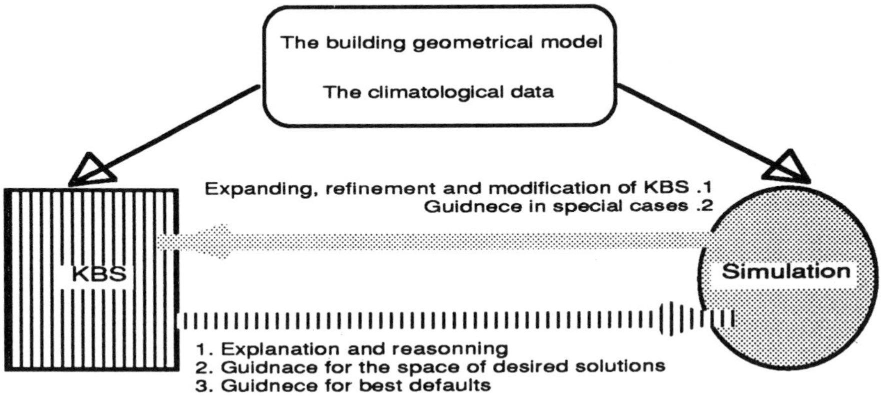

Figure 18.5. Advantages of combining KBS and simulation.

1. Provide energy-conscious decision making throughout all phases of the design process, using the same design tool.
2. Provide rigor to the heuristics used by early design phase expert systems, through determination of actual conditions as they develop in specific design cases. By testing the applicability of heuristic rules to particular design situations, it is possible to ensure that generalizations are not applied unduly, producing wrong results.
3. Provide context sensitivity to the selection of defaults by precise procedural approaches when the lack of information requires the use of defaults. As a result, the system is able to select default settings that are appropriate for particular design conditions rather than generalized ones, thereby improving the accuracy of the evaluation.
4. Provide a more explicit definition of the possible alternatives in all cases.
5. Provide an explanation facility, now available only in expert-system-based programs, to help the designer interpret the results obtained by accurate simulations.
6. Help derive new heuristics from the designer's own experiences.
7. Provide a learning tool to the inexperienced designer.

The integration of procedural approaches to energy evaluation of buildings with those of knowledge-based systems will avail the advantages and power of each method to the other method, as summarized in Figure 18.5. Furthermore, we strongly believe that this methodology is generalizable to other evaluation procedures and to the generation of design solutions as well. Therefore, other design processes that combine procedural methods with heuristic rules can benefit from the approach presented in this paper.

REFERENCES

Antony, S. (1987). "ARCADE: Architectural Computer Aided Design Evaluator," M.Arch. thesis, School of Architecture and Planning, State University of New York at Buffalo, Buffalo, N.Y.

ASHRAE (1972). *Handbook of Fundamentals,* American Society of Heating, Refrigerating, and Air-Conditioning Engineers, Atlanta, Ga.

Balcomb, J. D., R. W. Jones, R. D. McFarland, and W. O. Wray (1984). *Passive Solar Heating Analysis: A Design Manual*, LANL, ASHRAE ISBN 0-910110-38-7, American Society of Heating, Refrigerating, and Air-Conditioning Engineers, Atlanta, Ga.

Cross, N. (1977). *The Automated Architect*, Pion, London.

Gero, J. S. (1985). *Expert Systems in Design and Analysis*, Technical Report, Department of Architectural Science, University of Sydney, Sydney, Australia.

Gero, J. S., and R. Coyne (1984). "The Place of Expert Systems in Architecture," *Proceedings of CAD'84*, Butterworths, Sevenoaks, Kent, England, pp. 529–546.

Israel Ministry of Energy and Infrastructures (1984). *Introduction to a Master Plan for Energy Conservation in Buildings*, Technical Report (in Hebrew).

Jones, J. C. (1970). *Design Methods: Seeds of Human Futures*, Wiley, New York.

Kalay, Y., A. Harfmann, and L. Swerdloff (1985). "ALEX: A Knowledge-Based Architectural Design System," *ACADIA'85, Workshop Proceedings*, P. G. McIntosh (ed.), Arizona State University, Tempe, Ariz.

Kalay, Y. E., L. M. Swerdloff, and A. C. Harfmann (1987). "A Knowledge-Based Computable Model of Design," in *Expert Systems in Computer-Aided Design*, J. S. Gero (ed.), North-Holland, Amsterdam.

LBL and LASL (1982). *DOE-2 Reference Manual, Version 2.1.* Lawrence Berkeley Laboratory and Los Alamos Scientific Laboratory.

Maher, M. L., and S. J. Fenves (1985). *HI-RISE: A Knowledge-Based Expert System for the Preliminary Structural Design of High Rise Buildings*, Report R-85-146, Department of Civil Engineering, Carnegie Mellon University, Pittsburgh, Pa.

Mitchell, W. J. (1977). *Computer-Aided Architectural Design*, Van Nostrand Reinhold, New York.

Palmiter, L., T. Wheeling, and R. Judkoff (1982). *Solar Energy Research Institute-SERIRES: Residential Energy Simulator*, SERI Report, Solar Energy Research Institute, Boulder, Colorado.

Schmitt, G. (1987). "Expert Systems in Design Abstraction and Evaluation," in *Computability of Design*, Y. E. Kalay (ed.), Wiley, New York.

Shaviv, E. (1977). "A Model for Predicting Thermal Performance of Buildings and Its Use in the Design for Minimal Energy Consumption," *Building Center of Israel Quarterly*, **4:** 26–33, (in Hebrew).

Shaviv, E. (1979). "Building Design for Passive Energy Conservation," *PARC'79*, Berlin, pp. 135–144.

Shaviv, E. (1981). "The Influence of the Orientation of the Main Solar Glazing on the Total Energy Consumption of a Building," *Solar Energy*, **26**(5): 453–454.

Shaviv, E. (1989). "The Use of Simulation Models in the Schematic Design Stage for the Design of Energy Conscious Buildings." Proceedings of the 2nd European Conference on Architecture, "Science and Technology at the Service of Architecture," Paris, France, 1989. Edited by: T. C. Steemers and W. Palz. Commission of the European Communities, Kluwer Academic Publishers.

Shaviv, E., and U. J. Peleg (1990). "A Knowledge Based System for the Schematic Design of Passive Solar Buildings." ASHRAE Transactions V.ST.-96-2, American Society of Heating, Refrigerating, and Air-Conditioning Engineers, Atlanta, Ga.

Shaviv, E., and U. J. Peleg (1989). *The Influence of Geometrical Design Parameters on the Energy Consumption and Thermal Performance of Buildings*, Research Report No. 0220474, Technion Research and Development Foundation Ltd. Haifa, Israel (in Hebrew).

Shaviv, E., and G. Shaviv (1977). *A Model for Predicting Thermal Performance of Buildings*, WP ASDM-8 Report, Faculty of Architecture and Town Planning, Technion, Haifa, Israel.

Shaviv E., and G. Shaviv (1978a). "Modelling the Thermal Performance of Buildings," *Buildinq and Environment*, **13:** 95–108.

Shaviv, E., and G. Shaviv (1978b). "Designing of Buildings for Minimal Energy Consumption," *Computer-Aided Design*, **10**(4): 239–247.

Solar Conservation Study (1981). *A New Prosperity Building a Sustainable Energy Future*, Brick House Publishing, Andover, Mass.

Walton, G. N. (1983). "Thermal Analysis Research Program (TARP)," NBS Interagency Report (NBSIR) 83-2655.

PART SIX

HUMAN FACTORS EVALUATION

19

TOWARD ARTIFICIAL USERS

Edward Steinfeld

School of Architecture and Planning
State University of New York at Buffalo
Buffalo, New York

The evaluation and prediction of human response to building designs can be facilitated through computer applications. When selecting methods, two important issues should be considered: how to represent knowledge about human response, and how to introduce that knowledge into the design process. It is proposed that a modeling approach to knowledge representation and a conversational approach to knowledge integration are the most effective ways to develop evaluative and predictive tools. Taken together, they constitute "artificial users." Scenarios for implementing this concept are presented, based on extensions of existing concepts. It is proposed that artificial users would encourage a design process that gives greater priority to user perspectives within a reflective mode of design thinking.

INTRODUCTION

Human response to a design includes its perception, the ergonomics of its use, the impact of its intervention within human social systems, and its interpreted meaning. It is probably the most difficult aspect of performance to evaluate and predict before construction, but in the final analysis, it is the most important because a successful human response is the essence of successful design. Ultimately, it defines the meaning and value of built form. Currently, there are many methods that can be brought to bear in the analysis of human response to existing designs and somewhat fewer that are useful for evaluating and predicting performance during the design process. Computer-aided design tools are being developed to increase the power and efficiency of some methods.

The major problem in "evaluating and predicting human response" is not methodological; it is attitudinal. Currently, knowledge about human response is not used effectively. Only a small parcel of what we know ever reaches the building in most cases. What does find its way into practice is primarily through the imposition of regulations, client guidelines, or other rules that are often used blindly without a critical perspective. Computer technology can make a difference in utilization of knowledge because it provides the opportunity to manipulate information in new ways. By developing better forms to communicate knowledge more effectively through this new "design medium" [Gero 1985], perhaps our existing knowledge base can be applied more appropriately.

There are two major issues that need to be discussed: (1) how best to represent knowledge about human response, and (2) how best to introduce that knowledge into the design process. A method of representing knowledge is central to the development of any evaluative and predictive tool. Incomplete or inaccurate representations of knowledge lead directly to faulty results using the tool. As an example, if one evaluates the safety of a car using a checklist of "safe features" that does not include a passenger restraint system, an incomplete assessment will result. In general, there are two basic approaches to representing knowledge in evaluative and predictive methods. The first is the rule or norm. The checklist that purports to represent the essence of safety would be an example of this approach. The second is the model of behavior. An example parallel to the one above would be a manikin (or computer analog) that simulates the movement of a human body in the car during a crash.

In the field of architecture, knowledge about human response to buildings has been explicitly represented in design activities primarily through a normative process. Characteristics (i.e., goals) of the intended design are specified based on knowledge derived from experience with similar projects. This knowledge is usually codified from information obtained from precedents, the experience of the individual practitioner or firm, an outside "expert," the client, or an authoritative text. The normative approach is valuable and necessary to a certain extent. However, if used exclusively, it is very limited and perhaps dangerous at this time due to the relatively immature state of our knowledge on the subject. Our knowledge base is primarily of a general nature; application to specific cases such as a particular design context requires interpretation and judgment. A preferable approach is to represent such knowledge through models of human experience. This approach recognizes the divergent nature of each case and is more compatible with the imaginative mode of thinking so crucial for design tasks.

The integration of knowledge into the design process is influenced by the context of the process and the "style" of design practiced by the designer or design team. The context includes three elements: the social context (client, users, and other people or organizations involved in the process and affected by it), the physical context of site, and the time context (i.e., the stage of project development). By "style" I mean the behavioral style used to design, not the physical attributes of the product. The contexts of design projects put varying demands on the designer to search for and utilize information. In many cases, specific knowledge is often imposed on the designer by an outside agent (e.g., a building code or client). Different designers utilize information and search for it in highly variable ways as well. Thus it is obvious that the level and type of information used in any one particular project at any one particular time may vary significantly. To accommodate this variation, the integration of knowledge into the design process can best be accomplished through a "conversation" between the knowledge sources and the designer. It should be possible for each party to initiate that conversation. Furthermore, the conversation should, ideally, have all the characteristics of human conversation, including argumentative and negotiating modes.

These two ideas, the modeling approach to knowledge representation and the conversational approach to knowledge integration, combine synergistically in a concept that I call the *artificial user* (Fig. 19.1). The idea is to develop tools that convey knowledge about human response as if the user were standing next to the designer. From that vantage point, the user could be queried about the building or asked to simulate using it. He or she could also be watching the design develop with an informed eye and making unsolicited comments and criticism about it. The user could even suggest that the designer do a little role playing once in a while to study the design in his or her shoes. Of course, computer technology would allow much more sophisticated implementations than the singular tense would suggest. The plural is a more appropriate form for this idea since it would be most advantageous if there were a lot of users watching over the designer's shoulder all the time.

In the remainder of this paper, I endeavor to explain why this concept is the best approach to evaluating and predicting human response to building design and how it might be implemented. I first elucidate the limitations of the normative approach to representing human response knowledge and the advantages of the modeling approach.

Figure 19.1. An artificial user (Eduardo Paolozzi: Lepsius' Canon, 1960–1962. Published by permission of the artist.

Example 1
issue: Mounting height for controls
fact: A male wheelchair user of 50th percentile height has a comfortable forward reach of 55.5″.
rule: All electrical controls shall be mounted no higher than 48 in.
precedent: All the thermostats in previous projects were set at 54 in.
experience: "One day last winter, I was so cold that I was shaking. I kept adjusting the thermostat but, not being able to see the dial, I was turning it down instead of up and it kept getting colder. I thought there was something wrong with the heat so I called the maintenance man. Boy did I feel stupid and helpless when he told me what I had been doing wrong!"

Example 2
issue: Resident room design in a long term care facility.
fact: Everyone needs a personal territory.
rule: All single patient rooms shall be at least 125 sf. Double rooms shall be at least 200sf.
precedent: Typical patient room design.

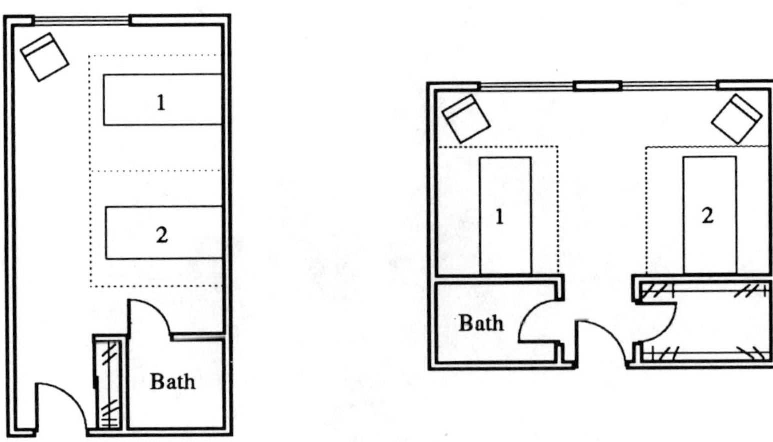

A: Deep Double Room B: Shallow Double Room

experience: "My last roommate was a real case. She had the bed near the window and wouldn't let me even sit in the chair over there when I wanted to watch the birds at the feeder."

Figure 19.2. Examples of knowledge.

Second, the need for the conversational approach to knowledge integration will be demonstrated. In the third section I present some speculations about how "artificial users" could be developed within the realm of computer-aided design using existing concepts and the impact they might have on the field of architecture.

KNOWLEDGE REPRESENTATION

Before going further, some definitions are necessary for the sake of clarity. In the literature on CAD, the definition of knowledge places an emphasis on facts and rules. Rules are sometimes defined as facts and sometimes as relations between facts. Consulting a few dictionaries, I realized that the more commonly used definition of knowledge puts the emphasis on awareness and experience, and rules are commonly defined as principles or guides for action. Moreover, it is important to clarify the definition of *precedent*, which is a special type of rule. The definitions I liked best were in my *Random House Dictionary*:

> Knowledge, n. 1. acquaintance or familiarity gained by experience. 2. the fact or state of knowing. 3. what is or may be known. 4. the sum of what is known. 5. the body of truths or facts accumulated by mankind.
>
> Rule, n. 1. a principle or regulation governing conduct or procedure. b. the customary or normal procedure, etc.
>
> Precedent, n. an act, decision or case that may serve as a guide or justification in subsequent ones.

Although knowledge can be defined as a body of facts and rules, that alone would be an incomplete representation. The differences between facts, rules, and experience are significant. They can be illustrated with the examples shown in Figure 19.2.

Each of the items of knowledge above can be useful in design. But each will be used in a different way. Most knowledge used by designers about human response to buildings is obtained in the form of experience or rules (including precedents). Experience is obtained primarily firsthand from the designer's own life or secondhand from a limited number of informants. The use of rules is a very efficient way to communicate knowledge to the designer. They serve as "how to" statements that demand little reflection. But they also provide little awareness about the issues behind them. Precedents are even more efficient. As soon as the designer accepts a precedent as a guide for a new design, it becomes a normative statement, a rule. The drawing or actual building embodies a complex set of rules about how the environment should be structured to accommodate human needs. These embodied rules include what is allowed as well as what is not allowed. Note that the personal territory provided in sketch A in Figure 19.2 is considerably different than in sketch B. In sketch A the inhabitant of bed 2 has the inferior territory because access to the windows is controlled by the inhabitant of bed 1. All circulation between the far end of the room, the entry, the closet, and the bathroom will also be an invasion of bed 2's territory. Sketch B, on the other hand, preserves territorial control.

If adopted, either in whole or in part, a precedent requires no decoding to uncover the embodied rules. Facts, although interesting, are not very useful to designers because they require interpretation. For example, the two "facts" in Figure 19.2, taken in isolation, given little assistance to the designer in decision making. Descriptions of experience raise awareness, stimulate reflection, and create a more complete mental picture of the issue at hand. That mental picture also has an emotional content that makes it difficult to ignore.

Designers and others (clients, officials, etc.) are more likely to understand a human response issue if they are exposed to it from the user's perspective. The best way to get that exposure is to have the experience directly, but that is rarely possible, particularly since the designer is often a very different type of person than the user. Without the opportunity for direct experience, access to a simulation of that experience is the next best thing. But to convey all the important information, simulations must be far more than purely visual walk-throughs of a building, as the examples above illustrate. A simulation is a model of experience. Its key attribute is the creation of a virtual world for the designer to visit, one where he or she can construct a mental image of what it is like to live in the kinds of places being designed. This will help the designer evaluate the success of the design by imagining (predicting) what life will be like in the eventual building. Virtual worlds help test ideas for action and allow reflection on their impact [Schön 1983].

There are many methods of modeling experience that are already in fairly common use (see [Clipson 1988]). Mathematical and gaming models, even computer based, exist that are used to simulate economic and social behavior. Empathic models have been developed to simulate the perceptual experiences of people who have physical, sensory, and cognitive processing limitations or the experience of living in an exotic place (e.g., zero gravity). Simulations of anthropometric characteristics in the form of manikins, either two-dimensional, three-dimensional, and computer animated, have been developed. In the social sciences and humanities, there are other forms of representation that are used to create models of experience. These include the imaginary worlds of fiction or poetry and the reconstructed worlds of history or participant observation. All these approaches have the shared characteristic of representing, as observable experience, the perspective of specific people in specific contexts.

KNOWLEDGE INTEGRATION

Why is it that current research-based knowledge about human response to buildings is not often applied in design? This question is asked *ad nauseam* by those of us who do this research. The form of information is clearly an important variable [Korobkin 1975, Seidel 1980]. Research that is reported with clear implications for design (i.e., rules) and illustrated with graphics are more "useful" to architects than the typical scholarly journal format. Architects are more likely to rely on their own experience, precedents, personally known experts, and professional journals for information rather than sources like academic research journals or books [Seidel 1980]. Professional attitudes toward the form of knowledge and the social nature of the design process are therefore critical factors affecting the integration of knowledge into design practice.

Schön [1983] argues that the dominant model of professional knowledge in the second half of the twentieth century has been technical rationality, the application of Positivist philosophy to professional activity. The view of professional practice supported by this model is "instrumental problem solving made rigorous by the application of scientific theory and technique" (p. 21). This view, Schön observes, is in contradiction of what professionals actually do in action. Professional practice is concerned as much with problem setting as with problem solving. Scientific theory and principles cannot be applied readily to many situations faced in the real world because there is often too much complexity and uncertainty surrounding them to use rigorous methods. Science is an activity that seeks to develop a convergent knowledge base, while professional practice deals with a real world that always has unique and unpredictable elements. Practitioners develop divergent thinking skills tailored to the unique requirements of their clients [Schön 1983].

Central to professional practice is *reflection in action.* This involves the use of tacit understanding, or know-how that has been developed through repeated experience. Confronted by the experience of a new or uncertain situation, skilled professionals allow themselves to reflect, which Schön likens to an "experiment" to generate new understanding [Schön 1983, p. 68]. Reflection involves surfacing and criticizing tacit understandings and results in reframing the situation and perhaps the role of the practitioner as well. Professionals, in fact, engage in conversations with the situations they are shaping. However, some professionals are better at it than others. Schön observes that many become locked into a view of themselves as technical experts and neglect the importance of reflection in action. They become skilled at selective inattention, junk categories, and situational control. To them, uncertainty is a threat rather than an opportunity and its admission is considered a sign of defeat.

Extending the line of Schön's observations, Archea [1987] has proposed that architects are not problem solvers but rather, puzzle makers. By this he means that architects are concerned primarily with bringing their own sense of structure to a given situation rather than responding to the situation as given. He proposes that architects view building design projects as unique opportunities to be discovered and exploited. Thus it is not always certain that they will solve problems; they may even create them. In his view, instrumental problem solving approaches are not consistent with the real strengths of professional practice. Thus architects are not likely to be particularly interested in conventional scientific knowledge bases. The architect's clients, regulatory bodies, and others involved in building decisions would be a more receptive constituency for the results of research on building performance.

Another way to look at architectural design, or for that matter, professional practice in general, is a dialogue between the pragmatic (i.e., technical) and the imaginative. Puzzle making is part of the imaginative side of the dialogue and certainly an important part, but much of architectural design involves dealing with pragmatic problems of structure, construction, habitability issues, circulation, cost, and so on. Although it is true that many of the pragmatic concerns previously handled by architects have now been delegated to their consultants, it does not mean that those professionals do not use imaginative thinking in their own work. Moreover, there are many technical issues that are still primarily the concern of the architect. Well-defined problem statements often precede design activity itself and certainly give inspiration to form making. Many examples of meaningful architecture begin with a pragmatic focus and use that to inform the imaginative process. Puzzle making, in other words, is not the exclusive domain of the architect, nor is problem solving the exclusive domain of the architect's consultants. The degree to which pragmatic and imaginative thinking are used in any particular project is probably determined by the proclivities of the actors in the process or the nature of the project context.

Now we might ask if it is even possible to consider evaluation and prediction of building performance as a feasible enterprise in practice? The concept itself may be too alien for the practitioner's mode of thinking because such analysis is generally conceived as a convergent, scientific activity. I believe it is feasible if we reinterpret that activity with respect to the goals of practice. Evaluative and predictive tools will be valued by practitioners *if they help them explore and structure the uniqueness of each situation.* But perhaps they will be valued even more by other actors in the design process, as Archea concluded.

The design of any significant building is an activity that involves many people—in most cases, the singular "designer" is obsolete. Clients and regulatory agencies are increasingly active in the process. Not only are many types of professionals involved, but the general public often plays a major role, even with respect to issues such as aesthetics. Two good examples are the recent public debate surrounding additions to the Whitney and Guggenheim Museums in New York City. Building design is almost always a social act. The process can be viewed as symbolic interaction. In this view the degree of involvement by the various participants reflects their relative power and influence. The building itself, since it embodies the values of the participants, reflects a negotiated order among them. Thus in addition to the interior dialogue of pragmatic and imaginative thinking, another, external dialogue is operative, involving communication and argumentation between the various actors.

The emphasis on imaginative thinking and interaction will vary considerably from project to project. Gutman [1988] reports on research that identified three types of architectural firms based on type of practice: strong idea firms, strong service firms, and strong delivery firms. Clients increasingly seek out the type of firm that will fulfill their objectives for a specific project. Where the "strong idea" firm is involved, we would expect that an imaginative approach (i.e., puzzle making) will be highly evident, and interaction, if it occurs at all, may be discounted. Where "strong service" firms are involved, we would expect that active communication would be the priority and the emphasis on imaginative thinking would be determined by the problem context. "Strong delivery" firms would probably be involved with projects that are routine or routinized and where neither dialogue nor imagination is particularly important; in fact, they may be considered a liability.

Given that design is a social act, the role of research on human response can be understood as an attempt to introject the perspective of users into the design process, where they have generally been underrepresented. Through development of knowledge about users, researchers hope to increase the attention given to this group as a participant in design. But given that researchers usually adopt a technical-rational mode of thinking, the attempt to influence the design process falls short. Users have to become part of the dialogue of design in the same way that other actors are involved—by being part of the project's social context or the design team itself. In this way they will become engaged in the conversation of design, not simply a subject of the conversation. Direct participation is usually, but not always, impossible since the design team cannot possibly ask everyone to join the discussion and, in fact, not everyone wants to because they have other things to do. But with computer technology it is possible to create sophisticated artificial representatives of users through new forms of dialogue and models of human behavior.

Many computer programs have been developed to simulate the designer's behavior [see, e.g., Shaviv [1987] and Flemming [1987]). At least one program simulates the designer while involving real building users as active operators [Kalay et al. 1985]. In fact, going back to the early stages of CAD, there has been a strong emphasis on creating artificial designers, but no one has yet considered the artificial user. Perhaps this neglect is consistent with the general trend in the profession to ignore the perspective of the user. It should be noted here that the idea of artificial users is different from participatory design. In the latter, the designer delegates decision-making power to the user. But the artificial user does not necessarily have to make design decisions. The role of the artificial user would be to answer the design team's questions, demonstrate their response to alternative designs, and interrogate the designer about aspects of the design. Various models could be developed to represent many different kinds of people. Of course, in the short run, it would be difficult to construct a full-fledged artificial "society" or even one representative. However, many concepts already exist that can be used to take steps in that direction.

Whatever tools are developed, they must be flexible and evolutionary. The knowledge base about human response to buildings is far from comprehensive and may be seriously flawed in many areas. The application of inadequate information about human behavior to building design has caused major problems in the past. For example, in the design of institutions, stairways, emergency egress, and low-cost housing, "knowledge" about human response to buildings has been explicitly applied. The result has been less than satisfactory in many cases, and in some, disasters have occurred. The most notorious example is the Puitt–Igoe Public Housing Project, which was partially demolished because it was so dangerous. This project was designed using a knowledge base derived from social research that within a short period of time was proven to be seriously flawed. Even though knowledge bases have been demonstrated to be inadequate and misleading, they continue to influence the design of contemporary buildings (see, e.g., Pauls [1980]). A primary reason for this lack of evolution in knowledge utilization is the codification of that knowledge into building regulations and standard reference books. Once this occurs there is great vested interest within the profession and among regulatory authorities to maintain the status quo.

A danger of computer applications is that they may, unwittingly, create a dependency on a *source* of information rather than encourage better ways of thinking. The mere existence and popularity of computer-based tools could suppress the introduction of new and better ideas and information, especially if they are developed and controlled by regulatory bodies. If knowledge in this field was mature and supported by predictive theory, the risk might not be so great. It is important to remember that we are talking about issues that affect health and safety of building occupants. Lack of flexibility and evolution in a knowledge base could result in unnecessary pain and suffering for building users.

SCENARIOS

Electronic Intelligent Bulletin Board: Discovery and Argumentation

Noble and Rittel [1988] described an *issue-based information system* (IBIS) that would allow the construction of interactive databases through which many participants can participate in an "argumentative" process, offering opinion and perspectives on issues

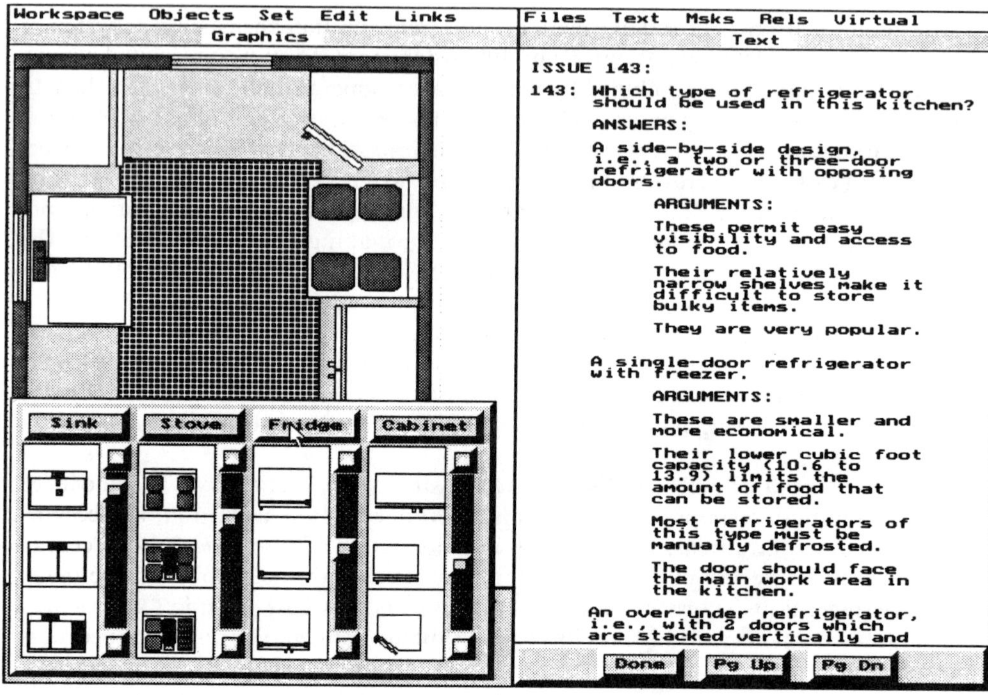

Figure 19.3. PHIDIAS output showing IBIS-like dialogue.

that evolve in the course of a design project. All discussion of the project can be documented through this system. Items are linked so that the development of discussion and relationships between issues, positions, and arguments can be analyzed. McCall et al. [1989] developed a CAD program integrated with an IBIS-like database for the design of kitchens. Such databases act like "critics," that analyzes the design in process and display criticism based on well-known design principles (see Figure 19.3).

Similar approaches could be developed within the framework of an electronic bulletin board. Resource information about human response to design could be posted onto such boards and a continual dialogue of opinion about the knowledge base could be maintained. The result might resemble a diverse and pluralistic *pattern language* [Alexander et al. 1977] with which researchers, designers, and others could interact. Case-oriented data could be linked to basic knowledge and opinion so that the knowledge in the system would become more and more divergent as more cases are entered. Information could be provided in a variety of modes: fact, rules (including precedents), and experience.

Imagine . . .

K. Solomon, the architect, received a major commission—the design of an outpatient cancer care clinic. He had never designed this type of building, but he knew that there were major human response issues involved in the project, so he convinced the client to subscribe to a new Compusource (an electronic information network) service for building design—"Engage." He and the client's representatives have been using the system to document all the discussion about the problem statement. They even recruited a panel of users from another clinic owned by the sponsor to engage in the dialogue. The preliminary design is now under way.

Solomon signs onto Engage to search for precedents. In the Ambulatory Health Care Area he finds 10 entries. Skimming through the Project Documentation File, he finds a similar clinic designed with a central plan surrounding a service core with nursing support. He likes the central plan idea but decides to check out experiences with it. The Dialogue File shows that there were no evaluations done on the building, but he discovers that several opinions of the core plan concept were recorded during the project's development. A psychologist thought that the plan would disorient people who were leaving the building after extensive chemotherapy. She felt that it would be best for the reception area to be visible from all cubicles. Solomon checks the file to see why the architect had not taken that advice. The architect and owner had responded that there would be a limited number of chemotherapy patients in the facility at one time so they could delegate the cubicles close to the reception area for them. Solomon moves to another window and checks the projected clientele for the new facility. He notices that there will be 67 percent chemotherapy patients, so he decides to abandon the central service core idea. He explores another central plan design with a skylit space in the center and the reception area at one end.

Later, he has to decide whether the cubicles will have doors or curtains. He moves to Issues Mode again and enters "privacy in treatment rooms." There is a four-star entry from the Knowledge Base which recommends that all treatment rooms have acoustical privacy. He trusts this suggestion because he knows that no recommendation can receive four stars unless a committee of experts on the subject have all agreed and it is based on systematic research. However, there is also a one-star entry recommending locks. He copies this recommendation into the project Dialogue File Agenda for later discussion with the project panel and the client.

Three months later it is time to decide on locks. By this time the panel and the owner have reviewed the recommendation and have left responses. The owner does not want them—too risky and, besides, they cost money. One of the users thinks they are necessary because, on several occasions, another patient walked in on him while he was undergoing treatment, and the resulting embarrassment caused him stress that provoked violent nausea. He also asks Solomon if he has included a call system, because patients are often left alone for a long time and sometimes need assistance. Solomon copies that response and suggestion to the Agenda for the next discussion with the client. It is also entered automatically into the Project Documentation. When the copy is sent, a message returns shortly, indicating that the client has read it; he agrees with the panel member's lock recommendation and instructs Solomon to include the call system as an alternate in the bidding.

Simulation Games: Awareness and Analysis

Sim City is a recently released interactive computer game that allows the player to take control over a simulated city. The game comes with "preconstructed" cities such as Rio or Tokyo, but players can build their own city. The object of the game is to manage events of urban life such as an attack by King Kong or an earthquake (see Figure 19.4). The player functions as the mayor, zoning board, budget director, and city planner. The

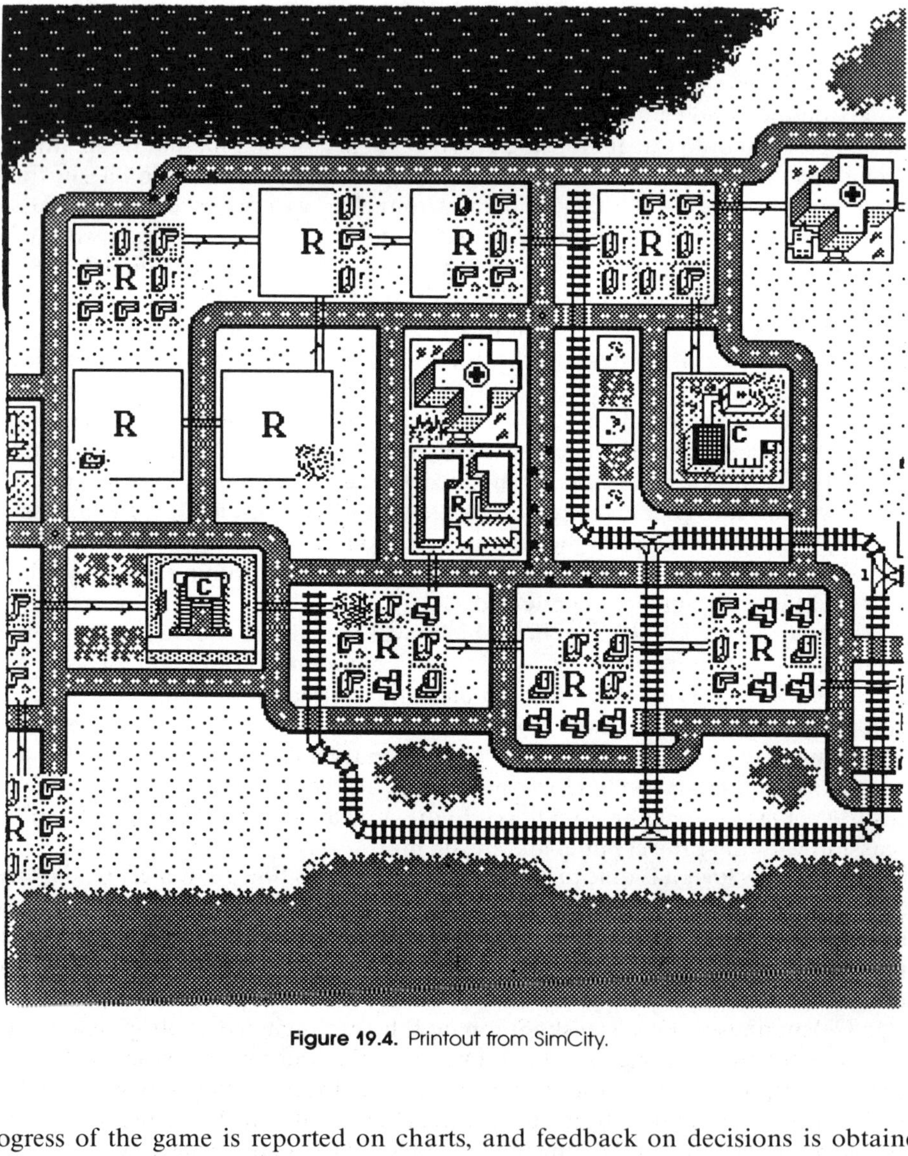

Figure 19.4. Printout from SimCity.

progress of the game is reported on charts, and feedback on decisions is obtained through such devices as public opinion polls. A reviewer writes: "As you play you learn a lot about how cities work, how mass transit and parks and airports and police stations all affect the urban calculus. But it doesn't feel like learning, it feels like fun" [Levy and Mello 1989].

The game format can be a highly effective way to develop awareness and even to communicate basic principles from research. Games may also be used as an exploratory tool in real design projects. Further, they could introduce simulation modeling tools that could be applied independently from the game itself. Games could simulate life in buildings of various types. Basic decisions about the nature of the building can be input that would change the simulated life experience, for example, living in a single as opposed to a double room. It could also be possible to experience life in a simulated building with different identities (e.g., a mentally impaired resident as opposed to a physically disabled but mentally alert resident).

Imagine . . .

J. Hasty has just taken a job with Enormous Associates, a firm with mostly institutional clients. She has been hired to work in the firm's Correctional Facilities Group but has never before designed such a building. Using SimShell, a program generation shell for gaming simulations, the firm's computer systems team has developed a game called SimPen which simulates life in a maximum security prison. This game is used by all the staff to become familiar with prison life. Players take the role of guards, warden, or prisoners and simulate a typical week within the prison. The game allows the player to "design" a prison, making a series of decisions about overall organization, size of living units, design of recreation space, security arrangements, cell design, and so on. A consulting sociologist has helped to incorporate knowledge about prison operations, including the impact of various design features into the simulation. The program is so effective that Enormous Associates has been hired to complete a study of prison environments to make the game even more comprehensive, so that it might be used in short-term training programs for corrections officers.

Hasty is using the game on her second day at work. She is running it with a predesigned prison. It is an ancient building that has been renovated and expanded several times over its lifetime. She has taken the role of a prisoner and is now starting her first "day." She studies the computer-assigned character and is surprised to discover that she is "Doc," a disabled man who uses a wheelchair and has tested positive to the AIDS virus. She knew that AIDS was prevalent among the prison population because prisoners participate in intravenous drug use and homosexual activity but she never had imagined that prisoners might be disabled. She soon finds that several areas of the prison are not accessible in a wheelchair. This restricts her character's involvement with some important aspects of prison life. He is forced to spend most of his time in the cell block area since not enough staff is available to supervise his passage through nonsecure areas in order to reach industrial work sites or the visiting area. The only accessible routes require leaving the building and reentering at different locations. This limitation turns out to be very serious since he soon has a visit from an AIDS counselor. Because of the lack of privacy, it leaks out to the other prisoners that he is an AIDS carrier. Later, his cellmate stages a protest about being confined in the same cell and using the same toilet. Doc receives a vicious beating from the cellmate when his request to be moved out of the cell is denied.

After the beating, Doc is visited by a Prisoner's Rights lawyer working on a task force to examine the plight of AIDS carriers integrated into the general prison population. The lawyer leaves some literature that describes various sides to the debate on segregation versus integration and summarizes the experience and policies of different states. . . .

After completing many rounds of the game, the program determines she should have learned everything there is to learn. Hasty is then asked to evaluate several prison designs. Plans of each design are displayed and she learns how to use a toolkit of checklists, simulations, and analytical models to evaluate the designs. One is a wheelchair-user manikin that can be positioned anywhere in the building and moved through the space to assess accessibility. Another is a program that identifies levels of visual access

Figure 19.5. Virtual reality system manufactured by VPL Research, Inc. (published by permission of the company).

and exposure in any space. In the case of prison design, it used to identify any hidden areas that might create a security problem. The program does not let her finish until she has mastered the use of the toolkit and completed a quiz on her knowledge of prison life. At the end of the quiz, she is referred to supplementary readings in the prison "library". Tomorrow, she will complete those and take another quiz. When she starts her first design assignment, she will be using the toolkit and the library on a routine basis.

Virtual Reality: Inhabiting the Model

A current exhibit at the National Air and Space Museum in Washington, D.C., features a film in which pilots discuss the use of contemporary flight simulation machines. These devices are airplane cockpits suspended within a structural framework by movable supporting rods. A hydraulic system adjusts the orientation of the cockpit and thrust forces applied to it in response to the cockpit controls. The crew in the cockpit sees the simulated environment through which they are traveling through the windows of the cockpit and gets audio and instrument feedback about the position and condition of the plane. They also feel the plane moving as the hydraulic system adjusts to their control. In the film, a jumbo jet pilot gives a vivid testimonial on the value of these systems by explaining that commercial pilots no longer have to train in a real plane before taking the controls of a new model; the simulator is so realistic that the pilots can move directly from using it to piloting the plane on their own! These flight simulation systems are extremely expensive, but the basic idea can be applied to architectural needs with much less equipment at significantly lower costs. As early as 1968, Sutherland developed a machine that gave the illusion of walking within visual models created through computer graphics. The machine took the form of a helmet that permitted the user to move around in space while seeing, through display devices in the helmet, a real-time stereoscopic view of an environment that did not exist in the real world.

Several companies are now testing virtual reality systems that can be used with CAD programs to allow the exploration of buildings under design. The most advanced of these systems include goggles with video screens for stereoscopic vision, binaural sound systems, and tactile feedback through a glove or complete body suit (see Figure 19.5). Users are completely immersed in an artificial world that they can explore at will and even feel [Heilbrun 1989, Tello 1988, Foley 1987].

Imagine . . .

Positive Space Design, PC, has just received a beta test model of the new Dr. Feelgood Virtual Reality System. Two architects in the firm, Chen Siyu and Jorn Johansen, are eager to try it out on a current project, a museum of science and technology in Duluth, Minnesota. Because of the long history of mining in the area, the architects have decided to integrate the building with the ground in some way. They have developed an earth sheltered scheme. At this stage of the design, they are particularly interested in how the building will be perceived during the approach and entry sequence. They want to achieve a feeling of integration between the natural forms of the surrounding landscape and the artificial forms of the building. The surroundings have been recorded on video and entered into their CAD modeling program and they have overlaid the design concept onto the context. They now specify the audio environment from a library of prerecorded background noises and use a modeling program to simulate acoustic performance of the building. A lighting model is also invoked by selecting and locating generic fixtures. Cloud cover is generated to reflect conditions at the site during early winter.

Chen is the first to get her hands on the system. It consists of goggles, earphones, gloves, and booties. She dons these and enters a small inflatable dome equipped with a wind generation system and a recording device to save the user's protocol of the experience. She "takes a walk" from the parking lot to the entry. She realizes that the nearby highway is a distracting influence and suggests that the perimeter of the parking lot be bermed to deflect some of the noise and hide it from view. She moves toward the entry. Actually, she just walks in place on a treadmill; the rate of her travel can be adjusted to her pace. She notes that the building is not prominent enough in the field of view. The sky is so huge in relationship to the building that her attention is diverted. In addition, she keeps losing visual contact with the entry as she negotiates through the cars in the parking lot. Therefore, she suggests extending the entry walk to a landscaped path through the parking lot. She finds that the overall distance from the edge of the lot to the entry is too far, particularly since the sky is threatening rain and the wind has picked up. She reaches the entry and passes through to the lobby. She notes that the change in light level is too abrupt and suggests increasing the size of clerestory windows. She notes that her view of the main exhibit gallery is too direct—it does not create the feeling of mystery they were trying to achieve.

Jorn then tries the system without listening to her protocol. Later, a comparison of their protocols demonstrates consistency in their criticism, with the exception of the distance between lot to entry. Jorn realizes that the difference in their experience was due to the fact that Chen is 4 ft 11 in. tall and he is 6 ft 3 in. They resolve to adjust the user mode to experience the building from other than their own perspective in future trials. Changes are made to the design, all the while using the system to check the impact. . . .

During the working drawing stage, they program the system to simulate a variety of users and take many tours of the building. The new Dr. Feelgood system has a physiological screen mode that allows them to simulate the vision and hearing of an older visitor. Using this mode, they discover that the increased size of the clerestory windows in the lobby has created too much glare, so they change their orientation. They also find that the entry canopy design casts shadows that create camouflage on the walking surface. They revise the canopy. Since children will be major users of the museum, they simulate children's stature and decide to change the signage system they had originally selected so that children could see the signs more easily. . . .

While the building was being designed, the Exhibit Systems Department of the firm has been working on the major exhibits of the building with the museum's curators. Before working drawings are completed, they merge the exhibit design models with the building design and do a final set of trips through the building. During these runs they adjust lighting, acoustics, and other details of the design. The client and staff also use the system to evaluate the design before accepting it.

CONCLUSION

Computer applications concerned with evaluating and predicting human response to buildings should not be focused solely on the automation and enhancements of traditional methods of knowledge application in the profession of architecture. This approach could cause results as pernicious as the institutionalization of flawed knowledge in earlier times. The contemporary posture toward knowledge applications is characterized by the practice of privatizing knowledge within a professional firm, a preference for mechanistic thinking as opposed to informed inquiry, a tendency to rely on personal experience rather than a generalized knowledge base, and reliance on limited authoritative sources of knowledge. The overall pattern can be described as a "resource conserving" approach to the search for knowledge. If applied to development and utilization of computer applications, it is likely to lead to a narrow and potentially dangerous application of knowledge with little outside control over the quality of information being used by professional firms.

On the other hand, computer technology holds the promise of new and more effective approaches toward the implementation of knowledge on human response obtained from research. These new approaches might improve our understanding of the subject and help apply knowledge in ways that have less chance of being misused in practice. They might also lead to advances in general design theory. The development of high-level visual reality, for example, may provide opportunities to reduce the cost of simulating environments to a point where behavioral modeling can become as integrated into design tasks as cost estimating is today. The use of computer networking coupled with computer graphics offers the opportunity to enhance communications with experts and rapidly develop consensus opinions on specific design problems for which existing research has no clear-cut answers. The use of interactive simulation games provides a powerful approach to educating practitioners about design issues in specific environments and the use of computer-based analytical tools.

To capture the potential of computer technology and avoid the detriments, both researchers and practitioners will have to adopt a new way of thinking and acting. Researchers must consider the adoption of computer-aided tools for communicating and organizing findings. Those who develop computer applications must avoid providing handy "drafting board" companions in a new guise and concentrate instead on developing tools that provoke inquiry and reflection, open up communication between practitioners, clients, researchers, and building users, and enable behavioral analysis of designs without making decisions for the designer. It will be critical to remember that there is no standard human response to an environment. Rather, there are many different responses based on the situation of the individual user. Tools that help designers to understand this perspective will be most beneficial to the profession and to society. They will help to reinforce the humanistic rather than mechanistic aspects of design practice.

> The feedback link through which man with his consciousness relates to his appreciated world is perhaps the most crucial one. It is here that the alternative between positive action and mere self-contained behavior is prepared. It is here that the whole rational framework for the justification of such action is welded together. But it is also here that the rational, the broader experiential and the plain irrational come together in greatest intimacy, touch and interpenetrate each other. It is the feedback loop of models and myths [Jantsch 1975].

REFERENCES

Alexander, C., S. Ishikawa, and M. Silverstein (1977). *A Pattern Language*, Oxford University Press, New York.

Archea, J. (1987). "Puzzle-Making: What Architects Do When No One's Looking," in *The Computability of Design*, Y. E. Kalay (ed.), Wiley, New York.

Clipson, C. (1988). *Simulating Future Worlds*, Architecture and Planning Research Laboratory, University of Michigan, Ann Arbor, Mich.

Flemming, U. (1987). "The Role of Shape Grammars in the Analysis and Creation of Designs," in *The Computability of Design*, Y. E. Kalay (ed.), Wiley, New York.

Foley, J. D. (1987). "Interfaces for Advanced Computing," *Scientific American*, October.

Gero, J. S. (1985). "On Overview of Knowledge Engineering and Its Relevance to CAAD," in *Computer-Aided Architectural Design Futures*, A. Pipes (ed.), Butterworth, London.

Gutman, R. (1988). *Architectural Practice*, Princeton Architectural Press, Princeton, N.J.

Heilbrun, A. (1989). "Virtual Reality," *Whole Earth Review*, Fall.

Jantsch, E. (1975). *Design for Evolution*. Braziller, New York.

Kalay, Y. E., A. Harfmann, and L. Swerdloff (1985). "ALEX: A Knowledge-Based Architectural Design System," in *ACADIA'85* Workshop Proceedings, P. McIntosh (ed.), Arizona State University, Tempe, Ariz.

Korobkin, B. (1975). *Images for Design: Communicating Social Science Research to Architects*, concept paper, American Institute of Architects, Washington, D.C.

Levy, S., and A. Mello (1989). "The Game Hall of Fame," *MacWorld*, December.

McCall, R., G. Fischer, and A. Morch (1989). "Supporting Reflection-in-Action in the JANUS Design Environment," in *Computer-Aided Architectural Design Futures '89*, M. McCullough, W. J. Mitchell, and P. Purcell (eds.).

Noble, D., and H. Rittel (1988). "Issue-Based Information Systems for Design," *Proceedings of ACADIA '88*.

Pauls, J. (1980). "Building Design for Egress," *Journal of Architectural Education*, **33**, Summer.

Schön, D. A. (1983). *The Reflective Practitioner*, Basic Books, New York.

Seidel, A. D. (1980). *Factors Influencing Conceptions of Information Quality in the Production and Utilization of Environment and Behavior Research.* Unpublished manuscript. Buffalo, New York: School of Architecture and Environmental Design.

Shaviv, E. (1987). "On Generative and Evaluative CAAD Tools for Spatial Allocation Problems," in *The Computability of Design*, Y. E. Kalay (ed.), Wiley, New York.

Swerdloff, L., and Y. E. Kalay (1987). "A Partnership Approach to Computer-Aided Design," in *The Computability of Design*, Y. E. Kalay (ed.), Wiley, New York.

Tello, E. R. (1988). "Between Man and Machine," *Byte*, September.

20

NEURAL NETWORK SIMULATION AS A COMPUTER-AIDED DESIGN TOOL FOR PREDICTING WAYFINDING PERFORMANCE

Michael J. O'Neill

University of Wisconsin–Madison
Madison, Wisconsin

Complex public facilities such as libraries, hospitals, and governmental buildings often present problems to the users who must find their way through them. Research shows that difficulty in wayfinding has costs in terms of time, money, public safety, and stress that results from being lost. While a wide range of architectural research supports the notion that ease of wayfinding should be a criterion for good design, architects have no method for evaluating how well their building designs will support the wayfinding task.

People store and retrieve information about the layout of the built environment in a knowledge representation known as the *cognitive map.* People depend on the information stored in the cognitive map to find their way through buildings. Although there are numerous simulations of the cognitive map, the mechanisms of these models are not constrained by what is known about the neurophysiology of the brain. Rather, these models incorporate search mechanisms that act on semantically encoded information about the environment. In this paper we describe the evaluation and application of an artificial neural network simulation of the cognitive map as a means of predicting wayfinding behavior in buildings. This simulation is called NAPS-PC (Network Activity Processing Simulator—PC version). This physiologically plausible model represents knowledge about the layout of the environment through a network of interconnected processing elements.

The performance of NAPS-PC was evaluated against actual human wayfinding performance. The study found that the simulation generated behavior that matched the performance of human participants. After the validation, NAPS-PC was modified so that it could read environmental information directly from AutoCAD (a popular microcomputer-based CAD software package) drawing files, and perform "wayfinding" tasks based on that environmental information. This prototype tool, called AutoNet, is conceptualized as a means of allowing designers to predict the wayfinding performance of users in a building before it is actually built.

INTRODUCTION

Numerous researchers have discussed the importance of being able to find one's way through the built environment [Carpman et al. 1984, Sommer and Aitkens 1982, Weisman 1981]. These researchers have found that when people have problems with wayfinding, there are costs in terms of time, money, and public safety. There are also less tangible human costs, such as stress resulting from disorientation in complex buildings. The very function of a building is impeded if wayfinding within it is a difficult task [Garling and Lindberg 1984]. For instance, a postoccupancy evaluation of an office building in New Zealand revealed that half of the sample of users had so much difficulty with wayfinding in the building that they required assistance from the staff [Beaumont et al. 1984]. In a hospital setting at the University of Michigan, a study found that in a one-week period, 70 percent of the staff had given directions to visitors, indicating the degree to which wayfinding was a problem there [Carpman et al. 1984].

Theory suggests that the physical characteristics of the environment that we use to find our way through the environment are stored in a mental representation of some kind [Kaplan 1976]. The psychological structure responsible for storing and retrieving spatial information about the environment is often referred to as the *cognitive map* [Kaplan and Kaplan 1982, Tolman 1948]. The accuracy of the information in the cognitive map can influence wayfinding performance [O'Neill 1986]. Although a wide range of architectural research supports the notion that ease of wayfinding should be a criterion for good design, architects have no method for predicting wayfinding within built or proposed buildings. In recognition of the need to understand the effects of the designed environment on wayfinding, this research evaluated an artificial neural network simulation of the cognitive map and used this simulation as the basis of a prototype CAD design aid.

MODEL OF WAYFINDING

Kaplan [1976] argues that people solve a complex wayfinding problem by breaking it down into a number of smaller problem-solving tasks. The solution to each of these smaller tasks is to locate a "subgoal" enroute to the ultimate destination. At each intermediate subgoal, one can reorient oneself and decide which direction will lead to the next subgoal, and so on until the destination is realized. These subgoals are often located at *choice points* in the environment. A choice point is a place within a building (such as the intersection of two corridors, or a corridor turn) at which a decision must be made as to which path to take.

Physical landmarks such as architectural details, color, or other memorable features of the environment are thought to play a role in identifying choice points and directions to follow [Weisman 1981]. The more distinct the place, the more easily it can serve as a cue to guide decision-making behavior. Kaplan [1976] also describes the importance of larger-scale landmarks, or *regions*, to the wayfinding process. Within a large building, a region can be made up of a series of landmarks that are experienced close together in time as one passes through the space. Just as it is possible to find one's way between individual landmarks within the small-scale environment, it is possible to find a distant destination by understanding the spatial relations between regions.

Based on the notion that wayfinding is a cognitive task, Weisman [1981] used Kaplan's theory of environmental cognition to develop general categories for physical features of the environment that could influence the wayfinding process within buildings. Of these, he found that complexity of floor-plan layout had the greatest influence on wayfinding performance. Other research corroborates the importance of plan configuration on wayfinding performance [O'Neill 1991]. The relations between choice points is one component of floor-plan layout. It is this environmental variable that is modeled by the simulation described in this chapter.

Representation of Environmental Knowledge

The wayfinding process is influenced not only by the features of the physical environment, but in the way that information is organized in the cognitive map [Kaplan 1976, Kaplan and Kaplan 1982]. Kaplan's cognitive map theory is based on the notion that the mechanisms of cognition did not arise by accident but were formed in response to environmental pressures, such as the need to find one's way through the environment under conditions of missing or incomplete information.

The organization of spatial information within the cognitive map is based on a principle of association called *contiguity*, which states that places experienced close together in time will become associated in memory. As one walks through a building, some places are experienced close together in time and others have little or no temporal relation. This is due in part to floor-plan layout and the choice of paths taken through the building. When a place within a familiar environment is perceived, representations of related places in that environment are experienced in no particular order. Due to the order and degree of experience with related places in the environment, the order of these associations is variable as well. This suggests many complex cross-associations in memory, which can best be represented as a network (rather than, for example, a list) of places. Kaplan's cognitive map theory uses a network model as a mechanism to represent environmental information. The operation of the network is based on Hebb's [1949] research on neural network mechanisms, which modeled the behavior of networks on the physiology of groups of neuronal cell assemblies in the brain.

During a wayfinding task in a familiar environment, if we have the start and destination points in mind, the cell assemblies representing those places in the cognitive map will be highly activated and pass their activity to other associated nodes (see Figure 20.1). The cell assemblies, or *nodes*, representing start and destination points are linked with varying degrees of strength to other nodes in the network that represent places. The strength of the connections between nodes is influenced by the floor-plan layout of the building and previous experience within. The activity passed between nodes is dependent on the strength of the experiential link between them. During a wayfinding task or search through the network, if a given node receives sufficient stimulation, it will become active itself. The activation of a node in the network during a search parallels the conscious recognition of a subgoal between the start and destination points. This mechanism, known as a *broadcast search,* allows a single place represented by the net-

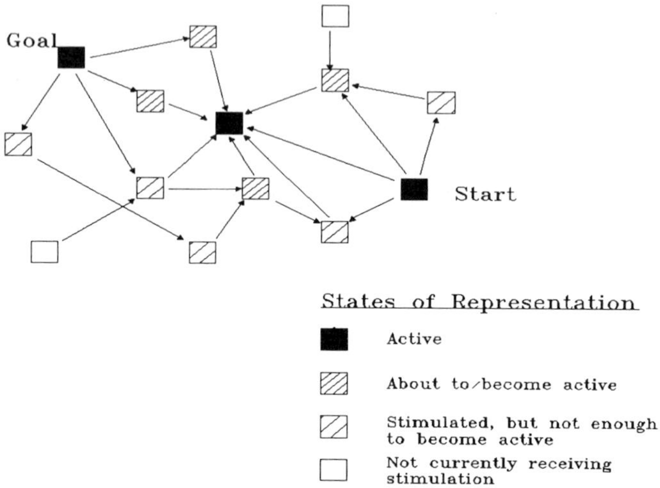

Figure 20.1. Diagram of associative, activity-passing network. (Adapted from [Kaplan and Kaplan 1982].)

work to be located and activated [Holland 1974]. This model of the cognitive map supports a number of capacities critical to the wayfinding task, such as the ability to learn, the ability to break a large problem down into smaller parts (use of subgoals), place recognition, prediction, and the ability to represent a range of concrete to abstract types of knowledge about the environment.

Learning. With experience in the environment, the connection strengths between places in the network can change. As new environmental information is encountered, new connections between places can be created. As the cognitive map alters its structure, it is capable of representing new knowledge about the environment.

Use of Subgoals. When the immediate solution is not apparent, the network must search for information to guide subsequent behavior. The network model uses the broadcast search mechanism, where the larger problem is broken into smaller subgoals, and the problem is solved a piece at a time, in small parts. The network is capable of representing the start and desired goal states (such as the start and destination points within a building) and uses its associative search mechanism to locate intermediate subgoals enroute to the destination.

Place Recognition. Place recognition is the ability to recognize places in the environment through perceptual mechanisms. In the network, lower-level *place nodes* represent more concrete, perceptual information about the environment. These nodes may be stimulated upon the perception of a specific place. Due to the web of connections between nodes in the network, these lower-level nodes can "feed upward" and activate more abstract, higher-level *regional nodes* that are compact representations of a series of individual places.

Prediction. The network encodes the experienced, sequential relationships between places. The *priming* of associated nodes allows the prediction of future places along a path. This system is capable of prediction, because the activation of a particular place node primes associated nodes for activation.

Abstraction. As we travel through the environment, we are often forced to rely on partial information to guide the wayfinding task. Often information about the immediate "next place" we are anticipating in the environment is missing. When information about a particular place is missing, we must shift to a larger-scale, more generalized representation in the cognitive map to determine the correct way to proceed. The cognitive map contains abstract, regional representations that encode the relations between groups of lower-level places. Abstraction is possible because experience is coded within a hierarchical structure in the network. Sensory material feeds into higher, more abstract layers, which are less closely linked to sensory events. When information is missing at the lower-level place nodes (e.g., an expected landmark is missing) the network can "jump up" to the more abstract layer that represents a much larger-scale portion of the environment. This structure allows the cognitive map to fill in the gap when a specific piece of information is missing. The use of abstraction within the cognitive map gives our understanding of the environment a feeling of continuity, even when the actual information we have is fragmented.

IMPLEMENTATION OF THE MODEL

The simulation used in this study, NAPS-PC (Network Activity Processing Simulator—PC version), was originally developed in the field of computer science to test the effects of different network models of knowledge representation on decision making [Levenick 1985]. For this research and application, the simulation was "ported over" from a mainframe to an MS-DOS environment and modified in numerous other ways. This implementation of NAPS-PC models an activity-passing, associative network-based representation of the spatial environment based on Kaplan's [1976] theory of the cognitive map (see Figure 20.1).

Kuipers [1983] suggests that the cognitive map can represent two distinct types of environmental information:

1. Metrical relations that indicate the positioning and distance of places in relation to each other
2. Topological relations that show the ordering of places and their connections to each other

A metrical relation is a type of information that allows people to understand the distance and direction of places in relation to their current location. Topological relations are simply information about the connections between places; whether it is possible to travel directly from one place to another, and what paths connect places in the environment. NAPS-PC can represent only topological information. The simulation can represent intersection and corridor-turn choice points within a building setting and the *paths* connecting them. The paths connecting places serve only to represent the experiential links between places and contain no information in themselves.

Major Components of the Simulation

The simulation works in two parts, a *builder* and a searcher. The *builder* creates the artificial cognitive map, and the searcher conducts wayfinding searches through the environment represented by the cognitive map. Before the builder can create a cognitive map, the layout of the floor plan must be translated into a format the simulation can use by assigning a letter of the alphabet to symbolize each intersection and turn within it. Since there are no rules for mapping the simulation onto reality, this was thought to be a reasonable approach to providing a functionally equivalent representation of the actual environment for the simulation to experience. An input file for a given setting is created by first describing the relations between all nodes in the horizontal direction and then between all nodes connected vertically. For example, consider the following simple environment shown on Figure 20.2*a*. An input file for this setting would be constructed as shown in Figure 20.2*b*. Note that all possible connections between places are made, and this list does not favor one particular path or route over another. Because the version of the simulation described here does not incorporate short-term memory functions, the ordering of the input path sequences is irrelevant. There are no memory effects that would make the final path sequences presented more strongly represented than the paths first presented.

From the information provided in the input file, the simulation creates a network representation of the relations between these places, the accuracy of which is influenced by level of experience, and complexity of the spatial relations between the places in environment. During the formation of the artificial cognitive map, the strength between nodes is increased each time the simulation passes from one node to the next. The link strength is increased relatively quickly when the environment is fairly novel, and more slowly as the link strength begins to approach its maximum. This was done as a crude approximation of the way human beings learn; with a steep initial curve and then in increasingly smaller increments. After the network representation is "built," a search can be initiated through the network by activating the start and destination nodes. The simulation will then search the network for appropriate subgoals making up a path connecting the start and goal nodes.

Mechanisms of the Network

The mechanisms for initiating and controlling activity in the network are constrained by general observations about the physiology of the brain.

Activity Propagation. Nodes are variably active, and this activity is specified through three sets of parameters: maximum and minimum activity, threshold for activation, and initial activity. These values can range from 0 to 100 and are thought of as percentages of a maximum value. Different nodes within the network can have different values. The

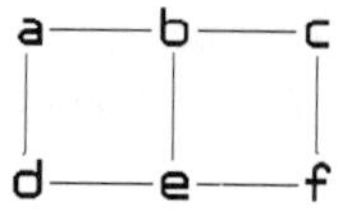

(*a*)

```
a b c
d e f
a d
b e
c f
```

(*b*)

Figure 20.2. Example of schematic floor plan and input file for NAPS-PC.

links between nodes are also variable in activity and can assume values between 0 and 100. The method used to propagate activity is very simple. For example, assume that three nodes, *A*, *B*, and *C*, are linked to node *X*. The links each have a value of 40. The activity levels are $A = 30, B = 40$, and $C = 50$. Their influence on node *X* is $(30 + 40 + 50) \times 40/1000 = 4.8$. If the activity of node *X* is currently 50, its activity at the next time step would be 54.8. If the activity threshold of node *X* was set at 75, it will be turned on, or activated, when this value is reached. When a node is activated, it means it has been selected as a subgoal during a search, and it will begin to act as a *feedback unit*, passing activity to other nodes in the network.

Control Mechanisms. To control activity within the network, the simulation uses a number of neurophysiological mechanisms, including fatigue, inhibition, and competition. Although there is no evidence from the work of neurophysiologists directly supporting the concept of fatigue, it is implemented in the simulation on logical grounds. Fatigue is a physiologically based construct, since in the human system the firing of nodes consumes energy, and the more active the system is, the more energy it will consume. In the simulation, each node has a fatigue value, which is linked to its level of activity. Fatigue is modeled through a formula that calculates a node's activity at each time step. Nodes that are highly active will fatigue more quickly, and fatigue makes a node more difficult to excite. Fatigue is required in an activity-passing system for a successful search to occur; otherwise, all nodes and connections become fully activated and no subgoals will be differentiated within the network.

Global inhibition is related to the concepts of caution and risk. If global inhibition is high in an organism, it could reflect a more cautious approach to evaluating potential solutions to a problem. At lower levels of inhibition, the decision-making behavior of a person could be characterized as being less cautious. Potential solutions to problems would be more readily considered. When global inhibition is set high, inactive nodes tend to stay inactive and the threshold for feedback is increased. This makes it relatively more difficult to come to a decision. While searching for subgoals, if no subgoal is found after a certain number of time steps, global inhibition is reduced. This reduces the feedback threshold and makes it easier for a subgoal to be selected. Within the simulation, global inhibition is either biased toward the higher-level representations, the lower-level ones, or is equivalent in both levels.

In NAPS-PC, the activation of a node is conceptualized as the focusing of the attention of the system. Nodes compete to become active, and very active nodes inhibit the activation of other, less active nodes. An analogy to competition in people's cognitive maps is the idea of competing thoughts, perhaps representing possible solutions to a goal. Once one or two very strong possibilities for solutions present themselves, the activity of other less attractive competing thoughts quickly dies down. In this way it is possible to have focused thought and concentrate our attention on the potentially feasible solutions and reduce the background "noise" from less likely competing solutions.

Wayfinding Task (Network Search)

To initiate a wayfinding task, the start and goal nodes are stimulated to their maximum possible activity and this activity spreads from each of these nodes until it intersects (see Figure 20.3). (Figure 20.3 should not be literally interpreted as a diagram of the cognitive map representation of NAPS-PC. This diagram is merely a visual aid in conceptualizing the notion of hierarchy in the network.) If any node other than start or goal exceeds the activation threshold within a certain amount of time, that node becomes a subgoal; otherwise, the search is inconclusive.

This is where the notion of hierarchy comes into play. The cognitive map can represent knowledge hierarchically, with higher-level, abstract nodes that essentially behave as compact representations of the lower-level data. Figure 20.3 shows how the simulation uses high-level nodes to represent groups of lower-level place nodes. The high-level regional nodes are represented by the spheres in Figure 20.3. Note how the lower-level place nodes are connected to the higher-level nodes. Each of the hierarchical nodes comes to represent an entire segment of sequentially linked places within the floor plan.

When engaged in a search within a large or complex environment, the activation between widely separated start and goal nodes may be so weak that it fails to converge on any subgoal. Figure 20.3 shows how the activation from start and goal nodes dies out before a lower-level subgoal is activated. As NAPS-PC gains experience (learns) in the environment, links from base-level nodes to the higher-level nodes are strengthened and the activation can jump up to the high level nodes. Once at this higher level, the search space (distance or number of items that must be searched) is effectively reduced because only the high level nodes are involved.

THE STUDY

In this section we describe the validation of the simulation against human wayfinding performance. The validation was accomplished by comparing the wayfinding performance of people in actual building environments with the performance of the computer simulation as it attempted wayfinding tasks in functionally identical representations of those built environments.

Settings and Participants

Three areas within the State University of New York at Amherst Lockwood Library were utilized as settings for this study. Figure 20.4 shows a schematic diagram of the layout of

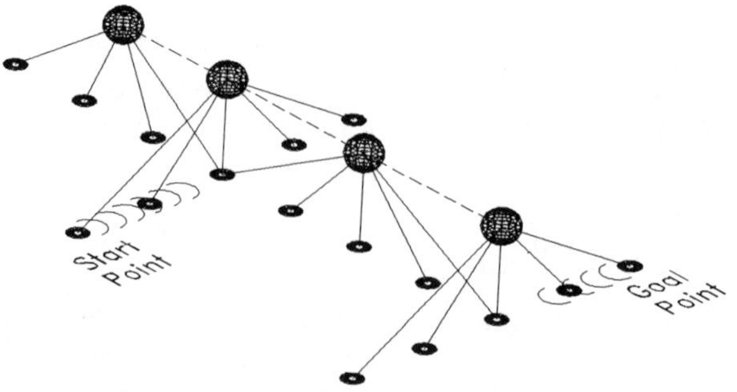

Figure 20.3. Schematic diagram of hierarchical structure in the cognitive map.

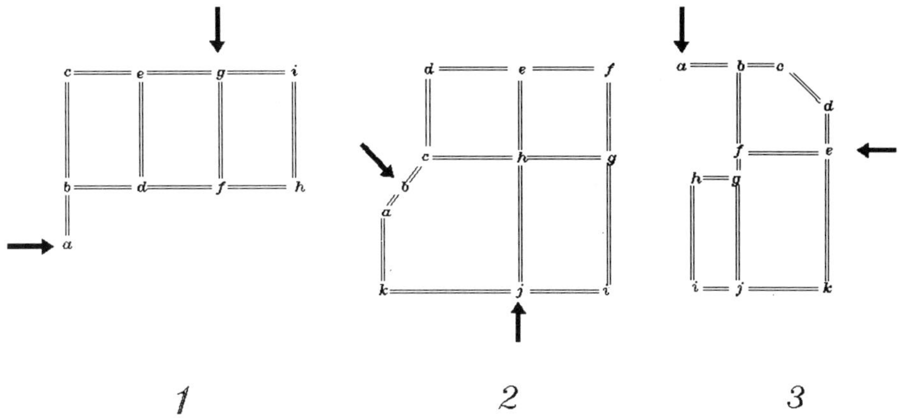

Figure 20.4. Schematic drawings of the three settings used in the validation.

the three settings and the relative locations of the corridor intersection and corridor turn nodes (shown by the letters of the alphabet). The arrows in this figure show the start and goal points for the wayfinding tasks, which are identical in the real and simulated settings. Of the 63 volunteers who participated in this study, slightly over half of the participants were male (60 percent), and the majority were undergraduate students at the State University of New York at Amherst in a variety of programs.

Calibration

Because the simulation first "builds" a cognitive map and then uses this existing representation to conduct searches, a pretraining technique was developed to match NAPS-PC's environmental knowledge with human knowledge of the setting. The participants had never experienced the actual environments. The pretraining involved a simulated walk-through of one of the three environments based on a series of black-and-white photographs. A separate experiment calibrated the level of environmental knowledge of human participants with the simulation so that the performance of the two groups could be compared from an equivalent baseline [O'Neill 1990] for the validation described in this paper.

Tasks and Measures

After pretraining, participants were asked to draw a sketch map of the environment they experienced during pre-training, and to mark the locations of two photographs representing start and destination points in that setting. They were then asked to imagine a path connecting these two points and draw that path on their sketch. This drawn-in path was used as a dependent measure. Each participant then performed the wayfinding task in one of the three environments. For this task, participants were brought to the start point of the environment and shown photographs of the start and destination points. After viewing the photographs, they were asked to find the goal point and indicate to the observer when they thought they had reached it. Once they indicated they had found the goal, they were walked back to the start point, shown the photographs a second time, and asked to find the goal point again. For this task, the path taken from start to goal point was used as a dependent measure.

The wayfinding task for the simulation was similar to that of the human participants. The simulation generated a cognitive map of the setting from the information provided, given the start and goal points, and then attempted to find a path connecting the start and destination points. The measure taken from the simulation was the "path" (series of choice points selected) between start and goal points.

Hypotheses

Because the calibration was made for first trial human performance only, hypotheses 1 and 2 reflect comparisons between the simulation and human data from the first trial. However, to illustrate trends in performance, in several cases simulation performance is compared to second trial human performance. Hypotheses 3 and 4 address general trends in simulation behavior and attempt to link them to human performance. With human behavior, additional experience brings with it increased speed and flexibility in wayfinding behavior. As the cognitive map develops, new connections between places become apparent, and alternative paths through the environment become available. With experience, the number of feasible paths taken through the environment should increase. If the simulation fails to generate alternative paths through the environment with experience, or if its performance deteriorates, it may indicate problems with the usefulness of the simulation as a predictor of wayfinding performance.

1. It is hypothesized that the paths drawn on sketch maps by human participants will be identical to the paths generated by the simulation.
2. It is hypothesized that the paths selected through the settings by human participants will be identical to those generated by the simulation.
3. It is hypothesized that the simulation will choose alternative paths as the level of environmental exposure increases.
4. It is hypothesized that the paths selected by the simulation across exposures will be the same paths taken by human participants.

Results

The first hypothesis states that the drawn-in paths connecting start and goal points on sketch maps will be identical to the paths selected by the simulation. The sketch map path task requires participants to rely on the information provided during pretraining, which is the same kind of node-and-path information that the simulation uses during its search. The sketch map task is an imaginary wayfinding task based purely on the knowledge representation. When participants draw a path on their sketch, there is no confound of learning from actual exposure to the environment. Participants rely solely on the information in their cognitive map and are not influenced by ongoing environmental cues as they may be in the actual wayfinding task. Figure 20.5 shows a summary of the paths drawn in for the three settings.

Figure 20.6 shows a summary of the paths chosen by the simulation in each of the settings at the calibration level of exposure. In setting 1, Figures 20.5 and 20.6 show that the second-most commonly drawn-in path (a) is identical to the path generated by the

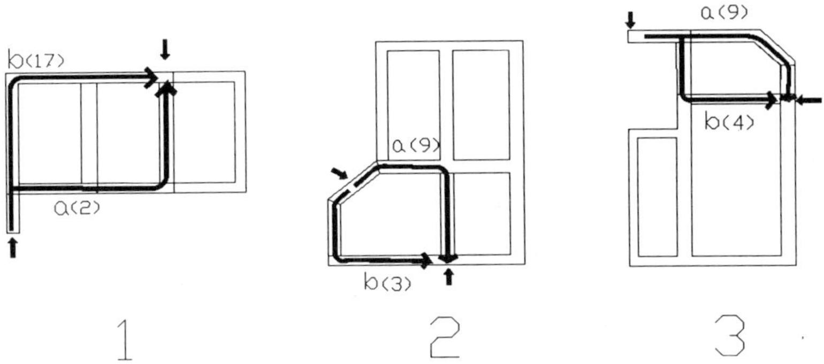

Figure 20.5. Summary of paths drawn by human participants on sketch maps connecting start and goal points in the three settings.

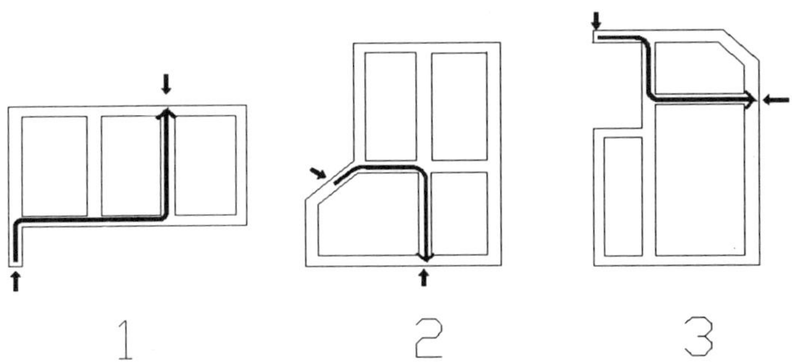

Figure 20.6. Record of simulation wayfinding behavior in the three settings.

simulation. These figures also show that the most commonly drawn-in path for setting 2 (path a) is identical to the path selected by the simulation. For setting 3, the less frequently drawn-in path (b) is identical to the path generated by the simulation. Although not always identical to the most commonly drawn path, all the paths the simulation generated were drawn by at least some people. This indicates that the knowledge stored in the cognitive maps of people may be similar to that represented by the simulation. At the very least, the simulated and human cognitive maps have the same functional consequences.

The second hypothesis predicts that the simulation will generate the same paths chosen by human participants in the actual wayfinding tasks. In Figure 20.7, the darker the shading, the more often a given segment of the plan was traversed. Examination of setting 1 in Figures 20.6 and 20.7 shows that the path taken by the simulation matches a less traveled path taken by people. In setting 2, human wayfinding activity is fairly diffuse in both trials (see Figure 20.7) but most people took the same path that the simulation generated (see Figure 20.6). In setting 3, initial human wayfinding activity occurred in all areas of the setting but is concentrated in the upper right of the floor plan (see Figure 20.7). With additional experience, activity is focused on a predominant path on

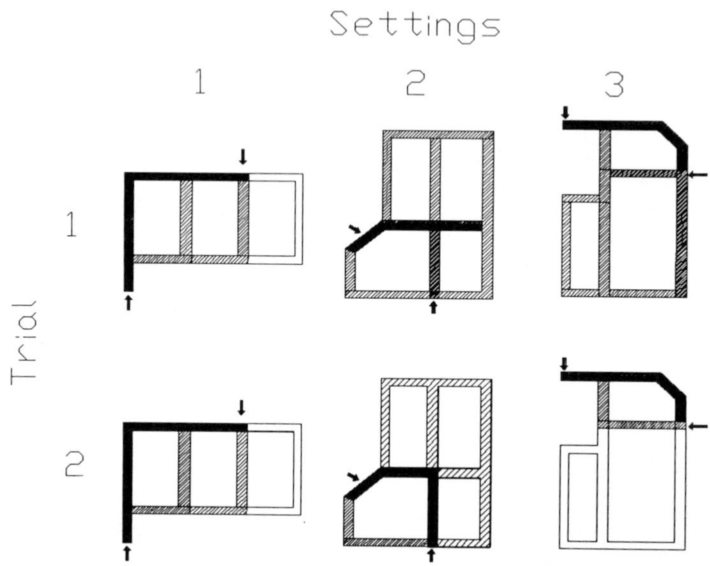

Figure 20.7. Human wayfinding performance across trials and settings.

the upper right, and a secondary path. This secondary path is identical to the path selected by the simulation (see Figure 20.6).

The results show that at the calibration level of exposure, the simulation is predicting some of the human wayfinding behavior. It is also obvious from looking at human wayfinding performance in the first trial (see Figure 20.7) that while there are trends emerging in terms of paths taken, wayfinding is fairly diffuse through the settings. In an effort to look at a wider range of human and simulation performance, the scope of comparison was increased to include both human trials through the environment. In addition, a wider range of experience within the environment was given to the simulation.

The third hypothesis suggests that the simulation will choose alternative paths as its level of experience within the environment increases. Figure 20.8 shows a summary of all paths selected by NAPS-PC. The frequency by which the paths were selected is indicated by the number in parentheses to the right of the path label. In setting 1 the simulation chose three alternative paths that exhaust the possibilities for reaching the goal point without actually going beyond it (Figure 20.8). In settings 2 and 3 the simulation chose the two most logical paths through the environment. These results clearly support the idea that the simulation has the flexibility to generate multiple alternative solutions to the wayfinding task. These solutions seem reasonable in that they are parsimonious and logical.

The fourth hypothesis investigates the degree to which these various solutions actually reflect the behavior of human participants in the actual settings. To test this hypothesis, the paths recorded for participants (see Figure 20.7) and those generated by the simulation (see Figure 20.8) were compared. In setting 1 the three paths generated by the simulation are identical to the paths chosen by people. In setting 2 the simulation generated a path that was most commonly taken by human participants.

While wayfinding behavior is rather diffuse over the whole of setting 3 in the first trial, by the second trial there are two paths. In setting 3 the simulation matched both paths taken by people. Because of the very high agreement in overall choice of paths between the two groups, it is apparent that the simulation does a very good job of predicting human wayfinding performance. This becomes especially apparent when comparing NAPS-PC to trial 2 human performance. After people got more experience in the environment, there was less wandering behavior and it becomes easier to discern overall paths through a setting. Here the similarities in behavior between the two groups are easier to see. Because of the variability of human behavior, any attempt to identify a single path as the only path people would take through an environment is impossible. Instead, comparing the routes most commonly taken by people with those generated by the simulation presents a fairer and more accurate form of comparison.

APPLICATION

In large or complex facilities such as governmental buildings, hospitals, libraries, or large office settings, designers have no way of evaluating the degree to which the building will support the wayfinding task. It is only after the building has been constructed that it may develop a reputation as being easy or difficult to understand and use. Wayfinding within settings such as these becomes especially critical during emergencies, such as during fire or other natural disaster. People may use these buildings only infrequently, which can contribute to a poor understanding of building layout and increase the difficulty of wayfinding. In addition, designers often have little incentive to give the floor plan a coherent organization, since the programmatic considerations of fitting in the many pieces of equipment, mechanicals, or other building functions takes precedence over design clarity.

Given these constraints, it would be desirable to predict the ease of wayfinding within a building before construction begins or interior modifications are made to existing facilities. Because the simulation is computer based, and because CAD is a method for design that easily accommodates iterative changes, it was decided to combine the two technologies to create a computer-based design tool that could interactively predict

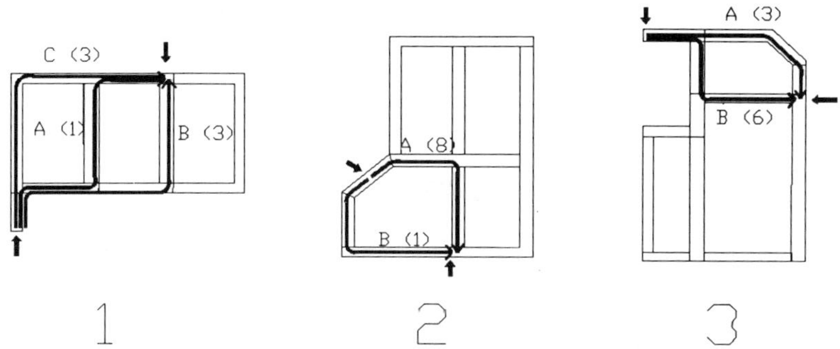

Figure 20.8. Summary of all paths taken by the simulation.

wayfinding within a building and provide the designer with feedback related to building performance on this dimension.

The CAD software chosen for this project is AutoCAD, a microcomputer based-package with a widely installed user base and an open architecture that encourages the development of third-party software. AutoCAD has its own programming language (AutoLISP), a variant of LISP that allows fully functional programs to run in an interpreted environment within the drawing editor. The memory requirements of the simulation program are small enough that it can be executed directly from within AutoCAD in the MS-DOS environment, or run as a background task in a multitasking environment. These factors made it possible to conceptualize an interactive relationship between the simulation and the CAD software.

NAPS-PC and the assorted LISP programs and data files were integrated into a new piece of software called AutoNet. This is a limited prototype system used to demonstrate the possibilities of integrating this simulation with CAD software as a means of assisting, rather than replacing, the decision-making process of a designer. This prototype tool gives the designer the capability of predicting wayfinding behavior between specific places within a building. Although this application is still in a very early stage of development, in this section we describe the use of operational portions of the software and describe the potential for actual use.

1. The user must edit an AutoCAD drawing of a building floor plan. Figure 20.9 shows an image of an AutoCAD drawing being edited and the pull-down AutoNet menu.
2. AutoNet automatically calculates the locations of intersections and corridor turns in the floor plan and inserts a symbol node at each of these locations. The user can add or modify nodes if desired. Figure 20.9 shows a portion of an AutoCAD drawing with nodes inserted.
3. The program calculates all possible connections between these nodes. The connections are constrained by the actual travelable paths between places. For example, if there are walls or other physical elements between two nodes, no connection is possible. Figure 20.10 shows the AutoConnect command highlighted and a portion of the drawing with the nodes connected.
4. The user selects one of three levels of environmental experience that the simulation should model. In this version the experience level can range from low to high. This is useful if the designer wants to model different user groups and building situations, such as an office building with a permanent staff versus casual users of a governmental building.
5. The program analyzes the connections between nodes and writes out a file containing these relationships in a format that the simulation can read.
6. The user invokes the simulation from within the AutoCAD environment, which reads in the description of the building generated in step 5 and creates a cognitive map representation based on the structure of the environment and the level of experience selected by the user.
7. By selecting the Wayfinding menu option, the user can select a start and destination point within the floor plan. In this way AutoNet can predict the wayfinding performance of people between any two selected points on the floor plan.

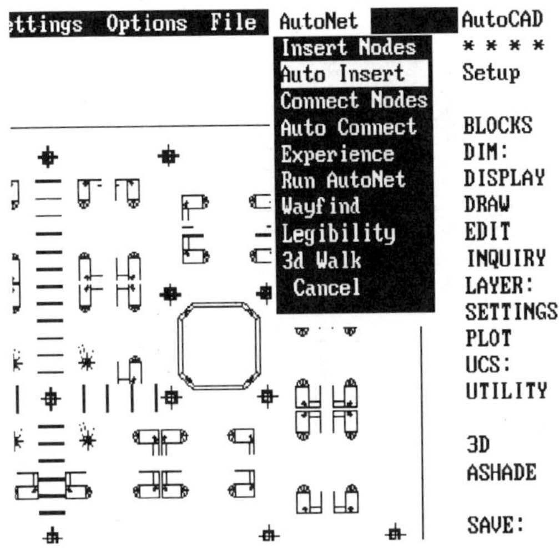

Figure 20.9. AutoCAD drawing with nodes inserted.

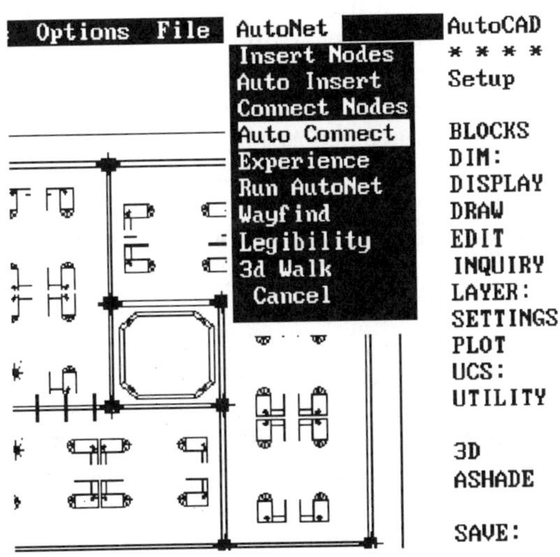

Figure 20.10. AutoCAD drawing with nodes connected.

8. Based on the level of experience selected by the designer and the layout of the floor plan, AutoNet predicts the path that would be taken (see dark arrows, Figure 20.11). The program will also generate alternative paths based on a wider range of experience to give an overview of the range of wayfinding behavior that can be expected for the setting. Figure 20.11 shows three paths predicted for users, with a start point in the upper left and a destination point in the lower right.

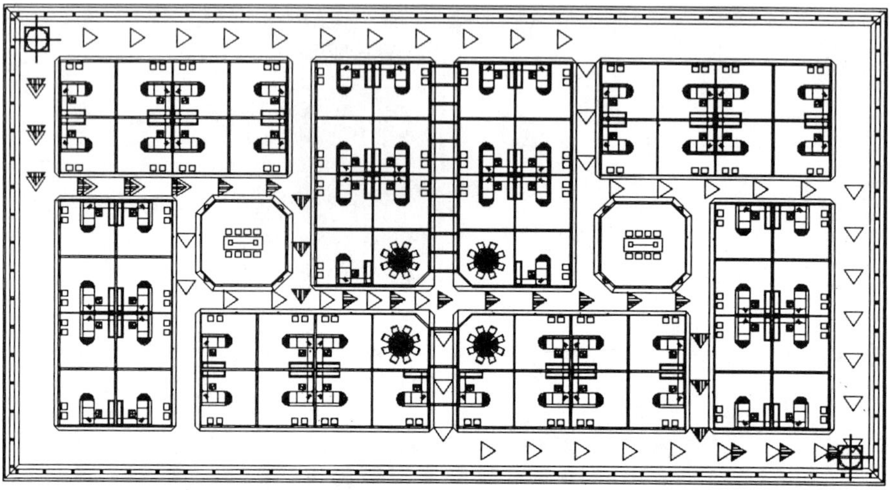

Figure 20.11. Floor-plan drawing with paths predicted by simulation.

In actual use the designer might find that the paths are concentrated near an area within the building that is zoned for private functions, or areas where noise and distractions from passers-by could disrupt quiet work. With this information the designer could alter the layout of the floor plan, relocate this function, or in an office setting use organizational techniques to reduce pedestrian flow through these areas. When more fully developed this tool has the potential to identify trouble spots within a building and to aid the designer in the placement of critical building functions or often-accessed areas. This tool could also be used to predict commonly taken routes through the building to exit points.

DISCUSSION

To provide answers to questions regarding environmental cognition and wayfinding behavior, a number of researchers have turned to the use of computer simulation techniques [Elliot and Lesk 1982, Kuipers 1983, Leiser and Zilbershatz 1989, Levenick 1985, Smith et al. 1982, Stahl 1984). An important characteristic of this type of simulation is that it replaces the human participant with a conceptual, dynamic, computer-implemented model of wayfinding behavior [McKechnie 1977]. In this section we describe several simulations that approach the problem of predicting wayfinding from two different perspectives. The first simulations discussed are concerned only with modeling the outcome variable, wayfinding, based on previously observed human behavior. These simulations are contrasted with work concerned with modeling the cognitive processes underlying the wayfinding task.

Stahl [1984] developed a simulation model called BFIRES that simulates emergency movement involved in the escape from a building during fire or other emergency. This emergency egress simulation explores decision-making mechanisms affecting route selection during fire emergencies. In this type of simulation, building floor-plan layouts are converted by the user into a "node and path" structure that is input into the simulation, along with information on user location, disabilities, and a number of other parameters. The program computes and prints the time to safety for each resident as well as the route selected. This type of emergency egress simulation is of limited use as a general model of wayfinding, because it makes no attempt to model human decision-making processes at the cognitive level. Actual occupant behavior is observed and formalized into sets of rules that reflect general characteristics of behavior. Ultimately, all behavior generated with this approach is predetermined. This stochastic approach is limited by the accuracy of the sets of rules that drive the simulation. In addition, this approach is situation specific, meaning that the rules of the simulation must be rewritten for each building analyzed. Although they are useful for predicting mortality rates and gross measures of occupant behavior, these types of simulation have no means of modeling cognitive processes or the impact of environmental features on wayfinding behavior.

Other simulations, such as the route-finding model developed by Elliot and Lesk [1982], are based on sets of rules that use an "economic man" rationale, where all players have perfect knowledge and always select the shortest route through the environment. As with the emergency egress work, this type of simulation does not contribute to an understanding of cognition or wayfinding behavior. It is useful as a heuristic for determining optimum routes through an environment. Other approaches attempt to simulate cognitive processes and wayfinding behavior. Kuipers' [1982] TOUR model simulates the cognitive map using two sorts of information: *views* of the environment and *actions* (such as rotate or travel) to be taken at each of these views. These view–action pairs are essentially elements of information in a database, where views of the environment act as *retrieval key(s)* [Kuipers 1982, p. 214] for subsequent actions. Kuipers' cognitive map model can provide exact information about the direction and distance between any two points represented. Metrical knowledge is modeled as lists of coordinates attached to places. Trigonometric computations of relative position vectors are calculated when locational information is desired. Given the inexact nature of human wayfinding and cognition, it is unlikely that people process environmental information in this manner. However, this model does recognize the importance of the network as a mechanism for representing spatial information.

Leiser and Zilbershatz's [1989] TRAVELLER simulation also uses a network representation and is constrained by empirical data from studies of wayfinding and spatial cognition. The model constructs knowledge through experience with the environment. Similar to Kuiper's TOUR simulation, this model uses *condition–action* pairs, which are production rules that link a specific place or node within the network with an action. This model recognizes that the wayfinding process is broken down into intermediate steps, or subgoals, and solved a piece at a time. The model emulates *spreading activation* through a network during a search, and information is distributed throughout the network. This simulation can build and store information about routes connecting places through the environment.

Like the emergency egress simulations, these models suffer from several shortcomings, the most important being that cognitive processes are modeled by specifying sets of rules that act on semantically encoded information. In both the TRAVELLER and TOUR simulations, knowledge is represented through a passive network, in which a set of rules act on an inert data structure. The links between nodes are sets of rules for action, whereas the links between nodes in NAPS-PC transmit activity. In NAPS-PC, the pattern of activation within the network represents information about the structure of the environment, while in the TOUR and TRAVELLER models, information about the world is represented by sets of rules that direct behavior.

This rule-based approach limits the way environmental information can be represented. It requires that information about the world be stored in a manner that allows the information to be manipulated by Boolean (if–then) rules of logic. This implies that environmental information is represented in a propositional, or wordlike format. If the simulation is thought to model the mechanism of cognition accurately, the theoretical implication is that perceptual and spatial information about the world is encoded into propositional descriptions. Although there are proponents of propositional representation (most notably Pylyshyn [1984]), rule-based simulations take the implicit stance that this is the mode of representation used in environmental cognition. Since most environmental information is perceptual in nature, models that force information to be stored in a wordlike format may have limited utility as explanatory mechanisms for the human cognitive map.

A simulation whose components bear little resemblance to the mechanisms of the human cognitive system will be useful at predicting behavior only within narrow domains. Ideally, simulation of wayfinding and decision making should include specification of the mechanisms of the system, not just reproduce behavior in limited circumstances. The mechanisms of the TOUR and TRAVELLER models do incorporate rules respecting the psychological constraints of the cognitive map, but are not physiologically plausible. Because of their reliance on a rule-based mechanism, these models fail to simulate what is known about the nature of the mechanisms of human cognition (see Hebb [1949], Kaplan and Kaplan [1982], Rumelhart and McClelland [1986]).

More important, these simulations cannot represent any aspect of the structure of the physical environment. NAPS-PC uses a functional representation of the environment (relations between places) to guide its behavior. The spatial relations between places are represented through an associationist, activity-passing network whose mechanisms are based on neurophysiological considerations and psychological constraints. The accuracy of the representation is influenced by the level of experience within the environment and the complexity of the relations between places in that environment. As in a natural system, the subsequent wayfinding behavior of the simulation is based on the accuracy of the constructed representation. It is this link between environmental features, the accuracy of the cognitive map, and subsequent wayfinding behavior that makes it possible to use the simulation to predict the effects of the built environment on wayfinding performance.

This research shows that it is possible to generate externally valid wayfinding performance from a simulation that relies on a structural representation of the environment, rather than an explicit set of rules, to predict behavior. The validation research presented in this paper clearly shows that NAPS-PC is a useful predictor of wayfinding behavior in different settings and at different levels of experience. Unlike other simulations of the cognitive map discussed in this paper, NAPS-PC does not rely on rules for wayfinding within a given environment. Rather, rules are emergent from the structure of the network representation. With additional experience, NAPS-PC "learns" more about the links between places in the environment. This learning occurs as experience strengthens the experiential links between nodes in the network. With experience, new connections (paths) between start and goal points are possible. This is similar to the learning process in people.

NAPS-PC does not pretend to be a completed, robust simulation of the cognitive map. It suffers from many limitations, not the least being that it can only represent topological relations between places in the environment. The simulation cannot represent distance or direction between places, and therefore cannot calculate shortest distance between places. The purpose of NAPS-PC is not to find least-distance algorithms or to calculate all possible paths to a goal point. Rather, the purpose of this simulation is to model the search mechanism of a natural system (e.g., human behavior) and the effects of the built environment on that behavior.

Future Research

NAPS-PC has a rudimentary learning mechanism that picks up information about the environment quickly at first and then more slowly with experience. Because NAPS-PC has no compensatory learning mechanism, it cannot throw out unneeded complexity in the representation as it learns an environment. Although not discussed in this paper, the time performance of NAPS-PC degrades due to level of exposure in the environment. In other words, with more experience in the environment, the search time from start to goal increases, rather than decreases, as would occur in a natural system. The cognitive map in a natural system has a mechanism for filtering out unnecessary complexity in the representation. NAPS-PC has no means of reducing repetitive incoming information. Some type of screening process for redundant information needs to be implemented that would be tied to the level of development of the cognitive map. A next step in this research is to develop an "output compensatory" learning mechanism that would force the output stimulation from every node to be a constant, regardless of how much environmental exposure the network has had. Currently, output from nodes is not regulated in this manner. The implementation of this compensatory mechanism should alleviate the slowing down of searches with experience and make NAPS-PC a more effective model of the cognitive map.

REFERENCES

Beaumont, P., J. Gray, G. T. Moore, and B. Robinson (1984). "Orientation and Wayfinding in the Tauranga Departmental Building: A Focused Post-occupancy Evaluation," in *Proceedings of the 15th Annual Conference of the Environmental Design Research Association*, D. Duerk and D. Campbell (eds.), San Luis Obispo, Calif.

Carpman, J., M. Grant, and D. Simmons (1984). *No More Mazes: Research About Design for Wayfinding in Hospitals.* Patient and Visitor Participation Project, University of Michigan Hospitals, Ann Arbor, Mich.

Elliot, R. J., and M. E. Lesk (1982). "Route Finding in Street Maps by Computers and People," *AAAI Proceedings*, pp. 258–261.

Evans, G. (1980). "Environmental Cognition," *Psychological Bulletin*, **88:** 259–287.

Garling, T., and E. Lindberg. (1984). "Postoccupancy Evaluation of Spatial Orientation and Wayfinding," in *Proceedings of the 15th Annual Conference of the Environmental Design Research Association*, D. Duerk and D. Campbell (eds.), San Luis Obispo, Calif.

Hebb, D. O. (1949). *The Organization of Behavior*, Wiley, New York

Holland, J. H. (1974). *Adaptation in Natural and Artificial Systems*, University of Michigan Press, Ann Arbor, Mich.

Kaplan, S., (1976). "Adaptation, Structure, and Knowledge," in *Environmental Knowing*, G. Moore and R. Golledge (eds.), Dowden, Hutchinson & Ross, Stroudsburg, Pa.

Kaplan, S. and R. Kaplan (1982). *Cognition and Environment: Functioning in an Uncertain World*, Praeger, New York (Ulrich's, Ann Arbor, Mich., 1989).

Kuipers, B. (1982). "The 'Map in the Head' Metaphor," *Environment and Behavior*, **14**.

Kuipers, B. (1983). "The Cognitive Map: Could It Have Been Any Other Way?" in *Spatial Orientation*, H. L. Pick, Jr., and L. P. Acredolo (eds.), Plenum, New York

Leiser, D., and A. Zilbershatz (1989). The TRAVELLER: A Computational Model of Spatial Network Learning," *Environment and Behavior*, **21**(4): 435–463.

Levenick, J. (1985). "Knowledge Representation and Intelligent Systems: From Semantic Networks to Cognitive Maps," Ph.D. dissertation, University of Michigan, Ann Arbor, Mich.

McKechnie, G. (1977). "Simulation Techniques in Environmental Psychology," in *Perspectives on Environment and Behavior*, D. Stokols (ed.), Plenum, New York.

O'Neill, M. J. (1986). "Effects of Computer Simulated Variables on Wayfinding Accuracy," in *Proceedings of the 17th Annual Conference of the Environment Design Research Association*, pp. 55–63.

O'Neill, M. J. (1990). "Evaluation of a Neural Network Simulation of Cognitive Mapping and Wayfinding Behavior," Ph.D. dissertation, University of Wisconsin–Milwaukee, (Abstracts International, 1990).

O'Neill, M. J. (1991). "Evaluation of a Conceptual Model of Architectural Legibility," *Environment and Behavior*, **23**(3): 259–284.

Pylyshyn, Z. W. (1984). *Computation and Cognition: Toward a Foundation for Cognitive Science*, MIT Press, Cambridge, Mass.

Rumelhart, D., and J. McClelland (1986). *Parallel Distributed Processing: Explorations in the Microstructure of Cognition*, MIT Press, Cambridge, Mass.

Smith, T. R., J. Pellegrino, and R. Golledge (1982). "Computational Process Modeling of Spatial Cognition and Behavior," *Geographical Analysis*, **14:** 305–325.

Sommer, R., and S. Aitkens (1982). "Mental Mapping of Two Supermarkets," *Journal of Consumer Research*, **9**.

Stahl, F. (1984). "Computer Simulation Modeling for Informed Design Decisionmaking," in *Proceedings of the 13th Annual Conference of the Environmental Design Research Association*, P. Bart, A. Chen, and G. Francescato (eds.).

Tolman, E. (1948). "Cognitive Maps in Rats and Men," *Psychological Review*, **55:** 189–208.

Weisman, G. D. (1981). "Evaluating Architectural Legibility: Wayfinding in the Built Environment," *Environment and Behavior*, **13:** 189–204.

21

PREDICTING WAYFINDING BEHAVIOR IN BUILDINGS: A SCHEMA-BASED APPROACH

Mark D. Gross

College of Environmental Design
University of Colorado
Boulder, Colorado

Craig Zimring

College of Architecture
Georgia Institute of Technology
Atlanta, Georgia

Postoccupancy evaluations of large buildings often reveal significant wayfinding problems caused by poor floor-plan layout. Predicting wayfinding problems early in the design process could avoid costly remodeling and make better buildings. However, we lack formal, predictive models of human wayfinding behavior. Computational models of wayfinding in buildings have addressed constructing a topological and geometric representations of the plan layout incrementally during exploration. We propose to combine this with a schema model of building memory. We argue that people orient themselves and wayfind in new buildings using schemas, or generic expectations about building layout. In this paper we give our preliminary thoughts toward developing a computational model of wayfinding based on this approach.

INTRODUCTION

Anticipating and avoiding potential wayfinding problems is an important concern in the design of large and complex buildings. Getting lost in buildings is not only frustrating but in emergency situations it can be a matter of life and death. Postoccupancy evaluations (POEs) in large institutional buildings such as hospitals often reveal significant wayfinding problems. Design guidelines that provide checklists of building characteristics help architects assess a design with respect to wayfinding [Passini 1984]. These efforts represent a good beginning, but we need predictive models that can help a designer anticipate and avoid wayfinding problems.

Most models for building assessment deal with physical factors such as thermal, structural, and acoustic performance of the building. For these physical domains, causal models exist—the computation to simulate the physical behavior of a building is fairly well understood. On the other hand, models for wayfinding assessment must deal with cognitive performance of human users. Relatively little is understood, compared with physical domains, about how people find their way in buildings. This makes constructing predictive models for wayfinding both interesting and difficult.

Some environmental cognition research has studied wayfinding in buildings. For example, environmental psychologist Gerald Weisman looked at factors that influence wayfinding in buildings [Weisman 1979, 1981]. Weisman found that plan geometry was most influential, followed by visibility of interior and exterior landmarks, signs and numbers, and finally architectural differentiation (e.g., clues provided by lighting, textures, or paint). As plan geometry is usually decided early in the designing, the relation between plan geometry and wayfinding is a logical first candidate for computer-based wayfinding models.

How do people remember information about the layout and organization of the built environment, and how do they use this information to plan routes? Central to our understanding wayfinding is how we structure spatial memory, the question of the *cognitive map*. Beginning with Lynch's original work [Lynch 1960], the cognitive map has been a focus of research in urban studies [Appleyard 1970], environmental psychology [Golledge 1987, Holahan 1986], and artificial intelligence (see below).

COMPUTATIONAL MODELING OF ENVIRONMENTAL COGNITION

Several attempts to construct intelligent wayfinding programs have focused mostly on large-scale (e.g., urban) space rather than on buildings. This section briefly reviews several computational models of cognitive maps and wayfinding. Three questions help organize our understanding of these various models: First, what structures are used to represent spatial information in memory? Second, how is spatial information acquired and assimilated into these data structures? Third, how is this information accessed and used in the planning of routes?

Kuipers' well-known TOUR model [1978] explored a simple map of Cambridge, Massachusetts, constructing an internal representation, or cognitive map as it learned new routes through the city. Five categories of spatial knowledge were represented in this program:

1. Routes (a series of motor actions)
2. A topological network of paths and places
3. Relative position with respect to a coordinate frame
4. Dividing boundaries
5. The regions they bound

The program learned routes by following directions through its simulated environment and it could solve wayfinding problems in its path network and answer questions about relative position and orientation.

Following up on the ideas in the TOUR model, Kuipers [1982] argues against the idea that our memory of large-scale spatial organizations is a *map in the head*, a sort of two-dimensional mental image that we inspect in our mind's eye. Though attractive, the "map in the head" metaphor fails to explain observed facts about human behavior and wayfinding. For example, it is possible to know two places and their local spatial relations with other places in their respective neighborhoods, without knowing their spatial relations with one another. But if the two places were located on our map in the head, we should be able to "read off" their spatial relation. Thus, Kuipers argues, the inability to represent such states of partial knowledge is a weakness of the map in the head as a structure for spatial memory. A second argument is that people can sometimes follow a route in one direction but not in the reverse direction. But if we were looking at a map in the head with our mind's eye, it should be equally easy to proceed in either way.

For these reasons, Kuipers suggests we should not promote the "map in the head" metaphor to the level of a theory of human spatial memory. Rather, Kuipers argues, we structure our memory of large-scale space by descriptions of routes leading from one place to another. A route description, in Kuipers' terms, is a sequence of linked associations between views and actions. (When you see view V1, take action A1, and you will see view V2.) A spatial memory structured this way accounts for the deficiencies in the "map in the head" structure discussed above. States of partial knowledge are easily represented, and routes may be known in only one direction. Kuipers shows how the elements of views and actions can be aggregated to define the concepts of place and path. He also shows how wayfinding performance using this memory structure degrades gracefully by providing useful partial information under limitation of computational resources. Kuipers concludes that spatial knowledge "consists of a number of distinct representations: metrical, topological, procedural, and sensorimotor," an assertion that appears also in his more recent overview article with Levitt [Kuipers and Levitt 1988].

Levitt et al. report a radically different approach to wayfinding. Their Qualnav (Qualitative Navigation) program [Levitt et al. 1987] uses distant landmarks (e.g., mountains) as reference points to navigate in a large-scale space. Lines drawn through each pair of landmarks divide the space into polygonal regions. Thus landmarks, connecting lines, and polygonal regions are the primitive elements of Qualnav's spatial memory. The navigator can identify which region it is in by examining relative positions of land along the horizon.

McDermott and Davis [1984] used a *fuzzy (Cartesian) map* to represent knowledge about large-scale space. Assertions about the relative positions of places are converted to regions (fuzzy locations) in the map. The map may have many frames of reference, so relative positions of objects in different frames may be known only vaguely, while within their own respective frames the objects are precisely located. As the program learns additional constraints on the locations of places by "assimilating fuzzy facts," it reduces the fuzziness of locations and remaps the relations between coordinate frames. An associated route-planning algorithm first connects origin and destination with a straight line on the fuzzy map, then circumnavigates any barriers that intersect this straight line.

Although most artificial intelligence (AI) work on cognitive maps deals with large-scale, urban space, a few studies have focused on smaller, building-scale space. Hiiro, Monnai, and Hara, for example, have exercised their wayfinding program on parts of Ginza station in Tokyo [Hiiro et al. 1989, Hiiro 1990]. O'Neill (Chapter 20) uses neural nets to model wayfinding in buildings. The TRAVELLER program also uses a neural net approach [Leiser and Zilbershatz 1989]. A key problem in building wayfinding is identifying rooms, halls, lobbies, and other spatial "places." W. K. Yeap's program explores a floor plan, constructing along the way *relative space representations* of rooms or partially contained places in the building [Yeap 1988]. The boundaries of each space are computed from the visible, occluded, and occluding vertices of wall edges in the plan, producing a stable representation of a place that does not fluctuate widely with small movements of the observer. As Yeap's simulated robot moves through doorways from one room to another, the program links local representations of the spaces to assemble a map of the entire floor plan.

Along similar lines, Kuipers and Byun's NX-Robot [1988] explores a floor plan and identifies a network of connected places and paths and incrementally constructs an internal representation of the plan to guide further route planning. A *place* is defined "as the point in an environment that is the local maximum of some distinctiveness measure defined over its neighborhood," and a *path* as "the control strategy . . . required to move from the neighborhood of one place to . . . another." The topological map represents the floor plan as a TOUR-like graph of connections, storing the local geometries of paths at each node, or place.

These efforts of Yeap and of Kuipers and Byun both focus on the important low-level task of constructing a consistent, absolute representation of a floor plan from local, incremental, and partial information gained during exploration. In the following sections we propose to augment this low-level plan recognition and learning process with a higher-level memory structure. In particular, we suggest that expectations based on generic knowledge, or schemas about building types, can guide learning and wayfinding.

SCHEMAS STRUCTURE MEMORY OF BUILDING LAYOUTS

People bring to any wayfinding task a large store of prior knowledge about typical building layouts. In many buildings, first-time visitors find their way easily; they know what to expect. For example, imagine visiting a shopping mall for the first time. You might recognize the overall plan layout and you might be able to locate a particular type of store, a drinking fountain, or a rest room. Our knowledge of building features and layouts enables us to orient ourselves quickly in an unfamiliar but typical building.

Frames, schemas, and scripts are often used in artificial intelligence models of representation and reasoning [Minsky 1975, Brewer 1987, Schank and Abelson 1977]. The idea is that we recognize a new situation as a specific instance of a general prototype; that particularities of the situation fill "slots" in the frame or schema, and that prototypes structure both recognition and memory. In case-based reasoning, a related approach, specific episodes, or cases, rather than generic or abstract schemas, are stored, recalled, and adapted to the situation at hand [Kolodner 1988].

Evidence from psychology suggests that *schemas*—prototypes or categories—structure people's memory of spatial organizations [Tversky and Hemenway 1983]. In informal experiments we found most people quite willing to sketch a hypothetical floor plan of a high-rise office building when shown only a photograph of the building exterior. Rumelhart and Norman [1975] found that students asked to sketch their dormitory drew balconies flush with the facade although the balconies in fact extend several feet. Presumably students were reconstructing their memory from a prototypical example of "balcony." In another experiment, Purcell [1986] found high agreement among subjects asked to rate the "churchiness" of each member of a set of photos of different church buildings. Mandler, following Bartlett, found that scenes were easier to remember if the elements in the scenes were in certain expected, normal relationships [Bartlett 1932, Mandler 1984].

Therefore, it seems reasonable that people organize their expectations and memory of building layouts using generic descriptions or schemas. Schemas describe typical building organizations at various levels of detail, from global plan organization to local arrangements of building elements or functions. For example, in a typical fast-food restaurant, rest rooms will be found in the rear, past the food-serving area, often on the right. At a global level, schemas describe overall plan topology and geometry and major building features ("parking garage in the basement levels"; "in a department store, the cafeteria will be on the top floor"). More local schemas describe typical partial organizations of building elements ("fire stairs at the ends and middles of corridors," or "coin-operated telephones will be near the rest rooms").

We suggest that our general knowledge of buildings—in the form of schemas—helps us to orient ourselves in unfamiliar settings. It tells us what building features to expect, and suggests possible overall plan organizations as well as local spatial patterns. A schema cannot specify precise metric distances or tell you whether to turn right or left, but it can suggest the presence of certain features and often a rough overall plan geometry and topology—in short, a sort of diagram [Ervin 1989]. For example, arriving in an airport for the first time, an "airport schema" (see Figure 21.1) suggests that baggage claim and ground transportation will be found in a central lobby area on the lower level, some distance from the arrival gate. A corridor will lead to the main lobby with rest rooms, snack bar, and newsstand along the corridor.

PRELIMINARY WORK TOWARD A COMPUTATIONAL MODEL

We are building a computational model of wayfinding in which expectations about layout inform exploration and route finding. The computational models we reviewed do not address this aspect. They focus instead on constructing an episodic memory of the plan based on exploration. We have built various pieces of a computational model,

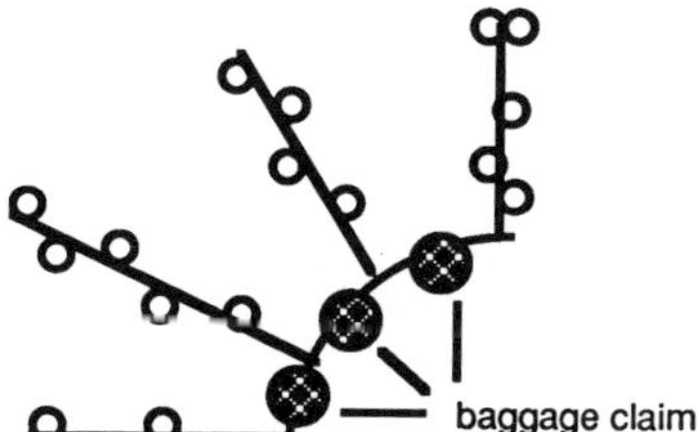

Figure 21.1. Airport schema diagram.

although we have not put the pieces together. We have experimented with two kinds of schemas: *place schemas*, which describe the features of particular kinds of spaces (e.g., kitchens, corridors), and *plan schemas*, which describe the topological connections between kinds of spaces in a plan.

Figure 21.2 illustrates the process of schema-based wayfinding, a conceptual model for our program. The building is put in as a labeled floor plan (A), in which major features and spaces are identified. A schema memory containing typical floor-plan layouts (B) is consulted in a matching and recognition step (C) that selects a particular schema (D) appropriate to the particular built context at hand. The schema is used to explore the particular built context (E) and to solve particular wayfinding problems. Exploration results in an internal representation or map (F) of the particular built context that can be used for future wayfinding tasks.

We have built several pieces of this conceptual model. We have built a drawing program that allows us to enter, label, and edit a floor plan. (Eventually, we would connect to a standard architectural drawing package.) Floor plan elements are displayed as simple geometric shapes but may belong to various classes. Material element classes include walls and windows; spatial element classes include rooms, lobbies, and halls. A catalog of space and material elements is available to facilitate drawing the plan.

We have built a "wayfinding turtle" that moves through the simulated floor plan executing simple motor actions (proceed, turn); the wayfinding turtle is also equipped with a simple vision primitive (isovist) that senses and reports nearbly floor plan elements. By drawing a route through the floor plan for the wayfinder to follow, and storing the sequence of isovists seen along the way, we can use the wayfinder to produce a route description: a sequence of (observation action) pairs. In addition to the vision primitive, the wayfinder can identify the class of space it is in: entrance lobby, private office, conference room.

We have experimented with *finding* spaces in the floor plan. A crude space-finding algorithm forms a polygon connecting the edges of walls visible from a point inside the plan (Figure 21.3). This method avoids problems caused by occluding surfaces and yields a simple and finite set of floor-plan spaces and their overlaps and connections can easily be identified. We also experimented with programming automatic guidance and exploration behaviors such as edge following and door seeking (see Kuipers and Byun [1988]) as part of plan exploring and space finding. The wayfinder wanders around the plan guided by local navigation rules, finding spaces and constructing an episodic memory of the plan.

We have also experimented with *identifying* spaces found by the space-finding algorithm. This involves recognizing features of a space that allow it to be classified—for

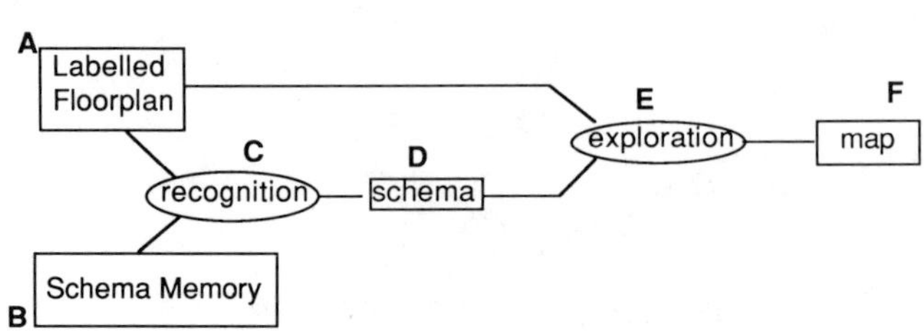

Figure 21.2. Schemas in the exploration process.

example, the presence of a stove and a sink in a room suggests that the room can be classified as a kitchen. We constructed a small library of *place schemas* that describe features of different classes of spaces. Features are of three types: *schema-compatible* features must be present in a space in order to match the schema; *schema-incompatible* features must be absent; and *schema-relevant* features, when present, increase the likelihood of a match. We wrote a classifier procedure that compares the features in a specific space with the library of place schemas and selects the schema that most closely matches the sample.

In the present experiment we simplify the problem of finding and classifying spaces by entering the floor plan with spaces already labeled. To simplify the movements of the wayfinder, we dispense with local navigation rules and limit the wayfinder's travel to paths between the access points of a space and its center. Figure 21.4 shows the wayfinder's paths through the floor plan of an apartment. White circles indicate centers of spaces; black circles indicate access points between spaces.

As in many of the computational models reported above, our wayfinder constructs a graph data structure representing the plan topology as it moves through spaces in the plan. In this *plan topology graph*, nodes represent spaces and links represent direct access between spaces. When the space-finding and space-identifying routines are connected, these will incrementally construct the plan topology graph as the wayfinder explores a plan.

A library of *plan schemas* contains topology graphs of various plan types. Although topology is not the only schematic aspect of a floor plan, topology can serve as a predictor of how a plan will function socially [Hillier and Hanson 1984]. In the plan topology graph for an individual floor plan, nodes represent individual spaces found and iden-

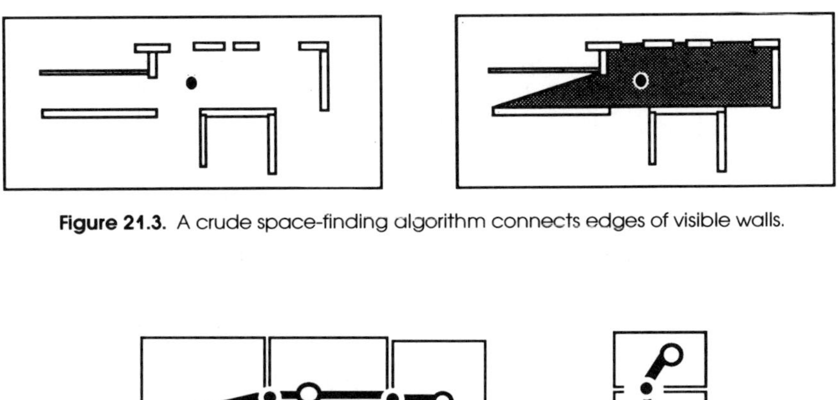

Figure 21.3. A crude space-finding algorithm connects edges of visible walls.

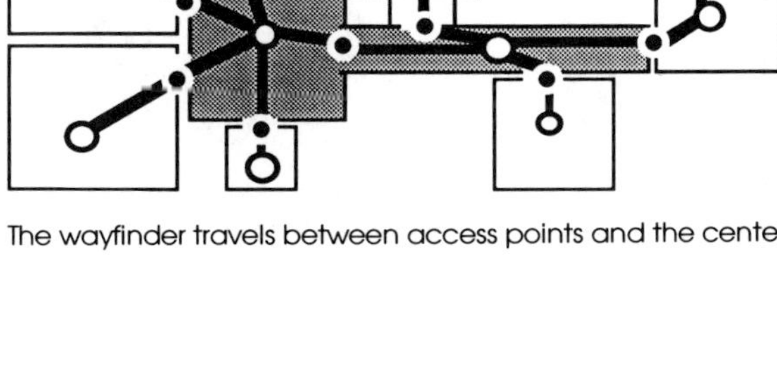

Figure 21.4. The wayfinder travels between access points and the centers of spaces.

tified. In a plan schema, however, nodes in the graph represent place schemas. Thus the topology graph in a plan schema represents a set of connections between a set of classes of places.

The wayfinder tries to match the topology graph of the floor plan it is exploring with a plan schema in the library. Figure 21.5 shows the plan topology graph after the wayfinder has moved through the entry (E) of an apartment and reached a space that it has identified as a hall (H). The wayfinder searches the library of plan schemas and finds a schema with a graph that matches the plan topology it has found in the floor plan so far (Figure 21.6). The schema predicts the topology of the floor-plan layout. However, at present no geometric information is supplied by the schema, so the wayfinder cannot predict which doorway is more likely to lead a particular room. However, according to the plan schema, the Diningroom must be traversed to reach the Kitchen.

The actual floor plan in this example differs somewhat from the schema (Figure 21.7). A second hall (H1) is attached to the first one (H), a porch is connected to the extra bedroom, an extra connection exists between the living and dining room and the kitchen, and the kitchen connects back to the second hall. As the wayfinder explores the plan, the topology graph instantiated from the schema is amended to match the actual floor-plan topology.

In summary, we have discussed the following steps of schema-based wayfinding:

1. Find enclosed spaces or rooms in the floor plan.
2. Identify each space by matching it against a library of *place schemas.*

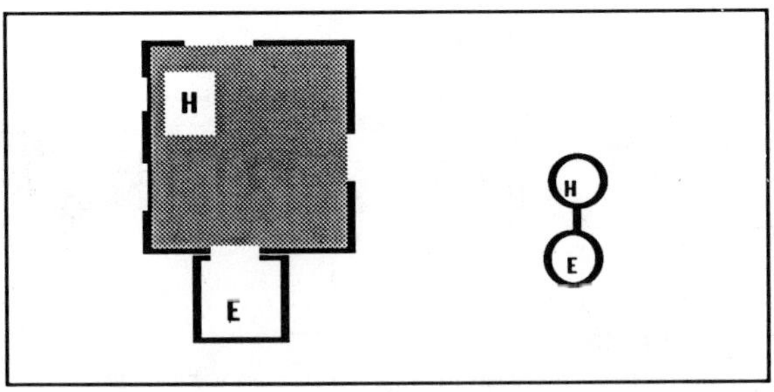

Figure 21.5. The wayfinder builds a topology graph as it moves through the floor plan.

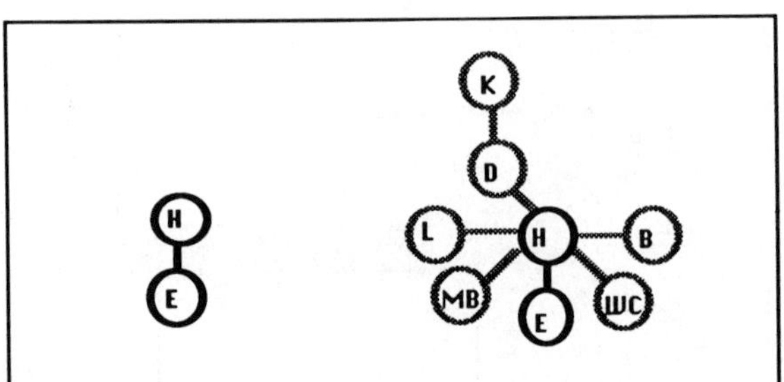

Figure 21.6. The wayfinder's partial plan topology graph matches a schema.

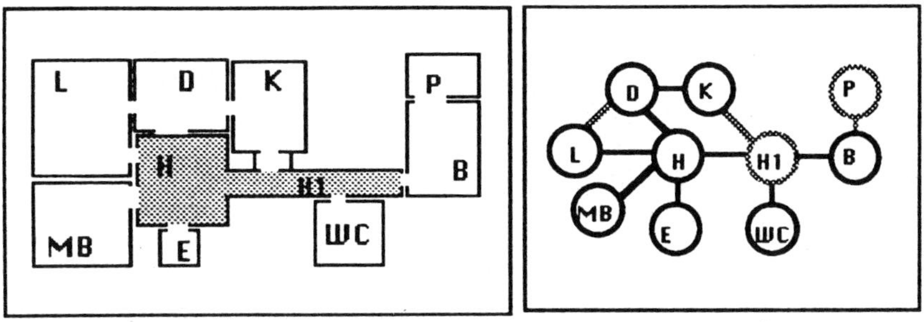

Figure 21.7. The actual apartment layout differs somewhat from the schema.

3. Construct a floor-plan topology graph.
4. Match the plan topology against a library of *plan schemas*, select a most-likely schema, and instantiate its topology graph.
5. Adapt the schema's topology graph to the floor plan.

We have built a computational model of each of these pieces of the process, but we have not connected all the pieces to work together. Although we believe that expectations about plan geometry plays an important role in wayfinding, we have limited plan schemas to plan topology. Local and global geometry should also be included in plan schemas. An approach to this might be to treat geometric relations the same as building features; thus in addition to particular spaces, the plan schema would also specify possible position relations.

DISCUSSION

This idea of schema-based wayfinding suggests several questions. First, how do we recognize a building as an instance of a schema? What features of a place allow us to recognize it as an instance of a schema? For example, do we recognize a building schema based on its use type (airport, hotel, office building), based on architectural style, or based on geometric aspects of the layout (such as building shape) that are immediately visible? How is our memory of building types indexed?

Second, once a schema is selected, how do we employ it in route planning and wayfinding in a particular built context? The schema contains only general information, based on the building type. But any specific instance will not match the type exactly. Specific metric information will be omitted, some information in the schema may not apply to the particular instance, or the instance may be an exception to its type. The route planner must be able to ignore the schema where the building is clearly different. It must also be able to distinguish minor discrepancies between schema and instance from major ones that indicate an incorrect choice of schema.

Third, as we begin to form a more specific map of a particular building, what role do schemas play in structuring this map? Do schemas provide a framework that is enhanced with specific information particular to the building, or is the mapping process carried out entirely separately from the schema?

Finally, how is the general knowledge of buildings that is stored in schemas acquired in the first place? That is, where do schemas come from? If schemas are the product of generalizing from prior specific experiences with buildings, this learning process must eventually be included as part of our model.

We have argued that people use schemas of typical building layouts to find their way in new but typical settings. We argued that if a building matches a well-known schema, it will be an easier building to wayfind in. That is why buildings that are similar in layout—airports, shopping malls—are easy to navigate. If ease of wayfinding were the sole design objective, we should make all buildings identical. Happily, wayfinding is only one among many criteria by which we judge a design, and other criteria militate for variation and adaptation to the site. Moreover, schemas need not specify the plan layout so precisely as to sustain no variation. On the contrary, variability within a type is essential to the idea of a schema. So our theory of schema-based wayfinding is not an argument for uniformity in design.

Nor would we argue for the universality of any particular schemas: knowledge about buildings is surely a matter of local culture. In many places, the courtyard house is a standard form; for the North American it is a novelty. Finding a rest room in downtown Tokyo can be a daunting task for an English-speaking visitor. And we would expect socioeconomic differences as well: not everyone will be equally adept at finding the health club in a high-rise office building. Certainly, however, schemas can be learned. It may be that differences in the schemas that different people hold will make it difficult to use schemas for predicting wayfinding behavior. On the other hand, it may be that people share enough cultural knowledge about how buildings are organized that schemas can help predict wayfinding problems caused by layouts that run counter to people's expectations.

CONCLUSION

Existing computational wayfinding models construct a representation of a floor plan from local observations gained incrementally during exploration. These models do not use prior knowledge about typical building layouts. We have proposed to augment these models with a schema model for wayfinding in buildings in which learning the layout of a new building is structured by expectations based on general knowledge about building types. In our proposed model, a library of schemas stores generic knowledge about typical building organizations. Schemas at the level of plan layout guide overall route planning through the building; schemas about more local organizations of floor-plan elements suggest local strategies. The wayfinding task is thus divided into three components: recognizing a schema that matches a particular place, applying the schema to guide route planning, and constructing a representation of the particular building. Our investigation is at a preliminary stage. We have built some pieces of a computational model and we look forward to reporting on a first round of experimental results.

REFERENCES

Appleyard, D. (1970). *Environment and Behavior*, **2:** 100–116.

Bartlett, F. C. (1932). *Remembering*, Cambridge University Press, Cambridge.

Brewer, W. F. (1987). In *Modeling Cognition*, P. Morris (ed.), Wiley, New York, pp. 187–197.

Ervin, S. M. (1989). "The Structure and Function of Diagrams in Environmental Design," Ph.D. dissertation, Massachusetts Institute of Technology, Cambridge, Mass.

Golledge, R. (1987). In *Handbook of Environmental Psychology*, D. Stokols and I. Altman (eds.), Wiley, New York, pp. 131–174.

Hiiro, M. (1990). In *Proceedings EDRA-21/1990*, Urbana, Ill.

Hiiro, M., T. Monnai, and H. Hara (1989). In *Summaries of Technical Papers of Annual Meeting Architectural Institute of Japan, E*, Architectural Institute of Japan, Tokyo.

Hillier, B., and J. Hanson (1984). *Social Logic of Space*, Cambridge University Press, Cambridge.

Holahan, C. (1986). *Annual Reviews in Psychology*, pp. 381–407.

Kolodner, J. (1988). *Proceedings: Case-Based Reasoning Workshop (DARPA)*, Morgan Kaufmann, San Mateo, Calif.

Kuipers, B. (1978). *Cognitive Science*, **2:** 129–153.

Kuipers, B. (1982). *Environment and Behavior*, **14**(2): 202–220.

Kuipers, B. J., and Y.-T. Byun (1988). *Proceedings of the National Conference on Artificial Intelligence (AAAI)*, St Paul, Minn., Morgan Kaufmann, San Mateo, Calif., pp. 774–779.

Kuipers, B., and T. Levitt (1988). *AI Magazine*, **9**(2): pp. 25–46.

Leiser, D., and A. Zilbershatz (1989). *Environment and Behavior*, **21**(6): 435–463.

Levitt, T., D. Lawton, D. Chelberg, and P. Nelson (1987). *Proceedings of the National Conference on Artificial Intelligence (AAAI)*, Seattle, Wash., Morgan Kaufmann, San Mateo, Calif., pp. 689–694.

Lynch, K. (1960). *The Image of the City*, MIT Press, Cambridge, Mass.

Mandler, J. (1984). *Scripts, Stories, and Schemas*, Lawrence Erlbaum, Hillsdale, N.J.

McDermott, D., and E. Davis (1984). *Artificial Intelligence*, **22:** 107–156.

Minsky, M. (1975). In *The Psychology of Computer Vision*, P. Winston (ed.), McGraw-Hill, New York.

Passini, R. (1984). *Wayfinding in Architecture*, Van Nostrand Reinhold, New York.

Purcell, A. T. (1986). *Environment and Behavior*, **18**(1): 3–30.

Rumelhart, D. E., and D. A. Norman (1975). *Explorations in Cognition*, W.H. Freeman, San Francisco.

Schank, R., and R. Abelson (1977). *Scripts, Plans, Goals, and Understanding*, Lawrence Erlbaum, Hillsdale, N.J.

Tversky, B., and K. Hemenway (1983). *Cognitive Psychology*, **15:** 121–149.

Weisman, G. (1979). "Way-finding in the Built Environment: A Study in Architectural Legibility," Ph.D. dissertation, University of Michigan, Ann Arbor, Mich.

Weisman, G. (1981). *Environment and Behavior*, **13:** 189–204.

Yeap, W. (1988). *Artificial Intelligence*, **34**(3): 297–360.

22

DESIGNING FOR PEDESTRIANS: A CAD-NETWORK ANALYSIS APPROACH

Foster Nichols, Jr.
Isabel J. Canete
Sagun Tuladhar

Parsons Brinckerhoff Quade & Douglas, Inc.
One Penn Plaza
New York, New York

Microcomputer techniques have been developed that combine CAD drawings with transportation network analysis software that uses spreadsheets and stand-alone programs activated from the DOS operating system. The CAD feature simplifies and improves the methods used to design pedestrian circulation facilities and evaluate the impact of new development on existing pedestrian flows. Through the use of customized software, the need for manual data entry is reduced, and the graphical display of analysis results in most intermediate steps in the process are automated. Three hypothetical case studies are presented, concentrating on proposed pedestrian circulation improvements at Penn Station, New York.

INTRODUCTION

Many large transportation centers and building complexes comprise a virtual maze of passageways, concourses, stairways, escalators, elevators, and other circulation elements. Where changes or additions are contemplated to an existing facility, the owner/developer may want to know whether existing pedestrian spaces will become congested or overcrowded following the proposed changes. In the case of a new development, it may be desirable to predict the movement of people through the complex to determine the required dimensions of pedestrian circulation facilities and optimize the location and layout of retail frontage during the design stage. In either case, a rational process is needed for modeling pedestrian flows, taking into account the physical layout of the facility and the predicted or observed origins, destinations, and volumes of pedestrian trips. The model needs to be sufficiently flexible to allow it to be used both as a design tool for sizing required facilities and as a means for assessing the impacts of proposed improvement programs (for environmental review or life safety code compliance purposes).

The methods described herein combine the well-established but heretofore separate techniques of computer-aided design and transportation network analysis. The result is an integrated set of software tools that includes AutoCAD drawings, AutoLISP routines, spreadsheets, and computer programs for building paths through a network, assigning pedestrian trips among alternative paths, and manipulating and transferring data between drawing files, spreadsheets, and other programs. The entire process runs on a microcomputer under the DOS operating system. Prior to the addition of the CAD element, the network analysis model required significant amounts of manual data entry to describe the characteristics of the network and produced only tabular output. The addition of CAD eliminates these drawbacks by allowing for network characteristics to be determined directly from a floor plan of the facility and the results of the analysis to be superimposed on the same floor plan.

EVALUATING AND PREDICTING BUILDING DESIGN PERFORMANCE

Computer-aided drafting and design techniques are well established for the development of architectural, structural, and systems designs for new buildings and the modernization of existing buildings. Automated routines can be used to draft repetitive elements, calculate quantities for cost estimating, construct databases for facilities management, and perform energy and lighting analysis. A logical extension of these applications is the use of computer-aided techniques for evaluating the interrelationships of people with the built environment. The pairing of CAD with transportation network analysis modeling permits such analyses to be carried out at a level of detail not possible with manual methods.

The interaction between computer-aided design and the network analysis model works in both directions. A conceptual building design or floor plan can be provided as one input to the model. Conversely, the results of the modeling effort can be used to improve the layout of space within the building. Three principal applications of these techniques were identified and are described in this paper, including a presentation of case studies.

1. Level of service impact analysis
2. Determination of fire and life safety code egress requirements
3. Retail location analysis

Level of Service Analysis

Level of service analysis is the most common use of pedestrian network modeling techniques. The level of service concept is described in the next section. This type of analysis is used in the design of buildings or group of buildings—especially transit stations, airports, rail and ferry terminals—or any other place with a heavy concentration of pedestrian activity in a relatively confined space. The techniques are well suited to the analysis of special-use facilities where bulk movement of pedestrians occurs, such as sports arenas, theaters, and convention centers.

Level of service analysis is also useful in environmental impact assessment. A pedestrian network model can be used to predict whether or not a proposed development project will cause congestion or significantly worsen the existing levels of service on adjacent sidewalks or in nearby transit stations. If an impact is found, the model can then be used to design appropriate mitigation measures (such as widened sidewalks or additional entrances and stairways to transit stations). The same techniques can be used to evaluate pedestrian improvements proposed by developers in exchange for zoning variances. In all of the cases above, the modeling process is undertaken iteratively to arrive at an acceptable facility design.

Fire and Life Safety Code Egress Analysis

A pedestrian network developed for a level of service analysis or to assist in building design can also be used to help determine whether or not a facility meets applicable fire and life safety codes for evacuation, or what improvements are necessary to bring a building into compliance with such codes. A different form of network analysis, employing many of the methods described above, can be used to trace the shortest egress paths to the nearest fire exits from all points within a building or terminal. The model calculates walking distances, average and maximum egress times, taking queueing into account, and compares them with applicable codes and standards. The designer can adjust the floor plan and rerun the model until code requirements are met.

Retail Location Analysis

Pedestrian network analysis can be used to determine the rent-generating potential of retail space within an office building complex, transportation terminal, or other large public spaces accessible to pedestrians walking from one place to another. As in the prior two examples, it can be used to optimize the design of a facility to maximize retail rent. It is particularly applicable where many alternative pedestrian routes exist through a facility. This method is geared to the analysis of concessions that serve impulse buyers or patrons who stop along their walking route from home to work, or from work to the bus stop or train station, and so on. Major retail establishments would generate trips in their own right.

The value of retail concession space (i.e., its rent-generating potential) is a function of total area (square feet), frontage (linear feet), and the number of pedestrians passing by the storefront. Maximizing the retail area and frontage often results in the fragmentation of pedestrian pathways, with the creation of many alternate routes. Consequently, if total pedestrian activity is a constant and the number of pedestrian links is increased, fewer potential patrons would use the network links adjacent to each storefront.

The model can be used to calculate and display the pedestrian volumes adjacent to all retail frontage in a facility; it also can provide a breakdown of who the users are (e.g., railroad riders, office workers). To complete this type of analysis, financial data would have to be added to the model, relating value (rent-generating potential) per square foot to daily pedestrian activity in front of the space. These data can be obtained from surveys and historical information.

Several classic examples exist, in New York City and elsewhere, of office buildings and transportation complexes that maximize retail space and frontages by elongating or fragmenting pedestrian routes. In many of these cases the pedestrian spaces are relatively underutilized and the adjoining concessions suffer because of insufficient demand.

NETWORK ANALYSIS PROCEDURE

The quantification and evaluation of pedestrian flows through a building or transportation terminal relies on network analysis techniques that are most commonly applied to vehicular traffic studies. The model encompasses all of the basic steps in the transportation systems planning process, as shown in Figure 22.1 [Nichols 1987]. The model analyzes pedestrian flows through a facility by tracking each possible pedestrian movement from entry points over a set of horizontal and vertical network links to exit points. Unlike vehicular traffic networks, the typical pedestrian network in a large building complex or terminal is three-dimensional, with corridors, entryways, and open spaces on several levels, connected by stairs, escalators, and/or elevators.

The major analytical functions of the model are to define a set of logical alternative paths through the network for each entry–exit point pair, assign the total number of trips between each entry and exit point among the alternative paths, and compute the resulting volume and level of congestion on each network link and travel times between all entry and exit points. The required computations are not mathematically complex, but the modeling process must handle effectively the creation, manipulation, storage, and retrieval of large quantities of data.

The model accepts as input a description of the network, including descriptions of links, entry points, and exit points; connections between links; link characteristics such as width or capacity; and a measure of travel time along the link (or impedance, which is equivalent to travel time plus a factor that describes the relative differences in the attractiveness of various routes—due to the presence of retail stores, the availability of escalators, or for aesthetic reasons). Also provided as input are actual or forecast peak pedestrian volumes (from each entry point to each exit point) and behavioral assumptions concerning pedestrians' selection of routes through the network.

The model determines the degree of congestion or queueing that occurs at points within the network. A "level of service" approach is used to describe the extent of congestion. This approach was developed in the field of traffic engineering and adapted by Fruin to pedestrian facility planning [Fruin 1971]. Six gradations in the level of congestion are possible, ranging from A to F. Level of service A represents unrestricted movement. At the opposite end of the scale, level of service F represents a very congested condition where pedestrians are reduced to moving at a slow shuffle or to waiting for unacceptably long periods in a queue. The level of service reflects the freedom of pedestrians to select their normal walking speed and bypass slow-moving and reverse-flow pedestrians at various pedestrian traffic concentrations. For escalators, train platforms, and lobbies or concourse areas, the level of service reflects the amount of queueing that occurs at the facility under various pedestrian loadings. Level of service C typically is

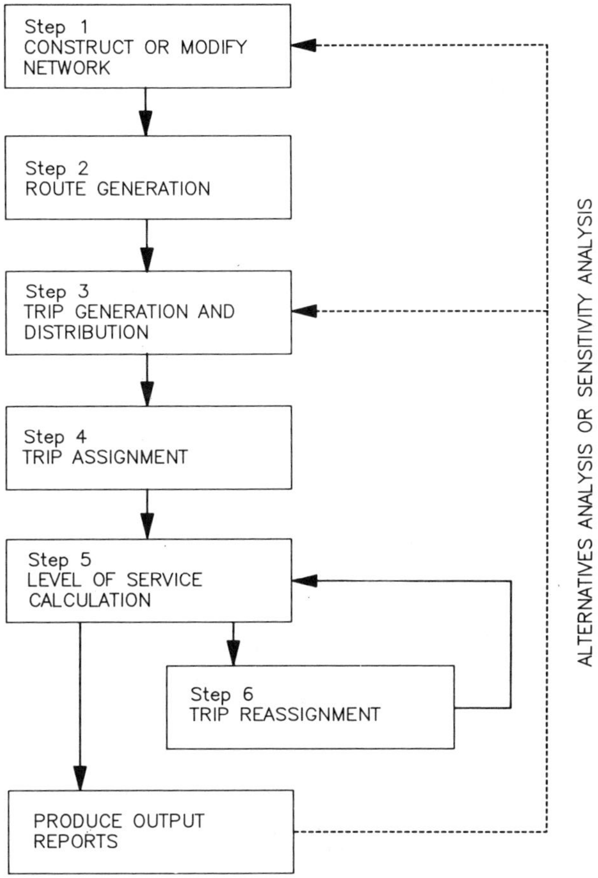

Figure 22.1. Components of the pedestrian level of service (PEDLS) model.

used as the design goal in transportation terminals and public areas heavily used by pedestrians.

The output of the model consists of peak hour and micropeak volumes, a calculated level of service at each link in the network, and travel times for origin–destination pairs. All of these results can be plotted on a network map, which is superimposed on floor plans or base drawings (one for each floor or level of the facility).

INTEGRATED CAD-NETWORK ANALYSIS SOFTWARE

When integrated with the network analysis model, computer-aided design automates many of the data-entry processes and allows for easy graphical display of analysis results. An AutoLISP routine has been developed for use within the AutoCAD environment. This procedure assumes that two-dimensional plans or three-dimensional models of the building design are available and can be loaded into AutoCAD. A customized menu for pedestrian network analysis is available for creating networks and displaying analysis results. The menu structure and available commands are as shown in Figure 22.2.

Once the two-dimensional plan(s) or three-dimensional model has been loaded into the AutoCAD environment, the pedestrian network can be created and/or edited. When adding links to the network, the following attributes must be provided by the user:

1. Link identification number (and optional text description)
2. Link direction (east, west, north, south)
3. Link type
   a. Corridor
   b. Stair
   c. Escalator
   d. Ramp
   e. Turnstile array
   f. Doorway
   g. Hybrid (combination of any of the above types)
4. Base link volume (optional)
5. Attraction factor

For each link, the AutoLISP routine calculates link length and travel time based on the specified walking speed. The routine also calculates link impedance, which equals link travel time, plus an attraction factor (a value added to the travel time to account for factors affecting the desirability of the link as a pedestrian walkway, such as the presence of retail concessions or aesthetic environment). The assignment of pedestrians over alternative paths through the network is based on the relative impedances of the paths.

Pedestrian flow capacities of the different link types are determined using the CAPACITY command. Capacities are defined in terms of effective width of corridors, stairs, ramps and doorways, and throughput of escalators and turnstiles. For example, the effective width of a corridor is determined by pointing at the two sides of the corridor and deducting specified widths of intermediate columns, trash cans, benches, and so on. Effective width data can be extracted from the network file and used as input to a spreadsheet that calculates capacity in terms of pedestrians per minute using a defined formula. Link attributes such as capacity and impedance are calculated by the AutoLISP routine, and these values can be displayed automatically on the network drawing.

From within AutoCAD, network drawing files can be SAVEd and RETRIEVEd and the program can CREATE ASCII files for use as input to the spreadsheet or to the PEDLS model. Files containing data from the Pedestrian Level of Service (PEDLS) model runs and other spreadsheet calculations can be IMPORTed for graphical display of results. For example, from using the spreadsheet, calculations indicate that a corridor with a width of 15 feet has a level of service of E. To improve the level of service to C, the spreadsheet calculations recommend a width of 24 feet. This type of data is saved in an ASCII file and the AutoLISP routine extracts the data and automatically displays them on the network drawing. Link impedances and volumes based on the results of the PEDLS model runs and level of service analysis results also can be extracted from ASCII files and displayed automatically on the network.

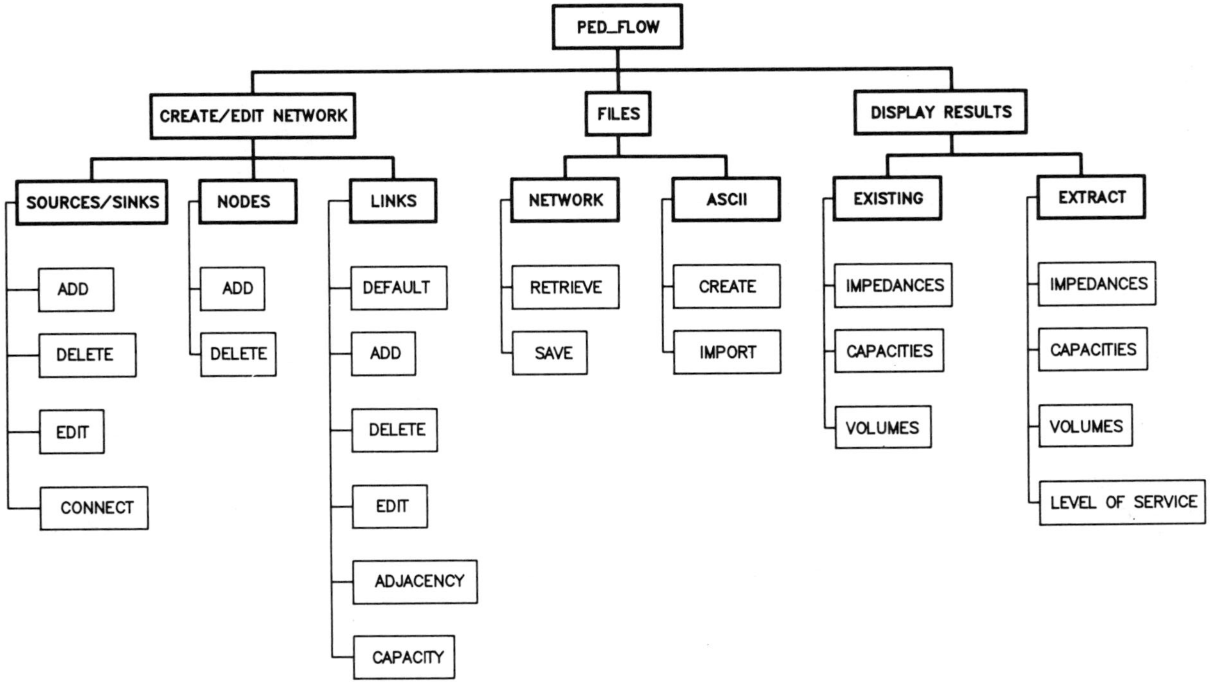

Figure 22.2. AutoLISP routine menu.

By itself, the CAD drawing provides a useful way to display dimensional information about the network. However, the power of using CAD to depict a network increases when the CAD drawing is linked with network analysis tools. The available network models comprise spreadsheets and/or other programs that can be activated from the operating system (in this case, DOS). The analytical techniques described in this paper use a pedestrian network analysis package called PEDLS [Nichols 1987], which includes a set of Lotus 1–2–3 spreadsheet models and a series of BASIC programs for generating paths through the network and assigning pedestrians to those paths based on relative travel times and other measures of impedance. The three computing environments—AutoCAD, BASIC programs activated from DOS, and spreadsheets—make up an integrated package, in which data can be sent to and from these environments. These input and output data are saved in ASCII format in a library that is maintained under the operating system or a shell or file management program (see Figure 22.3). The ASCII files include link information, route information, entry–exit point information, and descriptive information.

The sequence of steps that would be followed in a typical level of service analysis is presented in Figure 22.4, showing the dynamic interaction among the three computing environments. The methods that would be employed for an emergency egress analysis or a retail location analysis are similar.

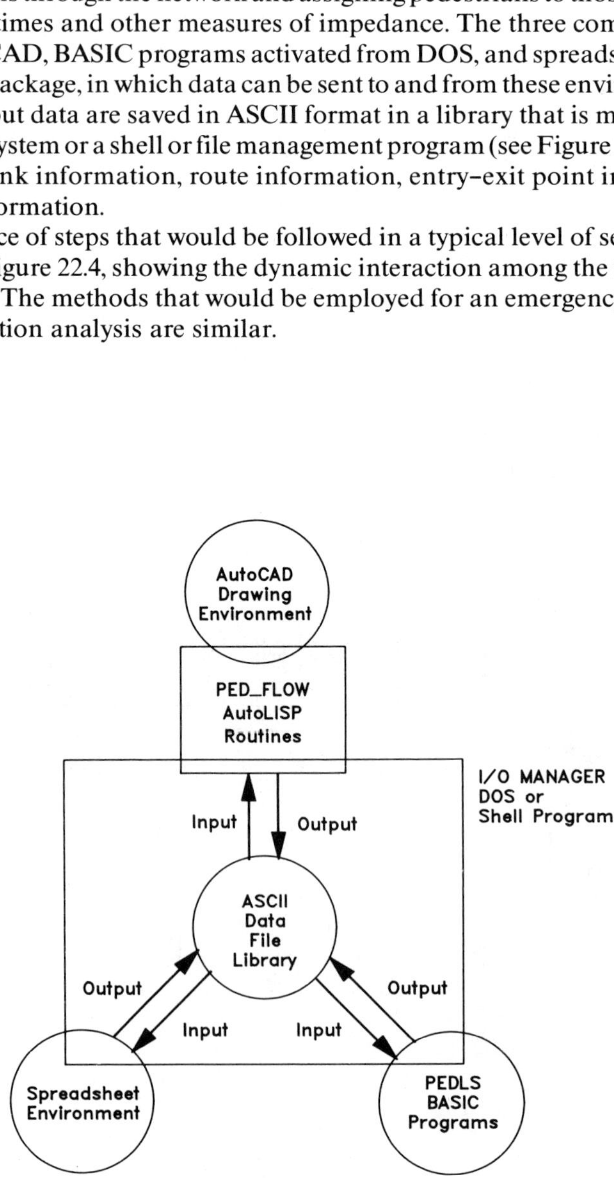

Figure 22.3. Computing environments.

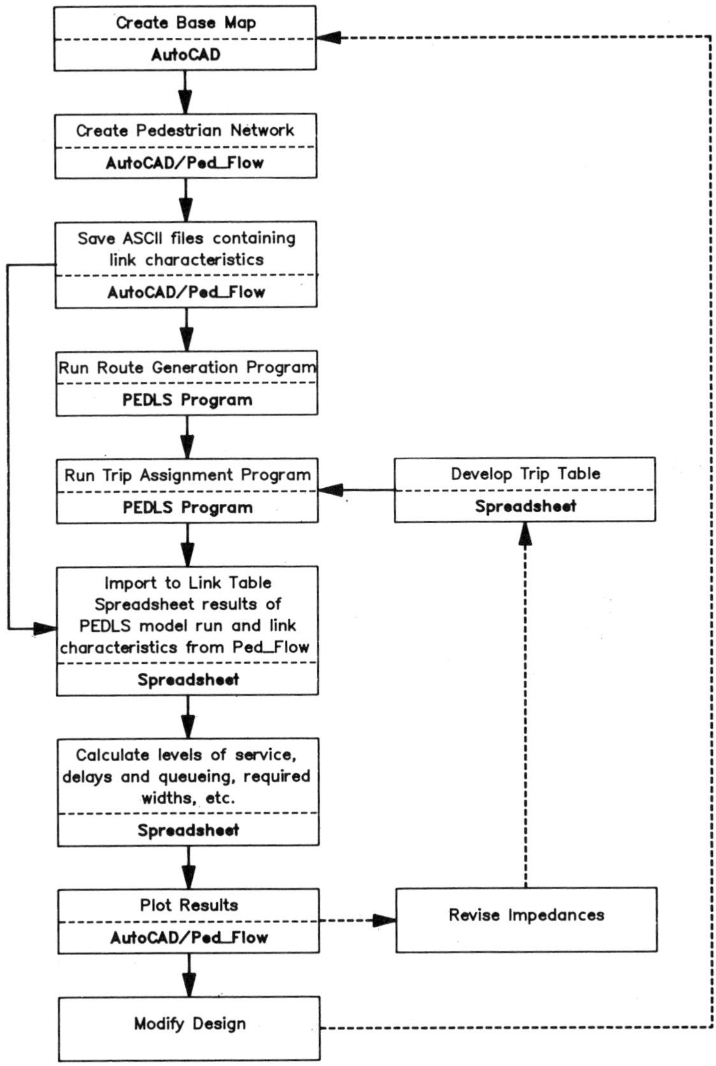

Figure 22.4. Level of service analysis methodology.

CASE STUDIES

Three case studies are presented below to demonstrate the usefulness of the CAD-network analysis approach. All the case studies involve Penn Station, New York, which is the nation's busiest railroad terminal, carrying upward of 200,000 daily riders into and out of midtown Manhattan. Penn Station serves the riders of the Long Island Rail Road, New Jersey Transit (NJ Transit), and Amtrak intercity trains, as well as two major New York City subway lines. The station comprises three subterranean levels, with 21 tracks and 11 platforms on the lowest level (Figure 22.5, platform level), a lower concourse level

(Figure 22.6, level A) providing access to all of the rail platforms and adjacent subway stations, and an upper concourse level (Figure 22.7, level B), containing most of the Amtrak passenger facilities and providing access to the street and selected rail platforms. There are five major exits to street level (Figure 22.8, level C), plus several additional exits through the subway stations and adjacent office buildings.

All three railroads have plans for improving pedestrian circulation through the station complex, in recognition of recent and projected future growth in rail passenger traffic. Amtrak and the Long Island Rail Road are in the midst of capital construction programs to improve the passenger flow capacity and pedestrian environment of the station. NJ Transit has longer-range plans for improvements, and this research and analysis effort was funded by NJ Transit as part of the initial planning for those improvements. The case studies below involve hypothetical capital improvements that might become elements of a plan to be considered by NJ Transit. Their use in this paper does not imply any endorsement, recommendation, or plan on the part of NJ Transit or the authors.

Each of the case studies involves a pedestrian network developed for Penn Station. The base network includes four vertical levels, several corridors, concourses, and walkways on each level, and connections between the various levels. The network has over 40 entry–exit (origin–destination) points, 400 discrete links, and 6000 possible pedestrian paths through the complex from entry points to sink points.

Case 1. Level of Service Impact Analysis

The objective in case 1 was to develop a program of pedestrian circulation improvements to Penn Station to shorten the time required to clear railroad passengers off of the crowded platforms on the south side of the station. The existing layout includes three or four points of egress from each of platforms 1 through 6, serving a single concourse area (see Figure 22.5). The hypothetical improvement plan includes three possible new concourses providing additional access to the platforms, as shown in Figures 22.5 through 22.8. The pedestrian flow model was used to predict the future peak flow volumes using the new facilities and to determine the required sizes and widths of new corridors, the required width of proposed stairways, and the necessary numbers and types of escalators.

The design criteria call for new pedestrian facilities at the lower concourse level (level A) to be able to accommodate the average flow over the peak 5 minutes at a level of service of C or better, a relatively crowded condition without any pedestrian queueing, which is acceptable for transit terminal design [Fruin 1971]. The peak loading condition, projected to occur in the morning between 7:30 and 8:30 A.M., was derived from historical peaking data and railroad ridership forecasts. The size of stairways leading up from the platforms is constrained by the width of the station platforms and is in the range of 5.0 to 6.0 feet. Queueing immediately following train arrivals is acceptable at these locations but not at the concourse level (level A).

In this example the analysis focused on a possible extension of an existing concourse to serve additional station platforms. The existing concourse serves four rail platforms and connects into an existing subway station at its northern end. The possible extension would serve four more platforms and provide a direct street exit at its southern end. An initial run was made using the existing network. Pedestrian volumes on each link were generated by the PEDLS model, and resulting levels of service were calculated in a spreadsheet and imported into the AutoCAD network drawing. Figure 22.9 is a com-

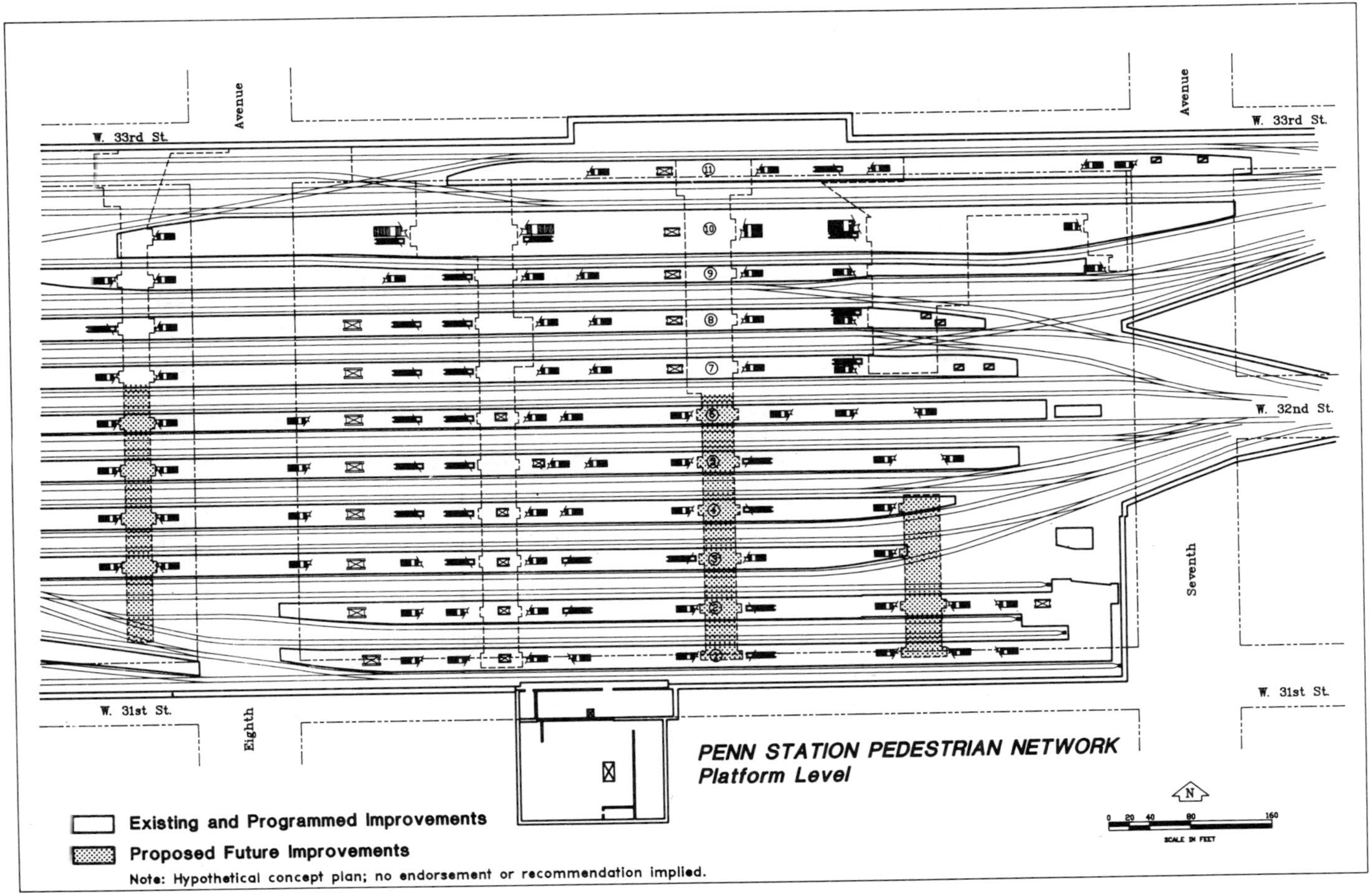

Figure 22.5. Penn Station pedestrian network: platform level.

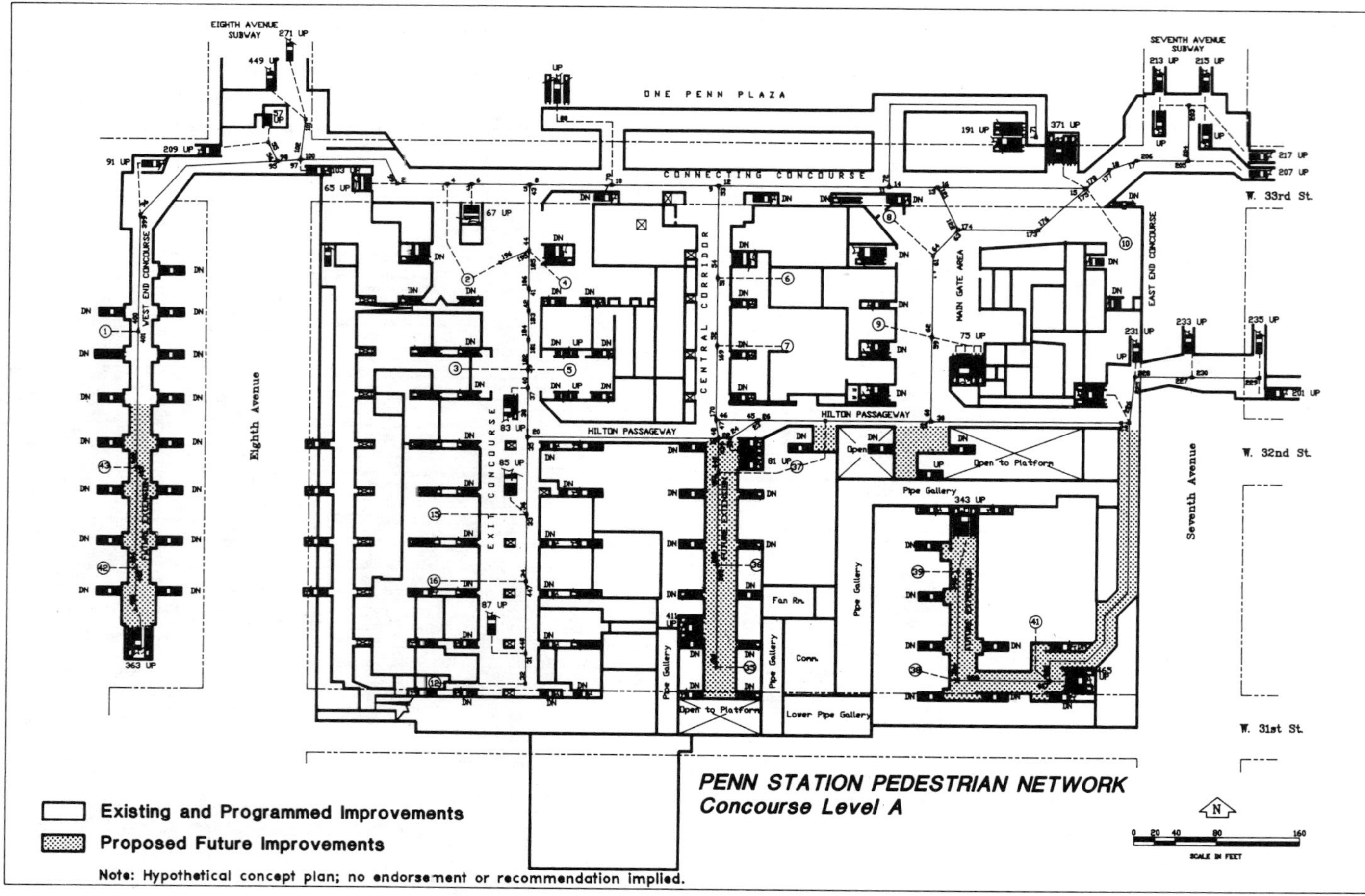

Figure 22.6. Penn Station pedestrian network: concourse level A.

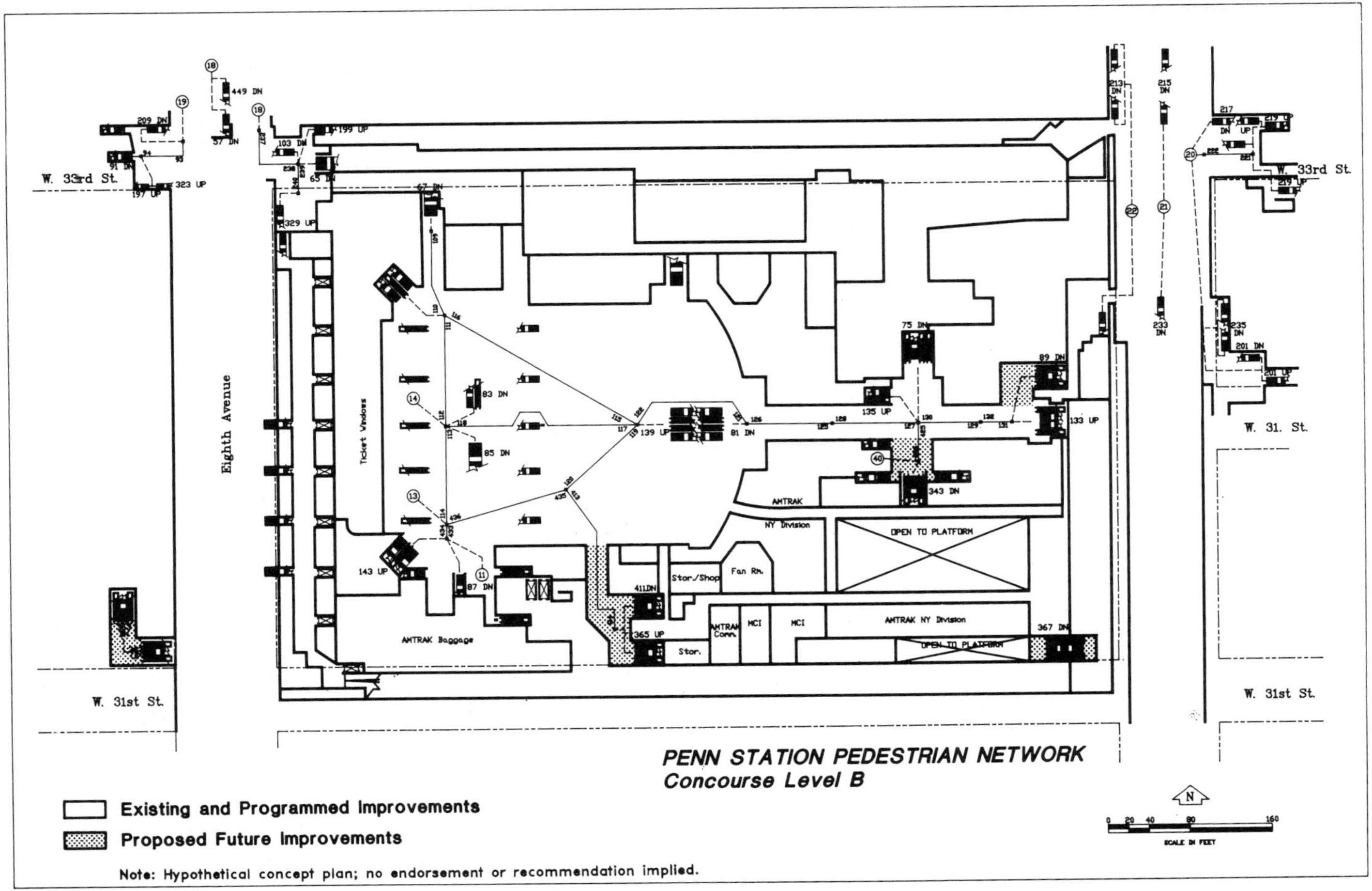

Figure 22.7. Penn Station pedestrian network: concourse level B.

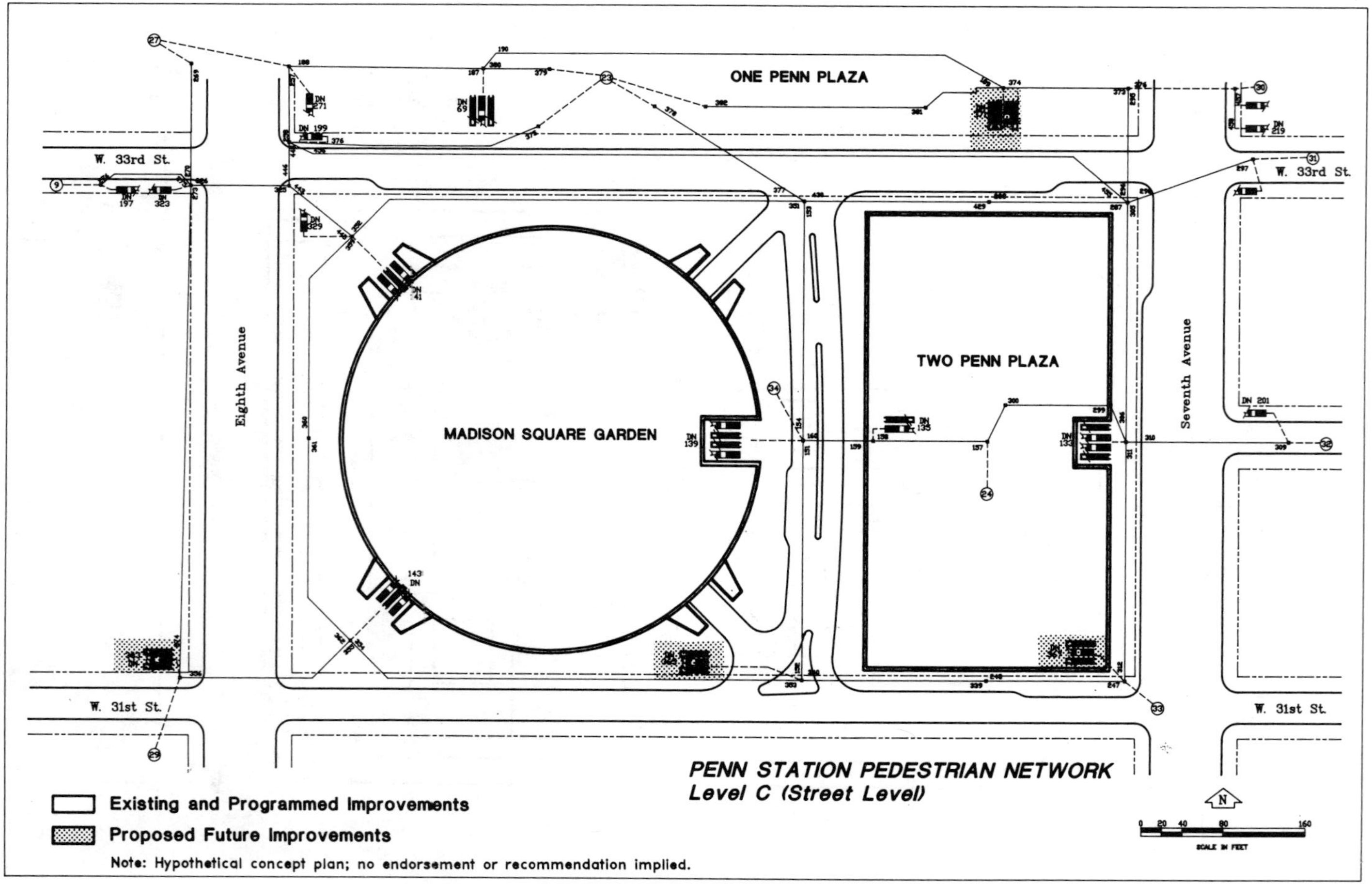

Figure 22.8. Penn Station pedestrian network: concourse level C (street level).

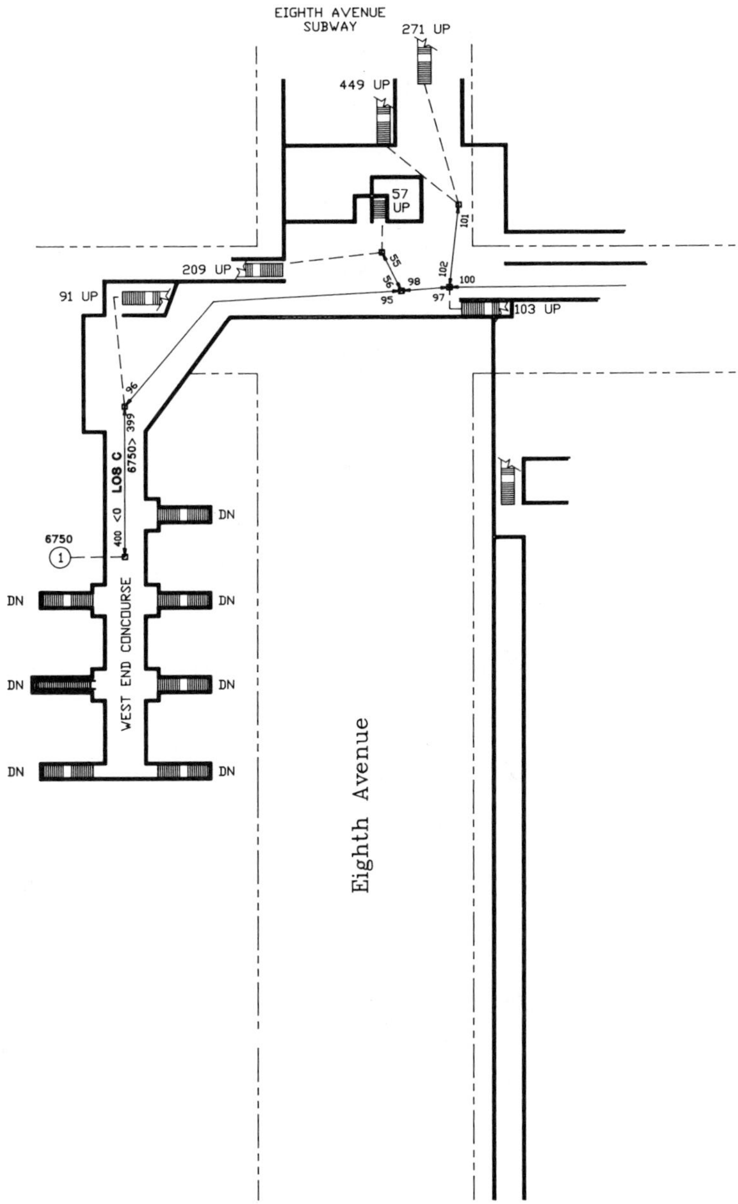

Figure 22.9. Pedestrian volumes and levels of service: existing plan.

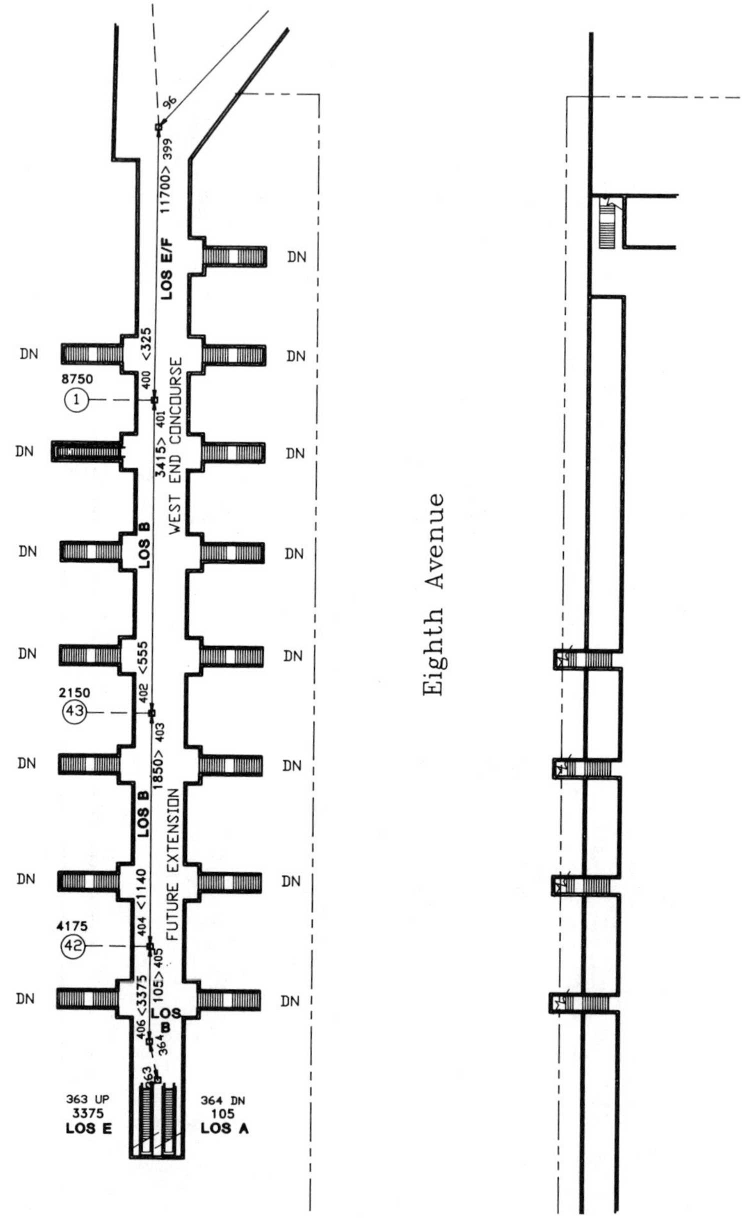

Figure 22.10. Pedestrian volumes and levels of service: initial plan.

posite AutoCAD plot of this concourse showing peak-hour pedestrian volumes and levels of service. The existing configuration is shown, generating a peak level of service C. With the extension, heavier peak passenger loads would be present. When the initial plan was analyzed (see Figure 22.10), the peak level of service was shown to be unacceptably high at the north end (level of service E/F) and at the southern escalator (level of service E). Through an iterative process of modifying the AutoCAD network map and trying different width/capacity values for the critical links, the recommended facility sizes were determined. To achieve an acceptable peak level of service, the northern segment of the concourse would have to be widened from 16 to approximately 22 feet, and a 6-foot-wide stair would have to be added at the south. Figure 22.11 presents this modified configuration, with improved peak levels of service at key locations, generated by the pedestrian flow model and displayed automatically on the AutoCAD base map.

Case 2. Fire and Life Safety Code Analysis

This case uses the same analytical tools to solve a different problem—that of testing the proposed facility improvements for compliance with pedestrian egress requirements of applicable building or life safety codes. The example discussed below addresses the requirements for emergency egress from the concourse levels of Penn Station. There are no specific codes directly applicable to commuter rail terminals. However, the National Fire Protection Association (NFPA) Code No. 130 for fixed-guideway transit systems offers guidance that can be used as a basis for design. One aspect of this code includes the requirement that in the event of a fire or other emergency, people at the most remote locations in the building must be able to reach a point of safety in 6 minutes or less. A separate platform clearance analysis indicated that the maximum egress time from the farthest spot on the rail platforms to the lower concourse level would be 4 minutes 20 seconds in the proposed plan. In this example, therefore, the goal would be to evaluate whether all passengers from the lower concourse level can exit to the street in 1 minute 40 seconds or less. The CAD-network model was used to determine whether or not the goal would be achieved for a given terminal configuration.

Both the pedestrian flow network and the loading assumptions were different in case 2 than in case 1. The network was modified in AutoCAD using the PED-FLOW software to create new network entry-point nodes at concourse level (at platform access locations and the extreme ends of concourse/concession areas in the station). Network exit points were added at each of the street exits from the station. The remainder of the base network (links and entry–exit points at platform and street level) was discarded. Otherwise, the base station plan was assumed to be the same as the one considered in case 1.

The assumed pedestrian loading was equivalent to the projected weekday peak station occupancy, which was based on an accumulation of the evening peak 15-minute passenger volume within the concourse areas (distributed over the various station areas on the basis of observed volumes), plus full trainloads of passengers on seven trains occupying station tracks. These volumes were distributed among the new network entry points using a spreadsheet and saved in an ASCII file. As in case 1, other relevant network information, derived from the AutoCAD map, were saved in other ASCII files, and the PEDLS software was used to calculate the shortest egress route and associated travel time to the street from each of the concourse entry points to the nearest street-level exit point. The routes and associated travel times were plotted on the AutoCAD base map.

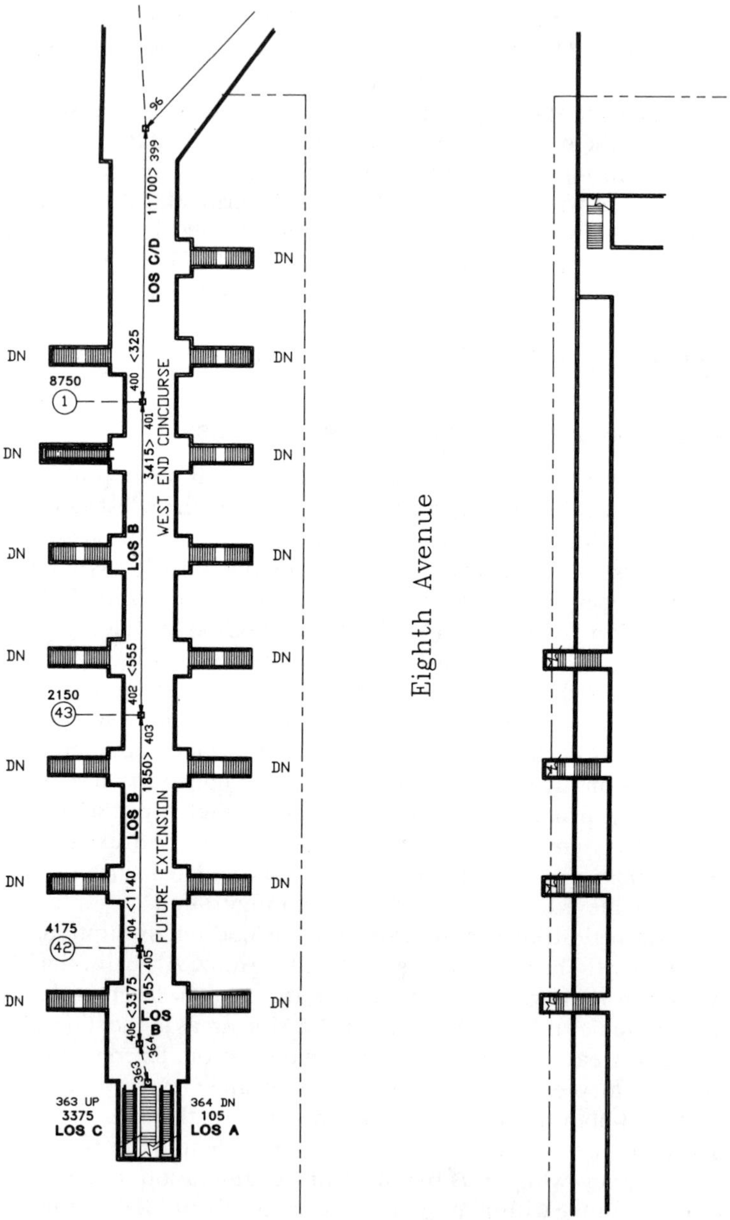

Figure 22.11. Pedestrian volumes and levels of service: modified plan.

An initial analysis was made of unimpeded walking times, which indicated that all concourse points could be cleared within 1 minute 40 seconds. The number and location of the proposed vertical circulation points are adequate, therefore, to meet the egress requirement. However, these unconstrained travel time figures do not account for queueing that may occur at bottleneck points due to the heavy volume of pedestrian flow. The PEDLS model was therefore used to calculate queueing delays resulting from the projected passenger loading. Where queues would be extensive and less-crowded alternative routes were available, the PEDLS program reassigned some passengers to the alternative routes to provide a more realistic representation of actual conditions.

At those locations where the actual egress times exceed the allowable maximum, the PEDLS model was iterated, with increased stair widths at critical locations and revised queueing and egress time estimates calculated. The amount of additional egress stair capacity required to bring the facility into compliance was then determined.

Case 3. Retail Location Analysis

A third type of analysis that may be undertaken using the CAD and network analysis software entails the assessment of the income-generating potential of proposed concession or retail space within a building complex. Concession space and frontage along heavily traveled corridors are more valuable than those in more out-of-the-way locations. If assumptions are made from past experience about the relative value of various types of retail space in different locations, the layout of a proposed new pedestrian facility (or modifications to an existing facility, as in the case of Penn Station) can be optimized to provide a mix of concession area and frontage that maximizes rent-generating potential.

The third case study entailed an investigation of the proposed Penn Station concourse improvement, with its new north–south concourses and exits to street level. It would offer new pedestrian routes into the main departure concourse of the station from points at street level to the south of the station, and would increase the available concession area and frontage.

The same pedestrian network as in case 1 was used, with the new concourses added. An estimate of total daily pedestrian traffic was made (incorporating the assumptions in the morning and evening peak period analyses) and converted into an entry point to exit point trip table for use with the PEDLS model. The model was run twice, once with the existing configuration, and once with the proposed configuration, to determine the aggregate volumes through the station under both cases. A comparison of these two runs demonstrated that the proposed concourse would generate relatively low levels of traffic and divert some pedestrian volumes from the existing spines. These factors, coupled with the estimated construction cost and an assessment of the relative value of the space, could then be used by the facility owner or a real estate consultant to determine whether or not this proposed enhancement to the plan is cost-effective.

CONCLUSION

The combination of CAD software with spreadsheet software and other computer programs offers a new methodology for conducting pedestrian flow analyses. The methodology can be used to design new facilities, evaluate proposed design changes to existing facilities, and optimize the layout of pedestrian circulation space. The technique makes it possible to perform analyses on large, complex networks at a higher level of detail. By eliminating many intermediate manual steps, accuracy is improved. Furthermore, the input–output flow of data and results is more logical and intuitive under this method. The initial investment required to develop a CAD base map and network can be offset by the ease with which multiple scenarios and configurations can be tested. Efforts to make the process more user-friendly are continuing, focusing on the links between programs, data file management, and the creation of a master "shell" program. Additional research will be needed to further improve and enhance the techniques discussed in this paper. However, the case studies referenced herein demonstrate the validity and reliability of this CAD-network analysis approach and reinforce its usefulness as a design tool.

REFERENCES

Fruin, J. J. (1971). *Pedestrian Planning and Design*, Metropolitan Association of Urban Designers and Planners, New York.

Nichols, F., Jr. (1987). "A Model for Analyzing Large Transportation Networks on a Microcomputer," *North American Conference on Microcomputers in Transportation*, Boston.

CONCLUSION: EVALUATING AND PREDICTING DESIGN PERFORMANCE

Yehuda E. Kalay

School of Architecture and Planning
State University of New York
Buffalo, New York

Evaluation can be defined as measuring the fit between achieved or expected performances to stated criteria. Prediction is the process whereby expected performance characteristics are simulated, or otherwise made tangible, when evaluation is applied to hypothetical design solutions. The multifaceted nature of design solutions precludes optimization of any one performance characteristic. Rather, a good design solution will strike a balance in the degree to which any performance criterion is achieved, such that overall performance will be maximized. In this conclusion chapter we discuss the nature of evaluation and prediction, their multilevel and multifaceted dimensions, and some of the approaches that have been proposed to perform quantitative and qualitative evaluations.

INTRODUCTION

Evaluation is a term that has many meanings and is subject to many interpretations. They range from measuring the performance of buildings (or other artifacts), to measuring the performance of the people who design them, and from interpreting "evaluation" as retrospective diagnosis, to dynamic "transformation" in the sense of mapping between different representational schemes. In between these extremes we find the usual meanings and interpretations of evaluation, as the counterpart of generating design alternatives in design processes, and as analysis of completed projects in their postoccupancy stages.

Evaluation can thus be understood as a process that compares what *has been* achieved to what *ought to be* achieved.* In the broadest sense, evaluation can be viewed as a *comparative process*, which is applicable both to *artifacts* and to the *processes* that lead to their articulation. Evaluation can thus be defined as *measuring the fit between achieved or expected performances to stated criteria.*

Evaluation is a proactive, operational term, not only because it requires effort to accomplish but also because it is a dynamic component of design, as well as other processes (e.g., medical diagnosis). Evaluation is called for when a given state of affairs is known to be, or is suspected to be deficient in some nonobvious ways (e.g., the patient is sick, but the cause for his illness is not known) and while there is still hope and desire to improve it. The purpose of evaluation, like the purpose of medical diagnosis, is to ascertain if anything could and should be done to improve that state and possibly determine what steps should be taken to do so. If the solution is known to be optimal (because it was, perhaps, generated by an algorithm that guarantees optimum), or if we are resigned to the fact that it cannot be improved (e.g., in the case of terminally ill patient), or if the steps that must be taken to improve the situation are obvious, then there is no need to evaluate.

The process of evaluation can only be applied to a given, specific set of performance characteristics, such as the form, composition, and location of a building, or the vital statistics of a patient. When evaluating hypothetical design alternatives where performances are not yet evident and measurable, evaluation must be preceded by *prediction.* Prediction is the process whereby the expected performance characteristics of a building (or another artifact) are simulated, hypothesized, imagined, or otherwise made tangible, and hence can be subjected to evaluation. For example, the rate of heat loss through a given building envelope must be predicted, often by way of simulation, before an evaluative procedure can determine whether this rate is acceptable. Similarly, a building model must be subjected to earthquake simulation before its nonelastic behavior can be evaluated.

In design, prediction and evaluation are so inseparably linked that they are often considered to be one and the same. Energy analysis by means of heat loss simulation has been considered for decades the undisputed method of evaluating the energy performance of buildings, as have been cost and structural evaluations. Moreover, even implicit evaluations are often confused with prediction, such as aesthetical appraisal, habitability, desirability, and other nonphysical measures of performance.

A discussion of evaluation and prediction must encompass two broad issues: their structure and their substance. *Structure* pertains to the division of labor between evaluative and generative design procedures. It raises questions regarding the partitioning of design processes into discrete *phases*, or *levels*, with their attendant conditions of uncertainty. Different methods of design raise different questions regarding the proper modes of performing evaluation and prediction, such as in constraint-based design, algorithmic design, precedent-based design, and postoccupancy evaluation.

In discussing the *substance* of evaluation we must consider the multifaceted nature of design, and hence the need for multicriteria evaluation, necessitated by the need to trade

*What ought to be achieved, known as constraints and goals, is itself subject to evaluation. To distinguish the process of evaluating constraints and goals from the process of assessing design solutions that purport to achieve them, researchers have termed the former *analysis*, and the latter *evaluation.*

off and balance the satisfaction of competing needs. We must also consider the differences between evaluating hard and soft criteria, lest the quantifiable considerations overwhelm the nonquantifiable ones. Finally, the various methods that have been developed for performing some quantitative and qualitative evaluations need to be explored, in relation to the particular structure and substance chosen for a given design process.

MODES OF EVALUATION

Evaluation can be applied at different stages of the design process, requiring different degrees of prediction. Only those components of a design solution need to be evaluated for which there are doubts concerning their degree of compliance with performance criteria. Hence the more aspects of a design proposal that are generated in compliance with stated performance criteria, the less need there is for evaluation.

The division of labor between generators and evaluators is pragmatic. It depends entirely on which performance criteria can be embedded easily in the generative procedures and which cannot. The more design aspects can be "designed right," in compliance with relevant performance criteria, the more efficient the design process is, since no effort is wasted on generating solutions that will later be rejected by the evaluative procedures. On the other hand, embedding compliance with performance criteria in the generators often relies on specialized, deep knowledge, which cannot encompass a global perspective of the candidate solution. Generators are therefore often criteria specific. In contrast, evaluators can relate information that is scattered in noncontiguous parts of the project, they can consider simultaneously multiple performance criteria, and they can heed points of view expressed on behalf of value systems that are outside the control of the designer, such as social, political, psychological, and economic ones. Furthermore, evaluators are able to cope with changing beliefs and value systems that are common in large projects whose design-built time is prolonged, and with post-occupancy changes over the lifetime of the project.

Multilevel Evaluation

The difficulty of designing large and complex artifacts such as buildings necessitates the use of abstraction and stepwise refinement as means to partition the process into discrete, manageable "chunks." It also means that some decisions must be made before others, hence before all the details that may effect them are known. How the process is partitioned, and what constitutes a chunk determine the informational needs of the designer and the evaluative and predictive procedures that he or she will use.

The design process can be partitioned vertically or horizontally. A *vertical* partitioning follows the levels of the abstraction hierarchy, from general to detailed. Also known as top-down (and bottom-up) partitioning, the levels of the hierarchy consider as chunks multiple design criteria in different levels of abstraction. The informational characteristics of such chunks are multicriteria, with differing degrees of vagueness and uncertainty. Conversely, a *horizontally* partitioned process considers as a chunk one aspect of the designed artifact each, such as structure or thermal comfort, from beginning to end. Also known as a *systems approach* to design, chunks in this mode of partitioning are informationally characterized by deep knowledge, yet they are mostly oblivious to the needs of other chunks.

Design tools that are applicable to these two modes of chunking are, necessarily, different. The vertical mode would tend to rely on broad-based design methods, such as constraint-based design and prototype refinement. The evaluative tools it uses range from expert systems in the early phases of the process, to postoccupancy evaluation at its conclusion. Horizontal chunking would rely on deep design methods, such as specialized algorithms. The evaluative tools it finds most appropriate range from specific simulations to highly specialized analytic procedures.

Designers often shift from one mode of chunking to another. They shift within one mode from top-down to bottom-up, and from one building system to another. Therefore, they make use of different design and evaluation tools at different stages of the process. Frameworks for availing different tools to designers, and facilitating dynamic shifts between them, are needed. The different approaches to partitioning design, and the computational frameworks that can facilitate creative and dynamic design process, constitute the first parts of this book.

THE DIMENSIONS OF EVALUATION

Proposed design alternatives, and realized design solutions, exhibit many performance characteristics and must meet many criteria. Each performance characteristic can, and often is, evaluated and compared to specific criteria, such as cost, thermal comfort, and structural stability. Furthermore, specific criteria often establish acceptability limits, by which the corresponding performance characteristics are expected to abide. For example, the total building cost should not exceed a certain budget, its envelop must maintain certain temperature and humidity levels, and it should withstand certain seismic forces.

Multicriteria Evaluation

Individual performance characteristics, nevertheless, rely on common factors and are therefore influenced by each other. For example, thermal comfort is effected by the form and materials the building is made of. Form and materials also determine the structural behavior of the building and its cost. Therefore, adjusting these common factors so that they satisfy individual criteria by which performance characteristics must abide often reduces the degree to which the common factors satisfy competing or even conflicting criteria. For example, a certain choice of form and materials may satisfy the structural criteria but induce unacceptable thermal conditions or cost too much. The designer must therefore consider all (or at least many) of the building's performance characteristics together, such that increasing the degree by which one is satisfied does not adversely affect the others.

Assessing the performance of a building is therefore a multicriteria evaluation effort where the predictable effects of each design proposal must be considered from multiple points of view. Since all performance criteria cannot be satisfied to the same degree, the relative importance of various performance criteria must be established as well as the marginal returns of relaxing some criteria in favor of others. For example, when the budget established for the purpose of building a single-family house requires choosing between a third bedroom and a garage, which of the two will the client prefer to have? And will the same choice be made when the trade-off is between a fourth bedroom and a two-car garage? Alternatively, will the client prefer to pay a little more and have both a fourth bedroom and a two-car garage? If so, how much more? Various approaches to integrating multiple independent evaluative procedures into one multicriteria evaluation system constitute Parts Two, Three, and Four of this book.

METHODS OF EVALUATION

Design performance criteria can be classified as hard and soft, corresponding to what is known as *quantitative* and *qualitative* evaluation, respectively. *Hard criteria* are typically measurable performances, such as thermal comfort, structural stability and cost. *Soft criteria*, on the other hand, are typically based on subjective perceptions, such as aesthetics, social, psychological, and political "fit," and are therefore difficult to measure. Yet both can be assisted by computational means using a variety of programming techniques. Furthermore, each can be done in different ways.

As examples of measuring hard criteria, energy evaluation was presented in Part Five of the book. First, a state-of-the-art algorithmic approach was presented, relying on deep knowledge. Then a knowledge-based approach was presented, which is applicable at the early phases of design. Finally, a combined method was presented, which attempts to bridge the gap between the two former ones. As examples of measuring soft criteria, wayfinding in buildings was presented. Again, several different methods were discussed: algorithmic, knowledge-based, and ones based on traffic simulations and observations.

CONCLUSION

Evaluating and predicting design performance was chosen as the subject matter of this book for two reasons:

1. Evaluation and prediction are central to design, and computers have been used to assist them ever since CAD became reality in the early 1960s.
2. New developments in the substance of evaluation, and in the methods that can be used to apply it in design, have emerged in the last decade that warrant a renewed look at evaluation and prediction.

In addition to exploring general theoretical questions concerning the nature of evaluation and prediction, the book also attempts to cover the topics of evaluation and prediction both in breadth and in depth. First, issues concerning the structure of evaluation and prediction in design, and frameworks for integrating computer-supported evaluations with other modes of design, were presented. They included postoccupancy evaluation, design by specification, automatic design procedures versus interactive ones, evaluation versus prediction, and evaluation at different phases of the design process. Second, issues concerning the substance of evaluation and prediction were presented. They included various aspects of, and approaches to, multicriteria evaluation. Finally, different approaches to quantitative and qualitative evaluation were discussed.

In addition to these specific topics, the theoretical foundations underlying most current approaches to design evaluation were examined, including the crucial roles in design of ambiguous and inconsistent representations and provisional beliefs, and models of design exploration based on nonmonotonic modes of reasoning. Many questions, of course, remain unanswered. They include:

- How can the results of prediction and evaluation of building performance be integrated into the design process such that they contribute to the generation of better buildings?
- How well can we assess the performance of buildings within given social, technical, political, and environmental contexts?
- How do changing design objectives (due to changing motivations of the various parties to the design, construction, and occupancy processes) affect evaluation?

Evaluation and prediction are not unique to design. They are commonplace in planning, economics, and other modes of intervention, where futures must be made explicit before their desirability can be judged. Therefore, it is expected that answers to these questions may benefit many disciplines. It is hoped that the discussions presented in this book will contribute to their exploration.